Recent Trends in Polymer Membranes: Fabrication Technique, Characterization, Functionalization, and Applications in Environmental Science, 2nd Edition

Recent Trends in Polymer Membranes: Fabrication Technique, Characterization, Functionalization, and Applications in Environmental Science, 2nd Edition

Guest Editors

Yan Wang
Gang Wei

MDPI

Basel • Beijing • Wuhan • Barcelona • Belgrade • Novi Sad • Cluj • Manchester

Guest Editors
Yan Wang
College of Chemistry and
Chemical Engineering
Qingdao University
Qingdao
China

Gang Wei
School of Polymer Science
and Engineering
Qingdao University of
Science and Technology
Qingdao
China

Editorial Office
MDPI AG
Grosspeteranlage 5
4052 Basel, Switzerland

This is a reprint of the Special Issue, published open access by the journal *Polymers* (ISSN 2073-4360), freely accessible at: https://www.mdpi.com/journal/polymers/special_issues/Y655347ITN.

For citation purposes, cite each article independently as indicated on the article page online and as indicated below:

Lastname, A.A.; Lastname, B.B. Article Title. *Journal Name* **Year**, *Volume Number*, Page Range.

ISBN 978-3-7258-7697-6 (Hbk)
ISBN 978-3-7258-7698-3 (PDF)
https://doi.org/10.3390/books978-3-7258-7698-3

Contents

About the Editors

Yan Wang

Yan Wang is a full Professor at the College of Chemistry and Chemical Engineering, Qingdao University. He received his Ph.D. in Chemical Engineering from the University of Bremen, Germany, in 2016. His research interests include chemical process simulation and optimization, the design of functional nanomaterials, electrohydrodynamic separation of micro-/nanoparticles, water environmental remediation, and lithium recovery from seawater. He has published more than 50 peer-reviewed papers in leading journals such as *Adv. Funct. Mater.*, *Chem. Eng. J.*, and *Carbohyd. Polym.* and holds several national invention patents. His research integrates theoretical modelling with experimental material design to address challenges in sustainable energy and wastewater treatment.

Gang Wei

Gang Wei is currently a full Professor at the School of Polymer Science and Engineering, Qingdao University of Science and Technology, China. From 2007 to 2011, he worked as an Alexander-von-Humboldt Fellow at the Friedrich-Schiller-University of Jena (Germany), and from 2012 to 2019 he worked as a senior researcher and group leader at the University of Bremen (Germany). Following this, from 2019.10 to 2025.05, he worked as a full Professor at Qingdao University, China. His research interests include polymer nanomaterials, supramolecular nanomaterials, hydrogels, two-dimensional materials, biomaterials, supramolecular self-assembly, and biomedicine. He has published 260+ papers in peer-reviewed journals, and the published papers have been cited more than 16000 times with an H-index of 72. He has also edited three books on protein/peptide nanofibers, peptide nanotechnology, and cellulose hydrogels (Springer and Elsevier) and has contributed nine chapters to other books.

polymers

MDPI

Editorial

Recent Trends in Polymer Membranes: Fabrication Technique, Characterization, Functionalization, and Applications in Environmental Science (Part II)

Gang Wei [1,*] and Yan Wang [2,*]

[1] School of Polymer Science and Engineering, Qingdao University of Science and Technology, Qingdao 266042, China

[2] School of Chemistry and Chemical Engineering, Qingdao University, Qingdao 266071, China

* Correspondence: wei-lab@qust.edu.cn (G.W.); yanwang@qdu.edu.cn (Y.W.); Tel.: +86-15066242101 (G.W.)

1. Introduction

Polymeric membranes have been widely applied in various fields such as adsorption, separation, energy conversion, chemical production, antibacterial, antifouling, and biomedicine due to their excellent porous and channel structure, as well as tailored separation performance, ease of operation, and low energy consumption. In terms of the polymeric membrane preparation technology, various methods such as phase inversion [1], interfacial polymerization [2], electrospinning [3], 3D printing [4], vacuum-induced filtration [5], template-assisted methods [6], and layer-by-layer assembly [7] have been widely applied. These techniques enable precise control over the membrane's pore structure, thickness, surface roughness, and mechanical properties.

The functionalization of polymeric membranes is crucial to enhance their performance and extend their applications. Through molecular conjugation, chemical linking, physical grafting, plasma treatment, and nanomaterial compositing, the main properties and functions of polymeric membranes, including hydrophilicity, antibacterial activity, antifouling capability, catalytic activity, and selective transport properties have been improved to a great degree [8,9]. For instance, the combination of metal–organic frameworks [10], two-dimensional materials (graphene, MXene, and others) [11], self-assembled nanofibers [12], multifunctional nanoparticles [13], or antibacterial agents [14] into the polymeric membrane structure can not only enhance separation efficiency but also impart antifouling, antibacterial, or photocatalytic functions, enabling multifunctional integration of polymers and other components together.

The design of stimuli-responsive polymeric membranes also allows the membranes to dynamically adjust their pore size and flux in response to environmental stimuli, such as the tiny changes in temperature, pH, enzymatic level, light, electric fields, and others, providing new strategies for water treatment, drug release, and environmental monitoring [15]. In the field of environmental science, polymeric membranes have been extensively used for seawater desalination, wastewater purification, surface antifouling and antibacterial purposes, air cleaning, and others [16]. Nanofiltration, ultrafiltration, and reverse osmosis polymeric membranes have been widely applied to remove suspended particles, heavy metal ions, organic pollutants, and microorganisms from water, achieving efficient separation and water purification [17,18]. By selecting different polymer substrates and tailoring the pore structure and surface properties of membranes, polymeric membranes can be specifically designed to meet particular requirements in environmental science.

https://doi.org/10.3390/polym18070895

This collection is an extension of our previous Special Issue [19] entitled *"Recent Trends in Polymer Membranes: Fabrication Techniques, Characterization, Functionalization, and Applications in Environmental Science (Part I)."* In this editorial, we summarize the studies on the fabrication, functional regulation, and applications of polymeric membranes that were published in this special issue, and only a part of the results is excerpted from them.

2. Overview of the Published Articles

Twelve articles, comprising 1 review and 11 research articles, were included in the second part of this Special Issue.

In a review article (Contribution 1), Sahu et al. summarized recent advances in the fabrication of polymeric membranes for liquid separation, systematically discussing material innovations, fabrication strategies, and industrial applications. First, the mechanisms of membrane separation, including molecular sieving, adsorption, and electrostatic interactions, were introduced, and enable selective separation with high permeability. The authors then discussed commonly used polymers, such as cellulose, polysulfone (PSF), and polyethersulfone, for the fabrication of industrial polymeric membranes. In addition, several emerging membrane materials were highlighted, including mixed matrix membranes (MMMs) incorporating 2D materials or metal oxides, thin-film composite (TFC) membranes, and biopolymer-based membranes. These advanced membranes exhibit enhanced separation performance due to surface functionalization strategies such as surface grafting, nanomaterial hybridization, and layer-by-layer assembly. Functionalized polymeric membranes have been widely applied in industrial fields, including desalination, dye removal, heavy metal ion adsorption, ion separation and recovery, as well as applications in the chemical and food industries. Finally, the authors analyzed the challenges and limitations of polymeric membranes in industrial applications, emphasizing issues such as stability, fouling, degradation, and selectivity. This article provides a comprehensive overview of recent advances in polymeric membranes for liquid separation and is highly valuable for researchers in materials science, chemical engineering, environmental engineering, and nanotechnology.

Polymeric membranes are commonly prepared using methods such as phase inversion, template-assisted synthesis, interfacial polymerization, and controlled swelling of monolithic films. These techniques enable precise regulation of pore structure, membrane thickness, and surface properties, thereby meeting diverse separation requirements. By rationally designing the polymer composition and preparation process, the separation selectivity, permeation flux, and antifouling performance of polymeric membranes can be significantly improved. The study by Balasankar and co-workers (Contribution 2) introduced a facile fabrication method of superhydrophobic poly(carbonate) (PC) and poly(methyl methacrylate) (PMMA) membranes through anodized aluminum oxide (AAO) imprint molds as templates. The hierarchical structure of the AAO template was successfully transferred onto the polymer membrane surface using a mold imprinting technique, resulting in micropatterned polymer membranes with high surface roughness and excellent hydrophobicity. The study demonstrates that this template-assisted imprinting method is simple, cost-effective, structurally controllable, and scalable, highlighting its promising potential for applications in self-cleaning, antifouling, and anticorrosion surfaces. Through interfacial polymerization, Burts et al. (Contribution 3) prepared a novel TFC nanofiltration membrane with PSF, polyaniline (PANI), and polyamide (PA) as the support, interlayer, and selective layer, respectively. The results showed that the thickness, uniformity, and polymerization conditions of the PANI interlayer strongly influenced the pore structure, surface roughness, water flux, and salt rejection of the resulting TFC hybrid membranes. Importantly, the study demonstrated that membrane performance can be

https://doi.org/10.3390/polym18070895

enhanced by tailoring the PANI interlayer, achieving both high permeability and improved selectivity. This interlayer-engineering approach combined with interfacial polymerization provides valuable guidelines for the design of functional polymeric membranes for water purification and industrial separation. In order to produce ultra-high-molecular-weight polyethylene (UHMWPE) ultrafiltration membrane, Basko and co-workers (Contribution 4) demonstrated a novel fabrication technique, called "controlled swelling of monolithic films". Under hot pressing, UHMWPE powder was utilized to create a monolithic film, which was then proceeded under controlled swelling to form UHMWPE film. After extraction and drying, the UHMWPE film was dried to produce a porous ultrafiltration membrane. This study indicated that the molecular weight of the polymer affected the structure and properties of the formed UHMWPE membranes. For instance, it was hard to form membranes with the low-molecular-weight polymer, but a high molecular weight promoted the formation of dense, porous UHMWPE membranes with excellent mechanical properties, which could be further used for high-performance filtration and separation.

In addition, advanced fabrication techniques such as electrospinning and electrospraying can be employed to produce polymer membranes with controlled thickness and pore size. Electrospinning is typically used to create nanofibrous membranes with high porosity and flexibility, whereas electrospraying is well suited for forming particle-assembled membranes with tunable microstructures and facile surface functionalization. For instance, the study by Zhang et al. (Contribution 5) focused on the production of nanofibrous membranes via gas-assisted coaxial electrospinning (GACES). Compared to traditional coaxial electrospinning, the GACES could significantly adjust the flow field by introducing assisted gas, which could increase the fiber deposition area by 3 times, improve thickness by 2.3 times, and reduce the average fiber diameter by nearly 37%. This advanced electrospinning technique provides a theoretical and experimental basis for large-scale, uniform preparation of coaxial polymeric nanofiber membranes. In a recent study (Contribution 6), electrospraying has been utilized for the synthesis of jicama (*Pachyrhizus erosus*) starch particles. The effects of the hydrolysis degree of jicama starch on the formation and structure of electrosprayed particles were investigated. The results indicated that acidic hydrolysis reduced the gel viscosity and surface tension while increasing the conductivity, thereby altering the particle formation behavior during the electrospraying process. Although this study is not directly related to polymer membrane fabrication, the electrospraying technique and the insights gained from this control study provide valuable guidance for the design of biomass-based polymer membranes and their subsequent functional regulation, showing great potential for applications in environmental and food science.

The functional regulation of polymeric membranes is essential for optimizing their properties and performances in environmentally related applications. Surface modification, plasma treating, and incorporation of nanomaterials further enable polymeric membranes to exhibit enhanced adsorption, antifouling, catalytic, or stimuli-responsive behaviors.

Polymer modification is an important strategy for tailoring the structure and performance of polymeric membranes. Through chemical functionalization, the introduction of additional active groups can endow polymeric membranes with higher selectivity and flux, as well as improved mechanical stability. Papchenko and co-workers evaluated the CO_2 capture performance of two commercial anion exchange membranes (AEMs), Fumasep® FAA-3-50 (Fuel Cell Store, Bryan, TX, USA) and Sustainion® X37-50 Grade RT (Dioxide Materials, Boca Raton, FL, USA) (Contribution 7). In these membranes, the polymer substrates were covalently functionalized with nitrogen-based cationic groups, where quaternary ammonium and imidazolium moieties were used to link with the polymer backbone, respectively. Through the CO_2/CH_4 adsorption tests, they demonstrated that both membranes exhibited good adsorption capacity towards CO_2, and Sustainion® had higher

https://doi.org/10.3390/polym18070895

adsorption ability than Fumasep®. However, Fumasep® showed higher CO_2 diffusivity, revealing promising uses for moisture-swing air capture, biogas purification, and other innovative processes. In another case (Contribution 8), the effect of the polymer backbone's side substituent on the performance of fabricated polymeric membranes was investigated. Comb-like polysiloxane was modified with side substituent (1-butanol, 1-propanol, and ethanol) and then fabricated to functional polysiloxane membranes. The obtained results indicated that the introduction of side groups could change the membrane's hydrophobicity, solubility, selectivity, and mass transfer ability. Therefore, it is effective to enhance the separation efficiency of membranes in organic removal and water purification, by adjusting the side chain and molecular structure of polysiloxane.

Additionally, plasma treating of polymeric membrane surface can enhance their surface properties and biological activity. For instance, Ge et al. (Contribution 9) prepared a polyvinyl alcohol/polylactic acid (PVA/PLA) membrane by electrospinning, which was then treated with plasma to introduce more polar groups onto the membrane surface and further improve the hydrophilicity and surface energy. It was found that the plasma-treated PVA/PLA hybrid membrane exhibited increased liquid-enrichment capacity of 350% and shortened the coagulation time to 258 s. Meanwhile, the treated membrane presented 79% higher hemostatic ability than the untreated membrane.

Finally, the incorporation of functional nanoparticles, nanofibers, porous materials, and 2D materials can effectively modulate the pore structure, surface properties, and active sites of polymeric membranes, thereby expanding their potential applications across various fields. Cai and colleagues (Contribution 10) demonstrated that hybridizing Nafion membranes with sulfonated porous aromatic frameworks (PAFs) significantly improves their high-temperature proton conductivity. The sulfonated PAFs serve dual functions: they adsorb and retain water within the membrane and provide additional SO_3H proton-conducting sites, which together stabilize proton transport channels under elevated temperatures. As a result, the PAF/Nafion composite membranes maintain superior proton conduction compared to pure Nafion at high temperatures. This study highlights the potential of sulfonated organic frameworks to enhance the performance of Nafion membranes for applications in energy conversion and environmental technologies. Guo and colleagues (Contribution 11) showed that functionalizing hydrophobic resin membranes with isothiazolinone (IS) and LDH–sodium pyrithione (SPT) composites effectively inhibits pipeline algal fouling. Chemical modification enhanced the hydrophobicity of the membranes, reducing initial algal attachment. Furthermore, IS and LDH-SPT acted as active anti-algal agents, suppressing algal growth and disrupting cell structures. Mechanistic analysis indicated a synergistic effect: the hydrophobic surface prevents adhesion while the chemical agents kill algae, together significantly lowering algae deposition and the risk of fouling in pipelines. In another study (Contribution 12), Liu et al. reported a strategy for preparing CuS nanoparticles through β-lactoglobulin (BLG) nanofiber templating and integrating them into MXene membranes to achieve photothermal antibacterial functionality. Their study showed that BLG nanofibers not only serve as templates to promote the uniform distribution of CuS nanoparticles but also enhance the structural stability of the MXene membrane. The composite membrane could rapidly heat under near-infrared light irradiation, killing bacteria through the photothermal effect while maintaining the membrane's integrity and water transport performance. The study provides new ideas for the biomimetic synthesis of MXene-based functional membranes in antibacterial, water treatment, and biomedical fields.

https://doi.org/10.3390/polym18070895

3. Conclusions and Outlooks

In summary, the studies collected in this Special Issue demonstrate that the rational design of polymer composition, microstructure, and surface functionality plays a crucial role in advancing the performance of polymeric membranes. The integration of emerging materials, scalable fabrication techniques, and multifunctional modification strategies is expected to further expand the capabilities of polymeric membranes in water treatment, energy conversion, gas separation, environmental remediation, and biomedical applications.

With increasing global water scarcity, environmental pollution, and energy demand, the importance of efficient separation technologies has become increasingly prominent, providing unprecedented opportunities for the development of polymeric membranes. Here, we highlight several potential directions in this promising research field. First, the design of polymer molecules at the molecular level is suggested. The regulation of polymer chain and functional side groups are useful for tailoring the properties and microstructure of polymeric membranes. The membrane permeability, selectivity, bioactivity, and chemical stability can be precisely tuned through the introduction of ionic, hydrophilic, hydrophobic, biological, and rigid aromatic groups. Second, it is necessary to develop biomimetic polymeric membranes inspired by biological systems, which have the potential to achieve exceptional separation performance while maintaining high selectivity. Advanced fabrication techniques, such as 3D printing and biomimetic synthesis strategies, could be highly effective in achieving this goal. Third, smart polymeric membranes with stimuli-responsive properties can be designed and synthesized for environmental science and other applications. This goal can be achieved by incorporating responsive polymers and nanomaterials into the membrane systems, to create permeability and selectivity in response to external stimuli such as temperature, pH, light, enzyme, or electric fields. Fourth, multifunctional polymeric membranes with integrated functions such as separation, catalysis, antibacterial, and antifouling abilities are recommended. A potential solution for addressing this challenge is the use of multifunctional nanoparticles and nanozymes. Such multifunctional polymeric membrane systems can significantly enhance process efficiency and broaden their applications in environmental science, sensing, and biomedicine. Finally, artificial intelligence (AI) and data-driven approaches are increasingly being applied to the design, synthesis, and performance analysis of polymeric membranes, which can accelerate the discovery of novel membrane materials and the optimization of membrane structures. Additionally, machine learning methods are helpful for elucidating the relationships between membrane structure, function, and performance, thereby significantly accelerating the screening and development of advanced polymeric membranes.

Conflicts of Interest: The authors declare no conflicts of interest.

List of Contributions:

1. Sahu, L.R.; Yadav, D.; Borah, D.; Gogoi, A.; Goswami, S.; Hazarika, G.; Karki, S.; Gohain, M.B.; Sawake, S.V.; Jadhav, S.V.; Chatterjee, S.; Ingole, P.G. Polymeric membranes for liquid separation: Innovations in materials, fabrication, and industrial applications. *Polymers* **2024**, *16*, 3240.
2. Balasankar, A.; Venkatesan, R.; Jeong, D.Y.; Oh, T.H.; Kim, S.C.; Vetcher, A.A.; Ramasundaram, S. Facile fabrication of hierarchical structured anodic aluminum oxide molds for large-scale production of superhydrophobic polymer films. *Polymers* **2024**, *16*, 2344.
3. Burts, K.S.; Plisko, T.V.; Penkova, A.V.; Yuan, B.B.; Ermakov, S.S.; Bildyukevich, A.V. Correlation between conditions of polyaniline interlayer formation and the structure and performance of thin-film composite membranes for nanofiltration prepared via interfacial polymerization. *Polymers* **2025**, *17*, 1199.
4. Basko, A.V.; Pochivalov, K.V.; Lebedeva, T.N.; Yurov, M.Y.; Zabolotnov, A.S.; Gostev, S.S.; Yushkin, A.A.; Volkov, A.V.; Bronnikov, S.V. Effect of polymer molecular weight on the structure

https://doi.org/10.3390/polym18070895

and properties of ultra-high-molecular-weight polyethylene membranes prepared via controlled swelling. *Polymers* **2025**, *17*, 2044.

5. Zhang, R.G.; Chen, X.; Wang, H.; Sun, J.F.; Huang, S.Z.; Zhang, X.Z.; Long, J.C. Study on deposition of coaxial electrospinning fibers by coaxial auxiliary flow field. *Polymers* **2025**, *17*, 396.
6. Serrano-Villa, F.S.; Morales-Sánchez, E.; Téllez-Morales, J.A.; Cuellar-Sánchez, V.; Farrera-Rebollo, R.R.; Calderón-Domínguez, G. Synthesis of jicama (pachyrhizus erosus) starch particles by electrospraying: Effect of the hydrolysis degree. *Polymers* **2025**, *17*, 2069.
7. Papchenko, K.; Kentish, S.; De Angelis, M.G. An assessment of anion exchange membranes for co2 capture processes: A focus on fumasep and sustainion. *Polymers* **2025**, *17*, 1581.
8. Grushevenko, E.; Chechenov, I.; Rokhmanka, T.; Anokhina, T.; Bazhenov, S.; Borisov, I. Effect of side substituent on comb-like polysiloxane membrane pervaporation properties during recovery of alcohols c2-c4 from water. *Polymers* **2024**, *16*, 3530.
9. Ge, X.T.; Zhang, L.; Wei, X.H.; Long, X.; Han, Y.C. Plasma surface treatment and application of polyvinyl alcohol/polylactic acid electrospun fibrous hemostatic membrane. *Polymers* **2024**, *16*, 1635.
10. Cai, K.; Yu, J.Z.; Tan, W.J.; Gao, C.; Zhao, Z.L.; Yuan, S.X.; Cheng, J.H.; Yang, Y.J.; Yuan, Y. The incorporation of sulfonated paf enhances the proton conductivity of nafion membranes at high temperatures. *Polymers* **2024**, *16*, 2208.
11. Zhou, R.D.; Zhao, Q.; Liu, A.; Xu, H.; Xu, Y.; Wang, W.B.; Han, J.R.; Liu, B.; Wang, M.; Di, Z.; Miao, L.; Liu, Y.; Guo, L. Inhibition and mechanisms of isothiazolinone and layered double hydroxide–sodium pyrithione with modified hydrophobic resin membranes against pipeline moss fouling. *Polymers* **2026**, *18*, 611.
12. Liu, Z.; Du, C.X.; Zhou, X.; Wei, G. Regulation of mxene membranes with β-lactoglobulin nanofiber-templated cus nanoparticles for photothermal antibacterial effect. *Polymers* **2025**, *17*, 1960.

References

1. Diwan, T.; Abudi, Z.N.; Al-Furaiji, M.H.; Nijmeijer, A. A competitive study using electrospinning and phase inversion to prepare polymeric membranes for oil removal. *Membranes* **2023**, *13*, 474. [CrossRef] [PubMed]
2. Lee, T.H.; Balcik, M.; Wu, W.N.; Pinnau, I.; Smith, Z.P. Dual-phase microporous polymer nanofilms by interfacial polymerization for ultrafast molecular separation. *Sci. Adv.* **2024**, *10*, eadp6666. [CrossRef] [PubMed]
3. Chinnappan, B.A.; Krishnaswamy, M.; Xu, H.Z.; Hoque, M.E. Electrospinning of biomedical nanofibers/nanomembranes: Effects of process parameters. *Polymers* **2022**, *14*, 3719. [CrossRef] [PubMed]
4. Thiam, B.G.; El Magri, A.; Vanaei, H.R.; Vaudreuil, S. 3d printed and conventional membranes-a review. *Polymers* **2022**, *14*, 1023. [CrossRef] [PubMed]
5. Chen, Y.; Yang, G.Z.; Liu, B.; Kong, H.; Xiong, Z.; Guo, L.; Wei, G. Biomineralization of zro2 nanoparticles on graphene oxide-supported peptide/cellulose binary nanofibrous membranes for high-performance removal of fluoride ions. *Chem. Eng. J.* **2022**, *430*, 132721. [CrossRef]
6. Ma, W.J.; Cao, W.X.; Lu, T.; Xiong, R.H.; Huang, C.B. Multifunctional nanofibrous membrane fabrication by a sacrifice template strategy for efficient emulsion oily wastewater separation and water purification. *J. Environ. Chem. Eng.* **2022**, *10*, 108908. [CrossRef]
7. Wang, J.X.; Qu, Y.; Liang, T.; Liu, Z.; Sun, P.D.; Li, Z.Y.; Wang, X.H.; Hu, Y.N.; Wang, L.; Wang, N. Fabrication of a graphene oxide-embedded separation bilayer composite nanofiltration membrane using a combination of layer-by-layer self-assembly and interfacial polymerization. *Environ. Sci.-Wat. Res.* **2022**, *8*, 1923–1937. [CrossRef]
8. Wang, X.Y.; Wang, R.R.; Xu, Y.Y.; Wei, G. Recent advances in biohybrid membranes for water treatment: Preparation strategies, nano-hybridization, bioinspired functionalization, applications, and sustainability analysis. *J. Mater. Chem. A* **2025**, *13*, 26967–27000. [CrossRef]
9. Nouri, M.; Poorkhalil, A.; Farrokhzad, H. Modification methods categorization of polymeric membranes for pervaporation: A review. *Polym. Rev.* **2025**, *65*, 425–452. [CrossRef]
10. Gowriboy, N.; Kalaivizhi, R.; Kaleekkal, N.J.; Ganesh, M.R.; Aswathy, K.A. Fabrication and characterization of polymer nanocomposites membrane (cu-mof@ca/pes) for water treatment. *J. Environ. Chem. Eng.* **2022**, *10*, 108668. [CrossRef]

https://doi.org/10.3390/polym18070895

11. Khosla, A.; Sonu; Awan, H.T.A.; Singh, K.; Gaurav; Walvekar, R.; Zhao, Z.H.; Kaushik, A.; Khalid, M.; Chaudhary, V. Emergence of mxene and mxene-polymer hybrid membranes as future- environmental remediation strategies. *Adv. Sci.* **2022**, *9*, 2203527. [CrossRef] [PubMed]
12. He, P.; Yang, M.H.; Lei, Y.; Guo, L.; Wang, Y.; Wei, G. Solvent-evaporation-induced synthesis of graphene oxide/peptide nanofiber (go/pnf) hybrid membranes doped with silver nanoparticles for antibacterial application. *Polymers* **2023**, *15*, 1321. [CrossRef] [PubMed]
13. Wang, L.X.; Zhao, L.Z.; Si, D.D.; Li, Z.X.; An, H.Q.; Ye, H.; Xin, Q.P.; Li, H.; Zhang, Y.Z. Polymeric membrane with nanohybrids of cu nanocomposites and metalloporphyrin-based nanosheets for enzyme-like catalytic degradation of congo red. *Sep. Purif. Technol.* **2024**, *331*, 125571. [CrossRef]
14. Li, X.L.; He, X.H.; Ling, Y.; Bai, Z.X.; Liu, C.C.; Liu, X.B.; Jia, K. In-situ growth of silver nanoparticles on sulfonated polyarylene ether nitrile nanofibers as super-wetting antibacterial oil/water separation membranes. *J. Membr. Sci.* **2023**, *675*, 121539. [CrossRef]
15. Huang, T.F.; Su, Z.X.; Hou, K.; Zeng, J.X.; Zhou, H.; Zhang, L.; Nunes, S.P. Advanced stimuli-responsive membranes for smart separation. *Chem. Soc. Rev.* **2023**, *52*, 4173–4207. [CrossRef] [PubMed]
16. Karki, S.; Hazarika, G.; Yadav, D.; Ingole, P.G. Polymeric membranes for industrial applications: Recent progress, challenges and perspectives. *Desalination* **2024**, *573*, 117200. [CrossRef]
17. Jin, Y.H.; Zhang, A.W.; Dong, G.Y.; Hou, J.W.; Zhu, J.Y.; Zhang, Y.T. Ultrathin membranes comprising polymers of intrinsic microporosity oligomers for high-performance organic solvent nanofiltration. *J. Membr. Sci.* **2025**, *714*, 123436. [CrossRef]
18. Yu, T.; Zhou, J.; Liu, F.; Xu, B.M.; Pan, Y. Recent progress of adsorptive ultrafiltration membranes in water treatment-a mini review. *Membranes* **2022**, *12*, 519. [CrossRef] [PubMed]
19. Wang, Y.; Wei, G. Recent trends in polymer membranes: Fabrication technique, characterization, functionalization, and applications in environmental science (part i). *Polymers* **2024**, *16*, 2889. [CrossRef] [PubMed]

https://doi.org/10.3390/polym18070895

polymers

MDPI

Review

Polymeric Membranes for Liquid Separation: Innovations in Materials, Fabrication, and Industrial Applications

Lalit Ranjan Sahu [1,2,†], Diksha Yadav [1,2,†], Debasish Borah [1,†], Anuranjit Gogoi [1,†], Subrata Goswami [1,†], Gauri Hazarika [1,2,†], Sachin Karki [1,2,†], Moucham Borpatra Gohain [1,2,†], Saurabh V. Sawake [1,†], Sumit V. Jadhav [3], Soumya Chatterjee [4] and Pravin G. Ingole [1,2,*]

1 Chemical Engineering Group, Engineering Sciences and Technology Division, CSIR-NorthEast Institute of Science and Technology, Jorhat 785006, Assam, India; lalitranjan00@gmail.com (L.R.S.); diksyadav96233@gmail.com (D.Y.); debasishbora051@gmail.com (D.B.); anuranjit@outlook.com (A.G.); subratagoswami098@gmail.com (S.G.); gauripriyahazarika98@gmail.com (G.H.); sachinkarkissk@gmail.com (S.K.); moucham1234@gmail.com (M.B.G.); saurabhsawake27013@gmail.com (S.V.S.)

2 Academy of Scientific and Innovative Research (AcSIR), Ghaziabad 201002, Uttar Pradesh, India

3 Government Polytechnic, Hol Tarfe Haveli, Nandurbar 425412, Maharashtra, India; svjadhav84@gmail.com

4 Defence Research Laboratory, Defence Research and Development Organisation (DRDO), Tezpur 784001, Assam, India; drlsoumya@gmail.com

* Correspondence: pingole@neist.res.in

† These authors contributed equally to this work.

Abstract: Polymeric membranes have emerged as a versatile and efficient liquid separation technology, addressing the growing demand for sustainable, high-performance separation processes in various industrial sectors. This review offers an in-depth analysis of recent developments in polymeric membrane technology, focusing on materials' advancements, innovative fabrication methods, and strategies for improving performance. We discuss the underlying principles of membrane separation, selecting suitable polymers, and integrating novel materials, such as mixed-matrix and composite membranes, to enhance selectivity, permeability, and antifouling properties. The article also highlights the challenges and limitations associated with polymeric membranes, including stability, fouling, and scalability, and explores potential solutions to overcome these obstacles. This review aims to guide the development of next-generation polymeric membranes for efficient and sustainable liquid separation by offering a detailed analysis of current research and future directions.

Keywords: polymeric membranes; liquid separation; material innovations; performance enhancement; scalability and commercialization; fouling mitigation

1. Introduction

Over recent decades, the membrane-based separation technology has increasingly captivated interest from researchers and industries globally, owing to its broad applications across numerous sectors. In recent years, commercial membranes have become widely available for water treatment [1–3]. Polymeric membranes have become a cornerstone in separation technologies due to their tunable properties, adaptability, and cost-effectiveness [4]. These membranes, primarily composed of synthetic polymers, are designed to selectively separate specific components from liquid mixtures, making them invaluable in various industrial processes [5,6]. In recent decades, advances in polymer chemistry and membrane fabrication techniques have markedly improved polymeric membranes' performance, durability, and application range. Their ability to be engineered for specific separation tasks has positioned them as essential tools in fields ranging from water treatment to food processing and chemical manufacturing [7–9]. Polymeric membranes are a sustainable solution for industrial applications due to their adaptive chemistry, structure, and performance attributes. Enhanced separation efficiency and the creation of advanced structures through innovative polymer chemistry have resulted in membranes with optimized pore sizes and free volumes [10]. With the rapid growth of modern industry, agriculture, and the global population, it has become evident that many countries and regions are facing water scarcity and severe water pollution [11–13]. With the increasing global demand for clean water and more sustainable industrial operations, the urgency for developing cutting-edge separation technologies has reached an all-time high.

Today, membrane engineering is a dynamic scientific field focused on developing, applying, and optimizing various membrane technologies. This field enables the precise determination of the most suitable membrane operations for specific industrial processes [14–17]. The widespread adoption of membrane technologies across multiple industrial sectors has further reinforced their selection as the preferred solution [18]. Researchers have investigated various composite and nanocomposite membranes for water purification, demonstrating significant potential and versatility. Nour et al. studied mixed-matrix membranes (MMMs) incorporated by the metal-organic framework (MOF) for efficient nickel and cobalt removal. They highlighted the excellent nickel and cobalt capture efficiency of the resulting MOF@PES MMMs, even in the presence of common ions, signifying future research avenues for enhancing the functionality of MOF-based MMMs [19]. Integrating MOFs within polymer matrices is an advantageous strategy for improving their performance. MOFs are recognized as an efficient and promising technology for heavy metals' removal from water due to several notable advantages, including microporosity, substantial water and structural stability, and the capacity for flexibility and adaptability, which are crucial for capturing and accommodating the targeted guest compounds [19–22].

In water purification, developing highly permeable antifouling membranes is a critical challenge for researchers, and significant progress has been made in this area [23]. Khoo et al. developed antifouling thin-film composite and nanocomposite membranes specifically designed with titania nanotubes (TNTs) [24]. More significantly, the produced TFN membrane demonstrated exceptional fouling resistance by obtaining a flux recovery rate of 85.77%, as opposed to the control membrane's 57.94%. Incorporating inorganic nanofillers has also played a crucial role in enhancing the antifouling properties, making these membranes essential for providing purified water in a world where clean water access is increasingly vital [25–27].

This review provides an extensive overview of the most current developments in polymeric membrane technology for liquid separation applications. It explores various types of polymeric membranes, their fabrication methods, and the materials used in their construction, focusing on innovations that improve membrane performance. The review addresses the key challenges polymeric membranes face, such as stability, fouling, and scalability, and discusses potential strategies to overcome these obstacles. The scope of this review encompasses both established and emerging applications of polymeric membranes, offering insights into their commercialization potential and future research directions. This review aims to inform and inspire continued innovation in polymeric membranes for liquid separation by synthesizing current knowledge and highlighting cutting-edge developments.

2. Methodology

This review aims to comprehensively analyze polymeric membranes, highlighting advancements that address the critical need for high-performance, sustainable liquid separation in various industrial sectors, such as desalination and wastewater treatment. This review article serves as a resource for understanding the recent innovations in materials, fabrication techniques, and functional enhancements to maximize polymeric membranes' separation efficiency and durability. This review encompasses the evolution of polymeric membrane materials and processes, with a detailed examination of advancements, such as phase inversion, electrospinning, interfacial polymerization, layer-by-layer assembly, and chemical vapor deposition (CVD). Emphasis is given to incorporating novel materials, such as nanomaterials and aquaporin, to improve membrane properties, such as the polymeric membranes' selectivity, permeability, and antifouling abilities. In addition, it explores strategies to enhance membrane performance, such as chemical and plasma treatment techniques, and investigates the roles of mixed-matrix, composite, and nanocomposite membranes. The primary function of this review is to guide readers in understanding the scientific principles of membrane separation, such as size exclusion, Donnan ion exclusion, hydrophobic–hydrophilic interactions, and electrostatic interactions. The review offers insights into advanced membrane functionality and performance in real-world applications by examining the intersection of polymer science with nanotechnology and surface engineering.

Moreover, it provides a critical perspective on the limitations of polymeric membranes, including chemical and thermal stability challenges and fouling. It presents potential approaches to overcome these issues through emerging innovations in membrane separation technologies. Ultimately, the intent is to inspire and inform the development of next-generation polymeric membranes that meet the rising demands for sustainable water treatment and industrial applications. This review supports the advancement of polymeric membranes by providing an extensive knowledge base that inspires new research pathways and practical applications for mitigating water pollution and enhancing industrial sustainability using polymeric membranes.

3. Fundamentals of Polymeric Membrane Technology

3.1. History of Membrane Technology

The history of membrane-based separation technology is rooted in extensive laboratory research that preceded its widespread industrial application. Initially, membranes were primarily a topic of scientific inquiry, with limited practical use. However, this changed dramatically in the 1950s, when the focus shifted to practical applications, leading to the rapid development of a significant membrane-based industry [28,29]. In the early 1950s, electrodialysis (ED), microfiltration (MF), and ion exchange membranes were employed and used on a lab scale in Europe to test drinking water safety [30]. However, polymeric membranes were first utilized for liquid material separation in the 1950s. Cellulose acetate, a cellulose derivative, became the primary material for developing polymeric membranes. In 1959, Sidney Loeb and Srinivasa Sourirajan developed the first practical polymeric membrane for seawater desalination at the University of California. They created an asymmetric membrane with a dense layer of cellulose acetate cast on a porous sub-layer for support. This membrane successfully turned saline water into potable water, leading to the construction of large-scale desalination plants to meet potable water demand [31,32]. With the addition of high-temperature thermoplastics, polymeric membranes became flexible, cost-effective, and applicable for various separations, including oil–water separation, protein separation, heavy metal ion separation, gas separation, dye separation, dialysis, and more. This versatility has led to a high demand for polymeric membranes in industries, such as dairy, fuel cells, packaging, petroleum, pharmaceuticals, sensors, textiles, medical, and energy sectors [33,34]. As a result, researchers have shown a deep interest in developing new polymeric membranes using various polymer materials for different purposes. Figure 1 illustrates the significant advancement of membrane technology.

With the advancements in polymer chemistry, many synthetic polymers were produced. Eventually, they became available to fabricate novel membranes with specific transport properties, flexibility, and exceptional thermal and mechanical stability. Based on the thermodynamics of irreversible processes, the transport properties of a membrane were illustrated by a comprehensive theory [31,35]. After the use of cellulose acetate polymers for membrane fabrication in the initial stage, soon, synthetic polymers, such as polysulfone (PSf), polyethersulfone (PES), polyamides (PA), polyacrylonitrile (PAN), polyethylene (PE), polyvinyl chloride (PVC), polyacrylonitrile (PAN), etc., were used as the primary material for the fabrication of synthetic membranes [31,36]. These synthetic polymers provide superior chemical and thermal stability and enhanced mechanical strength compared to cellulose acetate. Despite these advantages, cellulose acetate remained the dominant material

for reverse osmosis (RO) membranes until the advent of interfacial-polymerized composite membranes [31,37].

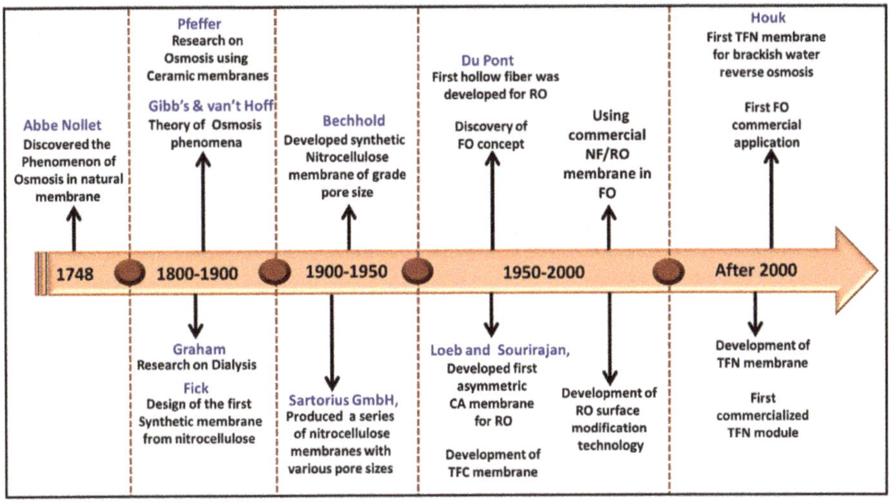

Figure 1. Significant advancements of membrane technology for liquid separation.

3.2. Membrane Separation Process

Membrane separation primarily relies on three key principles: molecular sieving, adsorption, and electrostatic interactions. The adsorption mechanism in membrane separation primarily relies on the hydrophobic interactions between the solute (analyte) and the membrane. This kind of membrane results in a higher rejection rate due to the smaller pore size of the membranes [38,39]. This indicates that the separation primarily depends on the size of the pore and solute. This advancement spurred the development of a range of membrane separation techniques, including ultrafiltration (UF), microfiltration (MF), nanofiltration (NF), forward osmosis (FO), reverse osmosis (RO), gas separation, dialysis, electrodialysis, pervaporation (PV), membrane contactors, and membrane reactors [38–40]. Figure 2 illustrates a schematic overview of various membrane separation processes categorized by their driving forces. Due to the low energy requirement compared to conventional thermal separation technology, membrane separation is often considered cost-effective and environmentally friendly [41].

Unlike many other separation methods governed by phase equilibrium relations, membrane separation techniques fundamentally rely on the relative rates of mass transfer. Transport takes place through different mechanisms, such as solution diffusion for liquid separation. The rate at which each component permeates through a membrane depends on its capacity to dissolve and diffuse within the membrane material, which dictates the membrane's selectivity. Transport through the membrane can be affected by factors such as pressure or temperature gradients, convection or diffusion driven by electric fields, or concentration differences [42].

3.3. Classification of Membrane Material and Structure

Membranes can be classified based on their structural morphology and the materials employed in their fabrication. Structurally, membranes fall into two main types: isotropic and anisotropic [43]. Isotropic membranes exhibit uniform properties throughout their structure and are composed of a single material type. They can be categorized as non-porous dense films, macro-porous, and electrically charged membranes. Macro-porous membranes feature pores ranging from 0.1 to 5 μm, enabling separation based on pore size. Non-porous dense-film membranes facilitate separation through diffusion driven by pressure and electrical gradients. Electrically charged membranes, also known as ion exchange membranes, separate ions according to their charge density, affecting transport rates via ion concentration and charge [43–45].

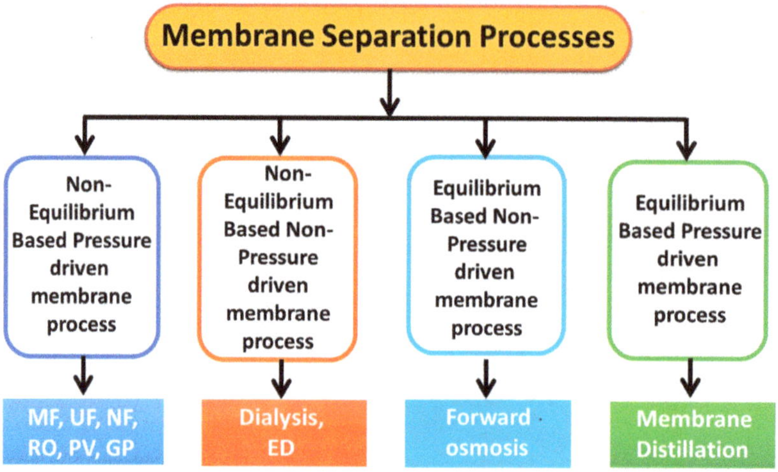

Figure 2. Different kinds of membrane separation processes.

In contrast, anisotropic membranes have varied structural and chemical properties and are classified into composite and phase separation membranes. Composite membranes are characterized by an asymmetric structure featuring a thin, highly cross-linked top selective layer (usually <1 μm) on a thicker microporous support. The properties of this top layer, including porosity, pore size, and thickness, control the transport rate of the feed mixture [43,46,47]. Phase separation membranes comprise multiple layers of the same material but differ in porosity, pore size, and thickness across layers.

Membranes are also categorized by the materials used in their fabrication, which can be inorganic or organic. Inorganic membranes are made from amorphous silica, carbon molecular sieves, graphene oxide, palladium alloys, zeolites, perovskites, etc. These membranes can be found in self-supporting structures or dense/porous multilayer supports. Polymers, such as cellulose acetate, polyimide, polysulfone, polyethersulfone, polycarbonate, polyvinylidene fluoride, and polydimethylsiloxane, are commonly used to fabricate organic membranes [43,48,49]. The membranes' classification by their structure and material is illustrated in Figure 3.

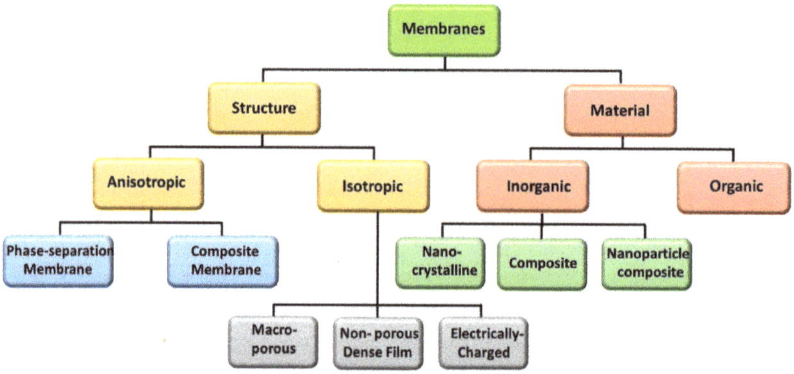

Figure 3. Classification of membranes based on structure and material.

4. Polymeric Membranes for Liquid Separation

4.1. Membrane Materials and Configurations

Membrane materials are either polymer or inorganic materials. The primary determinant in selecting materials for liquid separation membranes is the materials' ability to separate. Inorganic materials, such as metals or ceramics, are also used to create membranes [50]. Inorganic ceramic membranes have been developed from the oxides of titanium, zirconium, aluminum, and silicon [51]. Ceramic membranes are frequently employed for UF and MF because they are chemically resistant, thermally stable, and microporous [50]. Because of their stability, ceramic microfiltration/ultrafiltration membranes are especially well suited for applications in the biotechnology, food, and pharmaceutical industries, where the membranes must be repeatedly cleaned with strong solutions and steam sterilized. Unfortunately, the disadvantages of mechanical brittleness and high expenses have hampered widespread application. Metallic membranes possess highly refined porosity, with stainless steel being a frequently used material for fabrication [52].

Polymer-based membranes are widely used commercially due to their scalability and thermal durability. Synthetic organic polymers dominate the membrane market across UF, MF, NF, and RO filtration processes. RO membranes typically consist of a substrate material, such as polysulfone or cellulose acetate, and are then coated with a layer of aromatic polyamides [50]. Similarly, NF membranes are frequently constructed from blends of cellulose acetate or polyamide composites, although modified ultrafiltration (UF) membranes, such as sulfonated polysulfone, are also employed. These membranes can be assembled in different configurations, which refer to the arrangement and orientation of the membrane concerning the feed and permeate flow. The configurations are generally categorized into two primary geometries: planar (e.g., spiral wound and plate-and-frame) and cylindrical (e.g., hollow fiber and tubular) [53,54], as illustrated in Figure 4.

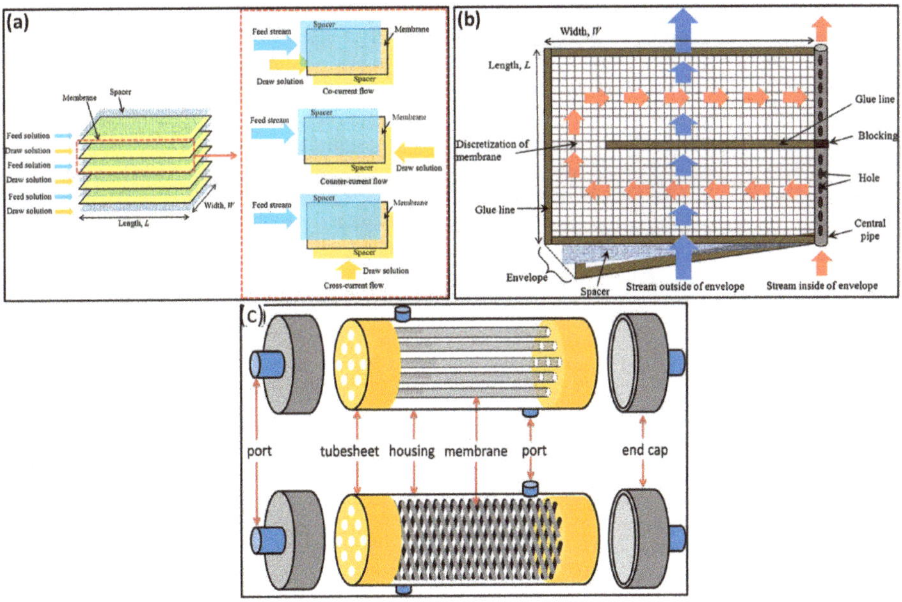

Figure 4. Schematic representation of planar, (**a**) plate-and-frame and (**b**) spiral wound [54], and cylindrical, (**c**) hollow-fiber membrane with parallel and crisscross arrangement [55], geometries.

Planar geometry: The plate-and-frame or flat-sheet module is a fundamental membrane configuration featuring a flat-sheet membrane, spacers, and two end plates. The feed mixture flows across the membrane surface, passing part into a permeate channel collected in a central manifold. Due to pressure limitations, these modules are predominantly utilized for MF and UF and exhibit a relatively low surface-area-to-volume ratio. There are two types: dead-end, where the feed flows perpendicularly, and crossflow, where it flows tangentially, reducing membrane fouling and extending membrane life, as shown in Figure 5 [51,53,56].

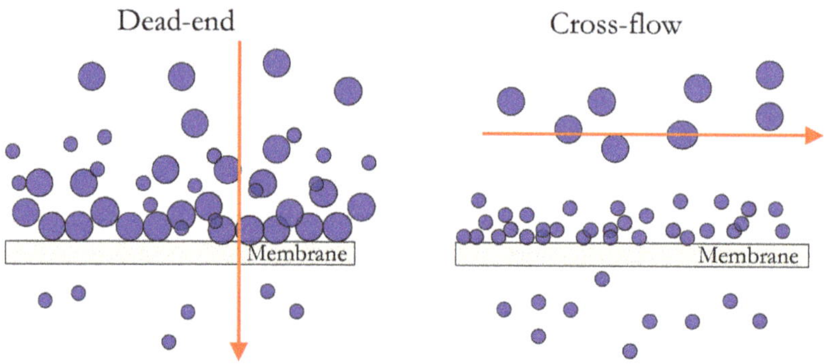

Figure 5. Representation of dead-end and crossflow filtration setups for membranes [55].

A prevalent design in industrial applications for NF and RO involves flat-sheet membranes arranged in a spiral configuration around a perforated central tube. This setup allows the feed to flow along one side of the membrane while the permeate is collected on the opposite side, moving inward in a spiral pattern. This design is favored for its efficiency, due to its elevated surface-area-to-volume ratio [53].

Cylindrical geometry: Tubular modules consist of a membrane encased within a cylindrical tube, through which the feed is directed and pumped. They are primarily used for ultrafiltration due to their better fluid dynamics [51,54].

Hollow-fiber modules utilized in seawater desalination typically comprise bundles of hollow fibers encased within a pressure vessel. These modules may include a shell-side feed configuration, where the feed flows around the exterior of the fibers and exits from their ends. In a bore-side feed arrangement, hollow-fiber modules can circulate feed through the fibers [50,56]. It is not often the case that hollow fibers used in membrane bioreactors and wastewater treatment are also utilized in pressure vessels. The permeate is extracted from one end of the fiber bundles suspended in the feed solution. Hollow-fiber modules offer key benefits due to their compact design, which allows them to design thousands of square meters of area in one cubic meter of module volume [51,56]. While flat-sheet and tubular modules are still in use, they are increasingly replaced by hollow-fiber and spiral-wound module designs.

Fabrication of hollow-fiber membranes (HFMs) is more complicated than flat membranes. The fabrication of an asymmetric hollow-fiber membrane, whose SEM images are shown in Figure 6, consists of the following steps: dope preparation, degas, metering, spinning, evaporation, coagulation, and solvent exchange. The bore fluid (mostly pure water) and polymer dope are subjected to a spinneret. The extruded polymeric dope solutions from the spinneret can pass through an air gap before the coagulation bath. They are further transferred over rollers to collect the hollow fibers [57].

Choosing proper spinning techniques results in better HFM fabrication. The most commonly used spinning techniques for manufacturing HFMs include dry-jet wet spinning, melt spinning, wet spinning, and electrospinning.

Wet spinning—Here, the air gap is absent; therefore, the extruded polymeric dope solutions from the spinneret are directly transferred to the coagulation bath containing non-solvent, where the dope solutions are precipitated and solidified. The just-spun fibers are collected over the roller, and before drying, a solvent exchange process is applied [58].

Dry spinning—Using air or inert gas in a warm air chamber, solid fibers are produced directly after extrusion from the spinneret by evaporating the solvent of the polymer dope solution, and further post-treatment is performed.

Melt spinning—It uses melted polymer in the viscous form as the polymer dope solutions. The resulting hollow fiber is obtained by forcing the molten polymer within the spinneret, where quick cooling and solidifying are carried out.

Electrospinning—It is a modern technique to fabricate ultrafine fibers. Unlike traditional techniques, it operates by utilizing electrostatic forces with the help of a high-voltage electric field to stretch a polymer solution or melt it into thin fibers, without the need for high temperatures and solvents [59].

Dry-jet wet spinning—It has the advantages of both wet and melt spinning for HFM production [60].

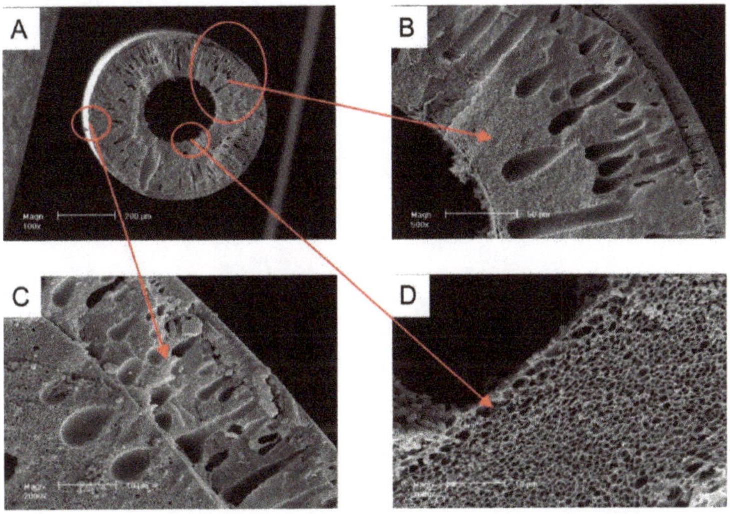

Figure 6. Cross-sectional SEM images of the dual-layer asymmetric hollow-fiber membrane [61]: (**A**) complete profile, (**B**) cross-sectional part, (**C**) outer layer, and (**D**) inner layer.

4.2. Mechanism of Liquid Separation

Various mechanisms are involved depending on the characteristics of polymeric membranes, including factors such as surface charge, pore size, chemical properties, and the liquid being separated. However, while size exclusion is the main separation mechanism, other mechanisms exist, such as electrostatic interaction, hydrophobic–hydrophilic adsorption, affinity-based separation, ion exchange mechanism, etc. [62].

Size exclusion (sieving/steric exclusion): In the size exclusion mechanism, the separation of molecules or particles is determined by their size relative to the membrane's pore size. Particles larger than the pores are retained on the membrane's surface, while smaller particles pass through. Membranes are characterized by their molecular weight cut-off (MWCO), which exhibits their cut-off value—the molecules of lower molecular weight than MWCO of the membrane will not be separated by size exclusion. Size exclusion works predominantly for microfiltration (MF) and ultrafiltration (UF) [63].

Certain ions in aqueous solutions form hydration shells—clusters of water molecules surrounding the ions. Although these ions carry charge, their hydrated form becomes too large to pass through the pores of a membrane. This separation process is known as the dielectric exclusion mechanism, a subtype of size exclusion [64,65].

Electrostatic interaction (Donnan ion exclusion): In this mechanism, charged particles are separated based on their charge when interacting with a charged membrane surface. Charged membranes either attract or repel ions based on their charge. They attract counter ions and repel co-ions. A positively charged membrane (cationic membrane) repels positively charged ions and attracts negatively charged ions, while a negatively charged membrane (anionic membrane) repels negatively charged ions and attracts positively charged ions. The Donnan ion effect, which is the electrostatic interaction, increases with the charge of co-ions and decreases with the charge of counter ions. This mechanism is found in NF membranes, where the separation of monovalent and divalent ions is required. It is also used in membrane bioreactors and certain types of electrodialysis [64,66].

A comprised representation of the size exclusion, Donnan ion exclusion, and dielectric exclusion mechanisms is represented in Figure 7.

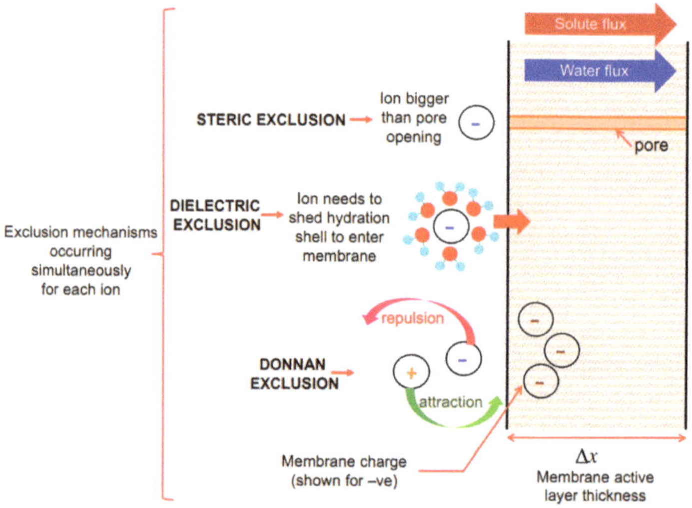

Figure 7. Diagram representing the size exclusion (steric exclusion), Donnan ion exclusion (electrostatic interaction), and dielectric exclusion mechanisms [65].

Hydrophobic–hydrophilic interaction: Hydrophilic and hydrophobic membranes interact based on their chemical affinity. Due to their non-polar nature, hydrophobic membranes repel water but attract and allow organic solvents and non-polar compounds, as represented in Figure 8. On the other hand, hydrophilic membranes attract water and repel non-polar molecules. This mechanism is commonly observed in pervaporation membranes, where water molecules can pass through for organic solvent recovery. It is also relevant in oil–water separation processes. This mechanism also operates in membrane distillation, a thermally driven process where a hydrophobic membrane separates the vapor of a liquid from a mixture of liquids [67,68].

Affinity-based separation (adsorption): In this type of separation, specific groups or ligands are attached to the surface of the membranes to act as receptors for certain types of molecules. This selective separation membrane is used to capture and collect valuable molecules, making it useful for obtaining an enantiomeric excess of one enantiomer from a racemic mixture and separating chiral molecules. Affinity membranes are mainly used in biotechnology and bioprocessing, especially for purifying proteins or antibodies. They are also important for sensor applications and medical diagnostics to capture specific biomolecules. Figure 9 shows how the membrane is utilized to capture boron, specifically by tailoring specific adsorption sites on the membrane surface [69,70].

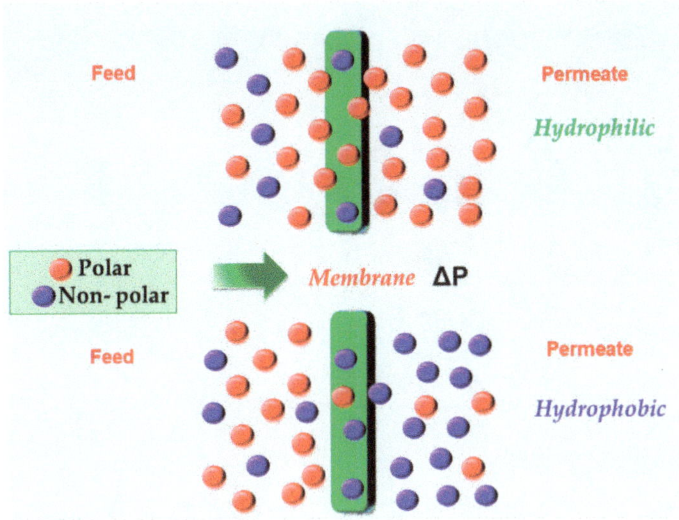

Figure 8. Schematic representation of hydrophilic and hydrophobic mechanisms (hydrophobic–hydrophilic interaction) of transportation [68].

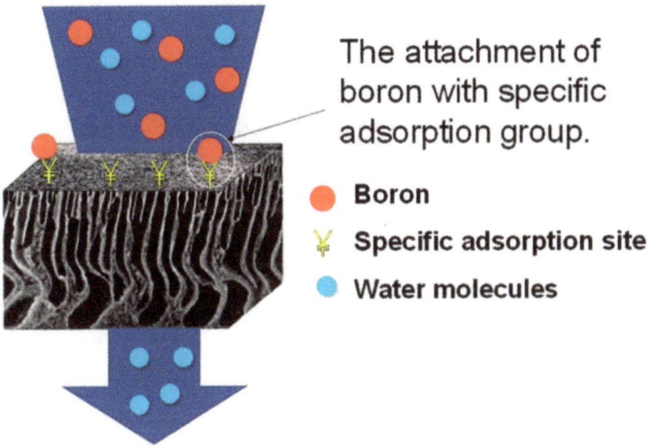

Figure 9. An illustration of capturing boron using an affinity-based adsorptive membrane [69].

Ion exchange mechanism: In the ion exchange mechanism, ions are exchanged between a solution and the membrane. The membranes are either anion exchange or cation exchange membranes. For example, a cation exchange membrane has a negative charge on the surface. When a solution passes through, positive ions are attracted to the membrane, allowing it to pass but repel the negative ions. The ion exchange membrane differs slightly from the ion exchange resin, as a counter ion on the resin surface replaces the charged ion. This advanced mechanism is used for demineralizing and softening water, among other applications. Diffusion dialysis is another important application. As represented in Figure 10, an anion exchange membrane (AEM) is used to collect hydrochloric acid and reject metal ions. This can apply to recycling and reusing acids while producing chemicals and collecting metal ions as residual products [71–73].

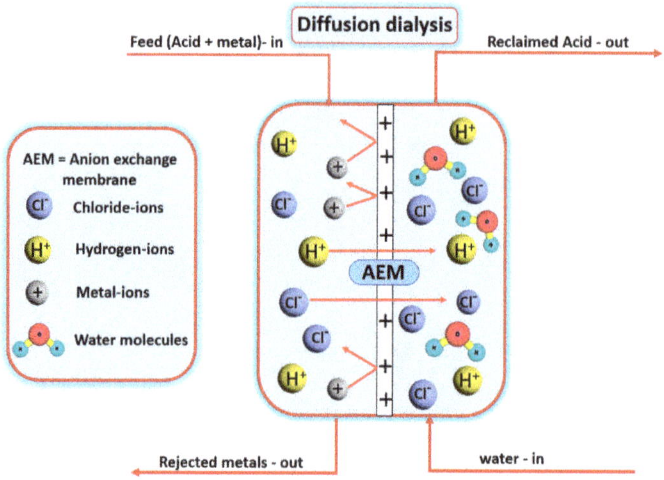

Figure 10. Diffusion dialysis using an anion exchange membrane (AEM) [73].

4.3. Fabrication Techniques for Polymeric Membranes

Fabrication of polymeric membranes comprises an intricate multiscale process controlled by different aspects, such as the phase formation kinetics, thermodynamic potential differences, and mass transfer rates. The fabrication process of polymeric membranes can be better understood through three crucial scales. At the molecular scale, the polymer molecules interact with the solvent molecules and go to the nanometer (nm) and nanosecond (ns) length and time scales, respectively. Meanwhile, at the mesoscale, different ordered structures and phase domains with lengths in the range of 10–100 nm are achieved over a 10–1000 ns time scale. At the manufacturing scale, membranes are prepared in the micrometers (μm) range, with pore sizes ranging from 10 to 1000 nm, taking several seconds or even hours for fabrication [74–79]. Researchers have employed various innovative fabrication methods, some of which are detailed below, to strengthen and enhance polymeric membranes' physical and chemical properties.

4.3.1. Phase Inversion

The most widely used method for producing microporous polymeric membranes is the phase inversion (PI) process, which was developed early in 1900. This method involves altering the physical state of a homogeneous polymer solution by coming in contact with other liquid/vapor phases or a temperature change, turning it into a solid phase. This process is used to create most polyamide membranes that are sold commercially [80–82]. Based on control parameters, phase inversion methods can be classified into four different types, namely, vapor-induced phase separation (VIPS), non-solvent-induced phase separation (NIPS) or immersion precipitation, thermally induced phase separation (TIPS), and evaporation-induced phase separation (EIPS) [83,84]. The idea behind the thermally induced phase separation (TIPS) method is that as the temperature drops, the solvent content usually decreases as well. Later, the solvent can be eliminated by freeze-drying, evaporation, and extraction after de-mixing [85]. The evaporation-induced phase separation (EIPS) method involves a solvent or a combination of volatile non-solvents to form the polymer solution. The solvent is removed through evaporation, forming a precipitate or de-mixing/precipitation. Solution casting is another name for this technique.

Vapor-induced phase separation (VIPS) is a technique wherein a polymer solution is exposed to an environment containing a non-solvent, often water. De-mixing or precipitation takes place once the non-solvent is absorbed. The polymer solution is immersed in a coagulation bath containing a non-solvent (usually water) in a non-solvent-induced phase separation (NIPS) or immersion precipitation process. Mixing and precipitation occur as the polymer solution solvents trade places with the non-solvent coagulation bath. The solvent and non-solvent must be miscible for this to occur [86]. Among these processes, the NIPS method is mostly used for fabricating polymeric membranes, which involve liquid–liquid de-mixing and the diffusion rate of solvent–non-solvent as the deciding factors for the membrane's overall structure and anatomy, such as a dense, spongy, porous, symmetric, and asymmetric nature [87]. The solvent and non-solvent used in this process diffuse across the interface, causing phase inversion, followed by the polymer solution's solidification (precipitation). Several conditions, such as humidity, temperature, coagulation bath composition, and solvent properties, must be checked during this process [83]. In a recent study conducted by Fareed et al. [88], they developed high-performance pervaporation membranes using an innovative coactive delayed phase inversion method. This approach involved the co-hydrolysis of a polyacrylonitrile (PAN) support membrane. Specifically, they substituted the conventional non-solvent, i.e., pure water, in the coagulation bath with a 1.0 M NaOH solution while sustaining a temperature of 50 °C. This modification facilitated a simultaneous, delayed phase inversion process and hydrolysis. Ren et al. [89] developed a functionalized ultrafiltration (UF) membrane by modifying the casting solution and coagulation bath by adding a radical initiator and functional monomers. The team illustrated the effectiveness of the UF membrane prepared by incorporating radical polymerization and a non-solvent-induced phase separation strategy using zwitterionic sulfobetaine methacrylate (SBMA) as a functional monomer.

4.3.2. Electrospinning

The electrospinning process is new and is used to create porous membranes for filtration and desalination, among other applications. A strong potential is applied between the grounded collector and the droplet of polymer solution. When the electrostatic potential is sufficiently increased, a charged liquid jet will occur to overcome the droplet's surface tension. These fibrous membranes are special because the aspect ratios, i.e., length or diameter of the fibers and nano/microfibers' shape, may be manipulated by adjusting the electric voltage applied, changing the flow speed and viscosity of the solution, as well as the environment in which the membrane is formed [90,91]. The ability to accurately manipulate the fiber size, morphology, and form has made electrospun fibrous membranes a preferred option for membrane filtration and membrane distillation (MD) applications. Research has shown that factors, such as the applied potential strength, polymer solution viscosity, ionic salt concentration, and solution feed rate, can significantly influence the nanostructure morphology and fiber diameter [92]. The four major components of an electrospinning system are typically a jet syringe, metal collector, spinneret (a fiber generator), and a high-voltage direct current (DC) power source (often 10–30 kV). There are two general categories for electrospinning: needle-based and needleless, as shown in Figure 11, based on the type of spinneret employed [93].

The viscous polymer solution is extruded from the syringe into the electric field while spinning through a spinning liquid thruster using a needle-based electrospinning method. This process charges the solution and, as a result, creates a repulsive interaction that counteracts the surface tension. The droplet changes into a jet after progressively taking the shape of a cone, called the Taylor cone. An unstable bending phenomenon is produced by the expelled jet. The polymer chain's entanglement prevents the charged jet from breaking as it approaches the low potential zone while evaporating the solvent. At the other end, the metal collector collects the nanofiber [94].

On the other hand, nanofibers can be rapidly and directly electrospun from an open fluid surface using a process known as needleless electrospinning. This process, often associated with needle electrospinning, shapes multiple planes concurrently from the spinneret without any effect of the capillary action. Here, the flight starts a self-assembly phenomenon on the fluid's free surface, making it difficult to regulate the spinning process in needleless electrospinning. This technique introduces a variety of spinneret shapes with varying production levels [95].

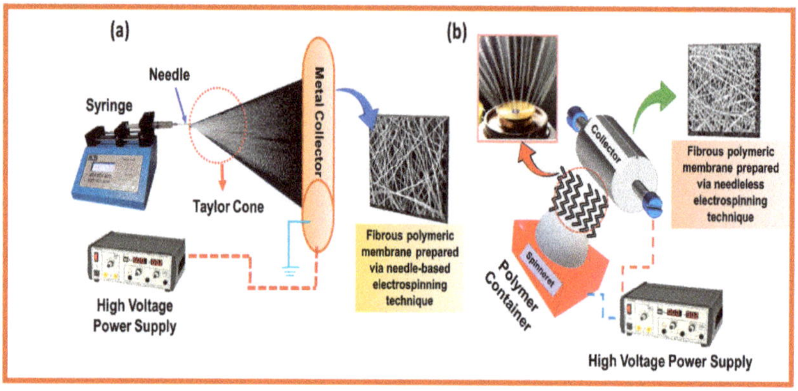

Figure 11. Diagram representing the electrospinning process for fabricating a fibrous polymeric membrane via the (**a**) needle-based electrospinning technique and (**b**) needleless electrospinning technique.

4.3.3. Interfacial Polymerization

The widely accepted polymerization technique, interfacial polymerization, has been used to make composite membranes based on polyamides since 1979, as shown in Figure 12. Interfacial polymerization is a process driven by reaction-diffusion that primarily occurs at the interface between two immiscible liquids containing multifunctional monomers, typically following the Schotten–Baumann reaction [96]. This method enables precise control in fabricating fibers, polymer films, and capsules, making it highly valuable for various applications. The polymerization typically occurs at the interface between the aqueous-phase monomer (activator) and the organic-phase monomer (inhibitor), with the organic-phase monomer immiscible in the aqueous phase. Consequently, the reaction is localized at the interface, where a thin polymer film rapidly forms due to fast polymerization. The presence of catalysts or initiators can accelerate the chemical reaction. However, as the polymer film develops, it creates a barrier between the two phases, restricting the diffusion of monomers within the organic phase.

Consequently, the reaction rate decreases over time and eventually halts as the film thickens. This is because the polymer thin film thickens or the diffusion barrier increases. This indicates that the interfacial polymerization-produced film thickness is self-limiting since it increases rapidly initially and then gradually slows down to achieve a maximum over time [97]. The interfacial polymerization (IP) reaction involves the reaction of amine monomers, such as piperazine (PIP; used for creating a semi-aromatic polyamide structure) and m-phenylenediamine (MPD; for a fully aromatic polyamide layer), with acyl chlorides, such as trimesoyl chloride (TMC), occurring at the interface of aqueous–organic phase on the surface of porous substrates, such as polyethersulfone (PES), polysulfone (PSf), and polyacrylonitrile (PAN). This reaction forms the active layer of thin-film composite membranes [98]. As a diffusion-limited process, the IP reaction results in a denser and thinner active layer when highly reactive monomers are used, compared to less reactive ones. M-phenylenediamine (MPD) and trimesoyl chloride (TMC) are the most commonly employed monomers for developing functional polyamide layers in nanofiltration and reverse osmosis membranes [99]. The thickness and morphology of polymer films formed during polymerization are influenced by several factors, including the monomers' solubility and diffusivity, reactivity, stoichiometry, concentration, morphology, temperature, additives, solvent media, etc. By controlling these features, one can create high-performance ultrathin polyamide membranes [100,101].

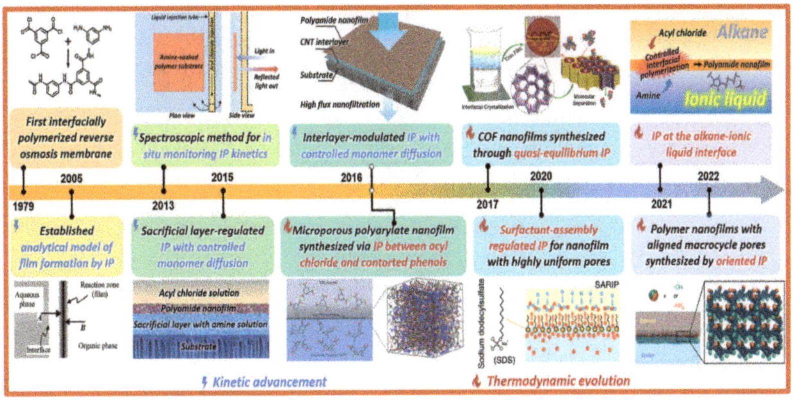

Figure 12. Significant advancements in the kinetic and thermodynamic aspects of polymer membranes created via liquid–liquid interfacial polymerization (IP) up to 2022 [100].

4.3.4. Recent Developments in Fabrication Methods

Polymeric membranes are essential for numerous uses, such as emerging organic micropollutant removal, purification of water, and biomedical applications. Recent advancements in fabrication techniques have aimed to improve membranes' mechanical strength, selectivity, permeability, and fouling resistance. In addition to the above-discussed fabrication techniques, many fresh and innovative processes are presently employed to manufacture advanced polymeric membranes, such as grafting, coating, layer-by-layer assembly, track-etching, incorporation of nanomaterials, the inkjet printing method, incorporation of aquaporin, as well as utilization of glassy polymers [102]. Numerous methods for producing polymeric membranes have been documented in the literature. For instance, since the early 1990s, diamond-like carbon (DLC) membranes have been fabricated using the chemical vapor deposition (CVD) technique [103]. Ultrafiltration freestanding DLC membranes were developed in 2012 using a parallel-plate plasma-enhanced CVD reactor [104]. The membranes exhibit excellent stability and remarkable mechanical strength in organic solvents, attributed to sp3 carbon networks within their structure. Thus, the CVD process offers a practical way to create membranes that may be adjusted in thickness. By fine-tuning the membrane properties for ionic and molecular separation applications, future advancements may enable the development of high-performance ultrathin membranes through the layer-by-layer (LBL) CVD technique [97].

Surface Modifications and Functionalization

To enhance membrane performance, especially concerning chemical stability, permeability, selectivity, and fouling resistance, modifying the surface morphology and functionalizing the membrane surface are essential approaches. By manipulating the membrane surface, these procedures can add new features or properties without affecting the overall properties of the membrane material. Researchers frequently utilize different techniques to modify the polymeric membrane's surface, which improves the surface physicochemical properties of the cross-linked polyamide networks and overcomes their limits, as shown in Figure 13. Adding nanomaterials, for instance, silver, silica, titanium dioxide (TiO_2), nickel oxide (NiO), aluminum oxide (Al_2O_3), iron oxide, GO, and CNTs, during the fabrication leads to nanocomposite polymeric membranes, significantly enhancing their properties [105,106]. Metal-based nanoparticles have been widely researched for their potential in water purification applications because of their remarkable adaptable characteristics, including high surface area, good selectivity, antifouling and antimicrobial properties, and increased hydrophilicity. Using metal complexations and coatings, a suitable surface modification can improve their adsorption affinity, selectivity, and reactivity even more [105]. Membrane surfaces can be functionalized with graphene oxide and carbon nanotubes to improve their mechanical strength, chemical resistance, and antifouling qualities. By creating targeted channels for water, these materials can also increase selectivity and permeability [106].

Moreover, plasma treatment techniques can introduce functional groups that strengthen chemical resistance or antifouling qualities. The plasma treatment activates the membrane surface, and monomers or polymers are grafted [106]. To create a multilayer structure, the layer-by-layer (LBL) method entails the alternating deposition of nanoparticles or polyelectrolytes that are oppositely charged on the membrane surface. With this method, one may precisely regulate the membrane's permeability, pore size, and surface charge [107]. The coating is another significant method to modify the surface of the polymeric membranes, which includes photocatalytic coating: applying a layer of photocatalysts, such as titanium dioxide (TiO_2), to membranes allows them to have self-cleaning capabilities when exposed to light, which lowers fouling and preserves the performance as time passes; hydrophilic coating: protein adsorption and biofouling can be minimized by coating the membrane surface with hydrophilic substances, such as polyethylene glycol (PEG) or zwitterionic polymers; antimicrobial coating: to stop the bacterial growth and biofilm formation, membranes can be functionalized with antimicrobial substances, for instance, silver nanoparticles, chitosan or quaternary ammonium compounds; dip coating: cross-linking agents (aqueous-phase and organic-phase monomers) are utilized to form covalent links between polymer chains to improve the thermal stability, mechanical strength, and chemical resistance of the membrane surface [108–112]. Chemical treatment is another widely recognized technique for modifying the surface characteristics of polymeric membranes by introducing different functional groups, such as amine, carboxyl, sulfone, and hydroxyl groups. Chemical treatment primarily involves hydrolysis, substitution, addition, and oxidation. When applied directly to the polymer solution, bulk modification techniques are straightforward and efficient. The backbone of aromatic polymers is frequently modified in bulk to increase surface hydrophilicity via amination, sulfonation, carboxylation, and epoxidation [108].

Mixed-Matrix Membranes (MMM)

Recently, researchers studying membranes have been experimenting with incorporating nanoparticles into traditional polymeric membranes. Significant advancements have been made in the last few years in creating nanomaterial-based polymeric nanofiltration membranes. These membranes demonstrate superior mechanical and thermal stability and effective antifouling characteristics. They also promise to address the traditional "trade-off" between solute selectivity and solvent permeability. The criteria that should be satisfied to fabricate sophisticated nanomaterial-based nanofiltration membranes are good nanomaterial compatibility with polymers, well-dispersed nanomaterials, and robust and stable nanomaterial–polymeric matrix interactions [113]. When nanofillers are incorporated into monomer/polymer solutions before membrane fabrication, the resultant nanofiller-blended membranes are known as mixed-matrix membranes (MMMs). Mixed-matrix membranes combine the advantages of inorganic materials' mechanical strength and functional qualities with the low cost and convenience of producing membranes. MMMs aim to leverage the mechanical robustness and functional qualities of inorganic materials while benefiting from the cost-effectiveness and ease of membrane fabrication typically associated with polymeric membranes. To overcome the shortcomings of polymeric membranes for gas separations, Zimmerman and his coworkers initially presented mixed-matrix membranes in the 1990s [114]. MMMs containing inorganic molecular sieves, such as silicalite and zeolites, are incorporated into the polymer matrix to give the target species preferential flow routes [115,116]. Figure 14 depicts the synthesis of a mixed-matrix membrane prepared using the electrospun technique for liquid separation [90]. The "percolation threshold", a volume fraction of filler material, is thought to be the point at which continuous paths of fast diffusion molecular sieves begin to form. Target molecules can now pass through the filler and the whole membrane cross-section [117,118]. Defects typically arise at the polymer–filler contact at a particularly high-volume fraction, reducing the selectivity. Due to their higher selectivity, targeted functions, and enhanced chemical, mechanical, and thermal stability, MMMs also provide the possibility of adjustable water treatment membranes.

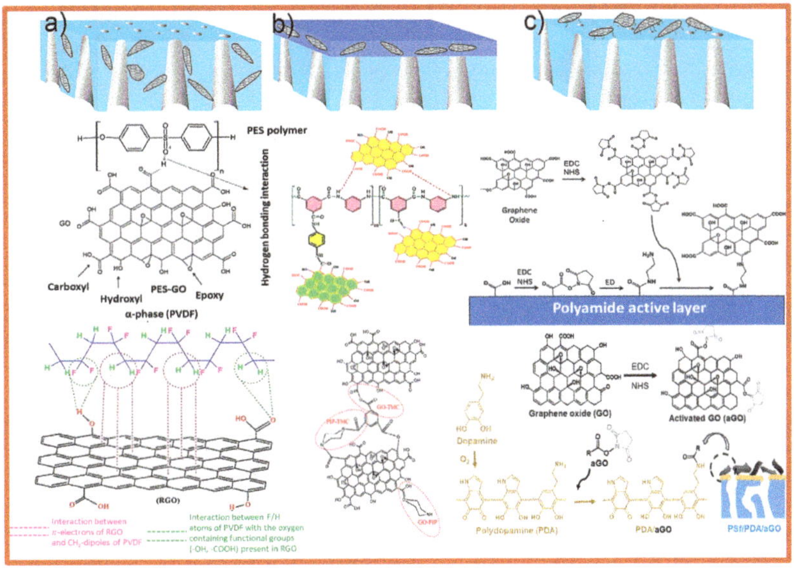

Figure 13. (**a**) Diagram showing the interaction of embedded carbon nanofillers with PES and PVDF polymers in UF membrane during the phase inversion process. (**b**) Interfacial polymerization process in NF and RO membranes between amine and TMC using carbon quantum dots (CQDs) and GO nanomaterials. (**c**) The surface modification of the membranes utilizing carbon nanofillers through covalent cross-linking [110].

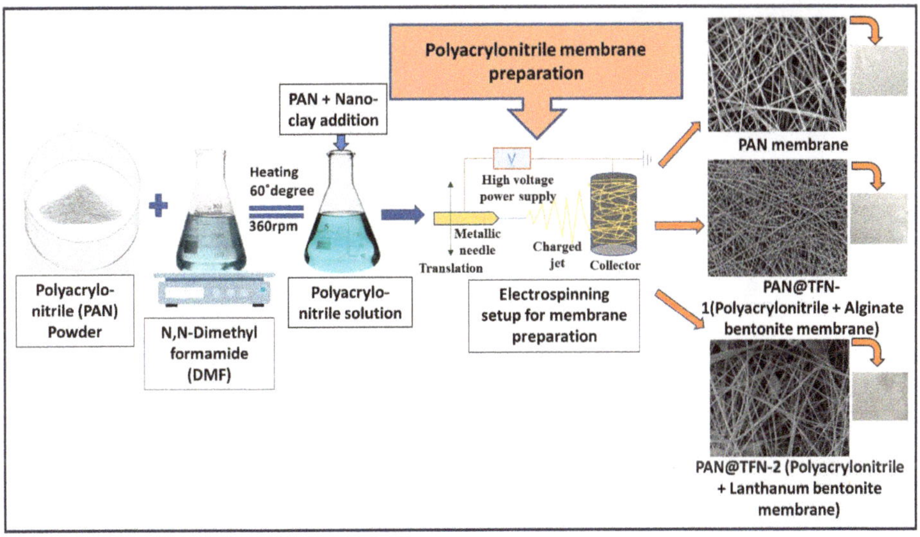

Figure 14. Electrospun nanofibrous MMMs' development via the electrospinning technique [90].

Composite and Nanocomposite Polymeric Membranes

Achieving favorable outcomes without risking the membrane's integrity requires balancing improved characteristics and defect formation [118]. A significant advancement in membrane separations was the fabrication of thin-film composite (TFC) membranes, which feature a porous polymeric support layer with an exceptionally thin "barrier" layer polymerized in situ. The terms "interfacial composite", "composite", or "TFC" membranes are frequently used to refer to these membranes generally, whereas layers incorporated with nanomaterials are referred to as thin-film nanocomposite (TFN) membranes. The primary advantage of TFC membranes is the ability to selectively tailor the chemistry of the upper selective layer and the porous support layer to enhance the overall functionality [118]. Phase inversion is typically used to create the porous layer, and IP or coating, such as dip, spray, or spin, is used to apply a cross-linked dense layer. The degree of polymer cross-linking is often increased by curing via heat, UV, and chemical treatment, which has a major effect on the thin film's stability, permeability, and selectivity [119]. TFC membranes, known for their remarkable salt selectivity and elevated water permeability, have been effectively engineered using polyurea for water treatment applications [120]. The vapor phase IP technique is becoming popular and successful in creating polymer-based TFC membranes or covalent organic framework (COF) membranes with an ultrathin separation layer [121].

Using additives, such as metal-organic frameworks (MOFs), layered double hydroxide, clay minerals, and nanomaterials, improved the separation efficiency. Membranes prepared by the addition of nanomaterials are regarded as nanocomposite polymeric membranes. Numerous investigations have been conducted on the impact of various nanomaterial types on the composition and functionality of TFN membranes up to this point, as mentioned above. Recently, researchers have been interested in the topic of polymer/clay nanocomposites because of the relationship between the filler and polymer, the effect of the nanoscale structure, the easily adjustable pore size of nano-clay particles, the distinctive chemical composition, and the high adsorption capacity [122]. Clay is classified as a two-dimensional nanomaterial with an anisotropic sheet-like morphology, in contrast to the previously described metal-based nanofillers for TFN. Aside from its hydrophilic nature, the most distinctive property of clay is the positive or negative permanent charge, which results from the isomorphous substitution in the crystal structure of the clay layers. Dong et al. [123] developed TFN reverse osmosis (RO) membranes by embedding a cationic clay, montmorillonite (MMT), and an anionic clay, layered double hydroxide (LDH), having dimensions on a hundred-nanometer scale and a high aspect ratio. Because of these nano-clay filler materials' special charged characteristics and hydrophilic nature, the properties of the TFN membranes were successfully altered. TFN membranes loaded with MMT and LDH demonstrated enhanced hydrophilicity, improved antifouling performance, and better desalination results. In another study by Selvan et al. [124], they utilized Kaolin clay and fabricated a zero-valent, iron-kaolinite, composite-based polyethersulfone (PES) membrane to separate As_2O_3 from water. The prepared nanocomposite membrane achieved a maximum removal efficiency of 50% for As_2O_3 from water.

Biopolymer-Based Membranes

In many industrial domains, membrane technologies have become a viable substitute for energy-intensive separation procedures. The production of separating membranes poses a sustainability problem that calls for an important shift away from petrochemical-based raw materials and toward more environmentally friendly bio-based sources. Hardian et al. [125] worked on the development of oil- and solvent-resistant NF membranes using blends of chitosan (received from shrimp farming waste) and cellulose (plant-based material). The membrane's molecular sieving capabilities were refined by varying the cellulose-to-chitosan ratio. Even at the highest temperature of 100 °C, the membranes demonstrated exceptional separation efficiency and chemical stability in harsh solvents, including polar aprotic solvents. The membranes demonstrated consistent separation performance after seven days of crossflow nanofiltration.

Jiang and his colleagues [126] employed the freeze-casting process and chemical vapor deposition to construct a novel superhydrophobic biopolymer aerogel membrane (T-SA/ligninx/rGO-MTMS) using graphene oxide (GO), sodium alginate (SA), and lignin. This innovative membrane significantly increased the oil removal efficiency, rising from 73.8% to 98.6%. Trimethoxymethylsilane (MTMS) undergoes silylation growth during production to provide a rough, hydrophobic surface for the ultimate protuberance. Even after 10 cycles of oil/water separation and oil sorption, T-SA/ligninx/rGO-MTMS showed promising cycling performance due to the ample protuberance and low-surface-energy reduced GO (rGO).

To create a more sustainable solution for the membrane fabrication industry, Tomietto et al. [127] explored a novel bio-based material combination for the phase inversion process, which included a polyhydroxyalkanoate (PHA) as a biopolymer and Cyrene as a green solvent. Epoxidized broccoli vegetable oil (EBO), polyethylene glycol (PEG), and polyvinylpyrrolidone (PVP) were all effective pore-forming agents for these membranes. Several phase inversion process factors were examined to determine the various membrane microstructure options and their ultimate performances. The use of dense membranes in pervaporation (PV) has proven successful in separating organic/organic azeotropic mixtures.

5. Applications of Polymeric Membranes

Polymeric membranes have numerous applications in various liquid separation areas, whether the requirement is purification or liquid separation. Desalination and cleaning brackish water were the initial applications; now, polymeric membranes are utilized in dairy, chemical, food, biotechnological, pharmaceutical, metal ion, beverages, metallurgy, dye, and other separation processes [128–130]. Developing polymeric membranes for water treatment and desalination is crucial for advancing global water sustainability, aligning with the United Nations Sustainable Development Goal 6. The SDG 6 and its subgoals are represented in Figure 15 [131].

Figure 15. Diagrammatic representation of SDG 6 goals and its subgoals [131].

5.1. Desalination and Dye Wastewater Treatment

Researchers have extensively investigated various types of polymeric membranes to achieve diverse requirements and applications. Jiang et al. used poly(m-phenylene isophthalamide) (PMIA) to prepare hollow-fiber membranes of the NF category via dry–wet phase inversion technology. They applied this membrane to the separation of Na_2SO_4 and $MgSO_4$ and achieved ~98% and ~94% rejection from their 2000 ppm solution. They have also shown that the membrane can effectively separate Janus green B and Chromotrope 2B dyes [132]. Li et al. fabricated a thin-film flat-sheet membrane utilizing PMIA on porous epoxy support for dye and salt separation. Although they achieved >99% rejection of various dyes, their membranes poorly separated the metal salts [133].

In a study by Usha et al., a microfiltration membrane made from polypropylene (PP) was employed to separate oil/water emulsions. Their modified membrane exhibited super hydrophilicity and underwater oleophobicity, which is great for an energy-saving gravity-driven separation [134].

PVDF hollow-fiber membranes were modified by Sairiam and his group using pulse inductively coupled plasma (PICP) and grafted by chloro-alkyl-silanes to enhance hydrophobicity. They applied this membrane for the ozonation membrane contractor process to decolor the azo dye DB 71 (Direct Blue 71). They reported that their membrane decolored DB 71 in 90 min and removed 63% and 20% COD and TOC, respectively [135]. Borban et al. developed electrospun nanofiber mixed-matrix membranes (MMMs) by incorporating lanthanum-bentonite (La-Bnt) and alginate-bentonite (Alg-Bnt) into polyacrylonitrile (PAN). The electrospun membranes demonstrated impressive performance, with water permeability reaching 18.8 $Lm^{-2} \cdot h^{-1}$ and rejection rates up to 99.08% for various micropollutants [90].

In another study, Gohain et al. developed thin-film nanocomposite (TFN) membranes via interfacial polymerization (IP) between an aqueous phase containing piperazine (PIP) and UiO-66-NH_2 and an organic phase containing trimethyl chloride (TMC) in n-hexane. These membranes were designed to separate phosphate and malachite green dye from water at low concentrations. The TFN membranes achieved excellent separation, with a rejection rate of 91.90% for malachite green (MG) and a flux of ~13.32 $Lm^{-2} \cdot h^{-1}$. In comparison, phosphate exhibited a rejection rate of ~78.36% with a flux of ~22.22 $Lm^{-2} \cdot h^{-1}$. The membranes also demonstrated a strong antifouling capability, with an antifouling tendency of ~84% and only ~14% irreversible fouling [136]. Figure 16a represents the permeation and rejection values obtained by their membranes (M1 and M2) against malachite green dye at 10 and 15 bar. To visually realize the performance, images of the permeate at each hour were taken, which can be seen in Figure 16b [136]. In another work, Gohain et al. developed MMMs by incorporating COOH-TiO_2-functionalized GO nanoparticles with cellulose acetate extracted from discarded cigarette butts and studied for dyes and salts' separation. The prepared membranes rejected up to 95% of dyes [137].

5.2. Metal Ion Removal from Wastewater

Nowadays, thin-film composite/nanocomposite and MMM types of polymeric membranes are prominently used for ion separation in the NF process. Due to their small size, metal ions are difficult to remove, and the Donnan ion effect (an electrostatic interaction mechanism) becomes more pronounced during their separation. In previous works, they successfully removed salt ions, especially heavy metal ions, which are required in industries for energy applications or removal in desalination [138,139].

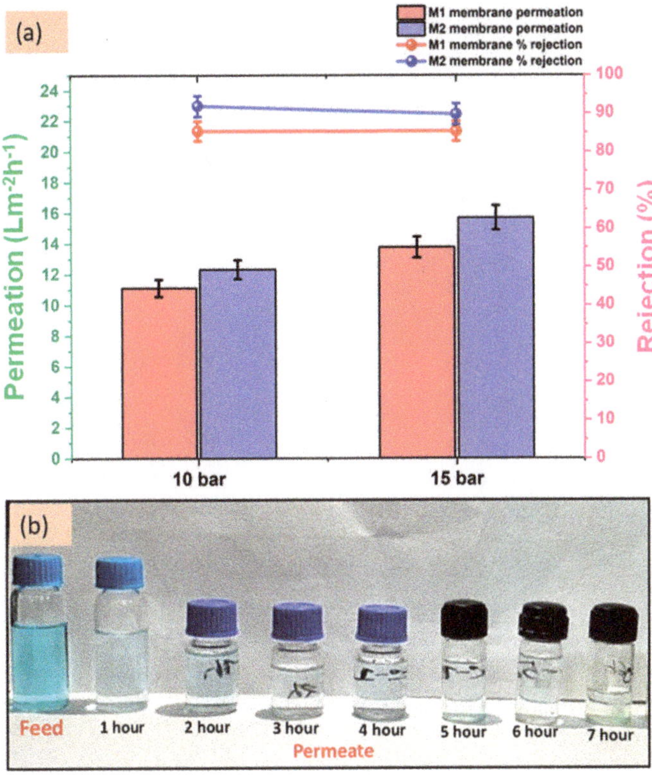

Figure 16. (**a**) Permeation and rejection graphs of M1 (without nanomaterial) and M2 (0.02 wt% UiO-66-NH_2) nanofiltration membranes for malachite green dye at 10 and 15 bar. (**b**) Images to compare the dye feed and obtained permeate per hour for up to 7 h at room temperature [136].

Incorporating COOH-TiO_2 nanofillers in the polyamide membrane prepared via the new vapor-phase interfacial polymerization (VP-IP) method on the polysulfone support provided rejection of Na_2SO_4, $HgCl_2$, $CuSO_4$, and $Pb(NO_3)_2$ salts up to ~87%, ~77% ~84%, and ~83% from their 2000 ppm contaminated aqueous solutions, respectively [140]. Using other functionalized nanomaterials, NHST and GO-NH-NH_2, rich in amine groups, delivered cationic thin-film nanocomposite (TFN) membranes. These membranes were developed simply by the IP, and dip-coating using NHST (amine-rich synthetic talc) as nanofillers for polyamide TFN membranes exhibited ~99%, ~95%, and ~94% rejection of Na_2SO_4, $MgSO_4$, and $CuSO_4$ salts with pure water permeation up to 38 $Lm^{-2}{\cdot}h^{-1}{\cdot}bar^{-1}$ [141]. TFN membranes were developed using another approach, integrating GO-NH-NH_2 nanosheets as nanofillers and indulging in the spray coating method, as shown in Figure 17a. The nanosheet addition influenced the prepared membranes cross-linking density, as evidenced by the SEM images in Figure 17b. Membranes with a higher concentration of the nanosheets exhibited denser cross-linked polyamide surfaces. Prepared R4-TFN and R3-TFN membranes with higher nanofiller loading appeared comparatively denser. These prepared membranes exhibited pure water flux up to ~47 $Lm^{-2}{\cdot}h^{-1}$ at 10 bar, and their separation quality was also good. They demonstrated ~96%, ~95%, ~94%, and ~92% rejection of Na_2SO_4, $Pb(NO_3)_2$, $CuSO_4$, and $CdSO_4$ salts, respectively [5].

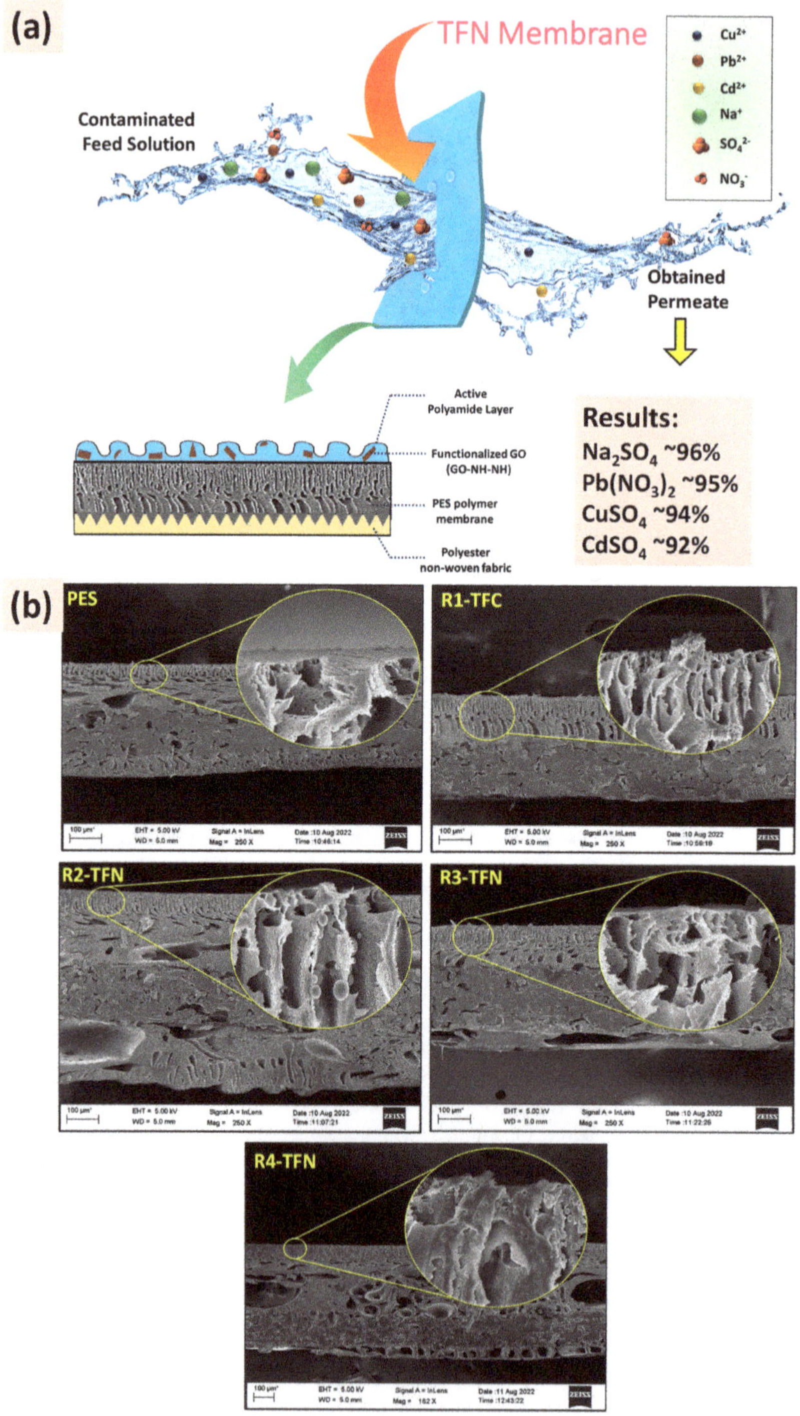

Figure 17. (**a**) TFN membranes prepared by the VP-IP method and (**b**) cross-sectional FE-SEM images of the prepared TFC, TFN, and pristine PES polymeric membranes [5].

Lithium-ion and polymer batteries are used for all kinds of electrical equipment, especially smartphones and laptops. Lithium extraction using membrane technology is far more efficient than other techniques, as articulated by Li et al. [142]. The applicability of NF membranes for the recovery of LiCl from seawater was first investigated by Wen et al. [143]. Zante prepared supported ion liquid membranes for extracting lithium from a complex aqueous solution. They displayed the separation of lithium from magnesium, cobalt, and nickel, and their lithium flux was 7×10^{-7} mol·cm^2·s^{-1}, with membrane stability up to two days [144]. Zhu et al. developed a composite membrane utilizing poly(vinylidene fluoride-co-hexafluoropropylene) (PVDF-HFP) as the host material reinforced with a nonwoven fabric. The prepared membrane exhibited an increased Na^+ ion transference number and excellent electrolyte retention, maintaining its performance even at 110 °C [145].

5.3. Ion Removal and Recovery in Fertilizer Production

Nitrogen and phosphorus are key ingredients for fertilizer production to increase crop yield. A high quantity of ammonia is generated from pharmaceuticals and other waste found in water, which needs to be recovered to enhance nitrogen fertilizer production. Atsbha et al. [146] utilized commercial hollow-fiber membrane contactors (HFMC) to remove and recover ammonium from pharmaceutical wastewater at bench and pilot scales. The recovered ammoniacal nitrogen can be used to make ammonium-sulfate fertilizer. Composite membranes with active and support layers effectively recover ammonia during wastewater or potable water treatment. Shahgodari et al. [147] recovered ammonia from wastewater using a thin-film composite forward osmosis polymeric membrane by determining the permeability coefficient of NH_3. They found a recovery of 15% of NH_3 and, simultaneously, 35% of water at a pH of 11.5. A study by Kurama et al. [148] revealed that 96.9% of ammonium ions were recovered from potable water using RO membranes.

Phosphorus removal and recovery from wastewater has great potential in the fertilizer industry, particularly in producing phosphorus-based fertilizers through membrane processes. In a study by Kannan et al. [149], a high level of phosphorus was recovered in the form of calcium phosphate for making fertilizer using an anaerobic membrane bioreactor. Results obtained show variability in phosphorus removal due to the high carbonate alkalinity of the permeate, which was precipitated as calcium carbonate. This was overcome by aeration and the addition of acid, leading to 97% phosphorus removal. In a wastewater treatment study, Santos et al. [150] utilized two commercialized polyamide NF membranes, NF90 and NF270, to remove and recover phosphorus from surface water. They found that NF270 had a higher water flux than NF90; however, NF90 had a superior removal efficiency (>90%), with more than 96% of phosphorus recovered from surface water. In another study, Gerardo et al. [151] recovered nitrogen and phosphorus from dairy farm sludge using crossflow membrane filtration.

6. Challenges and Limitations

Despite notable advancements in polymeric membranes, some challenges still need to be addressed. One prominent issue is membrane fouling, which occurs when microorganisms and particles accumulate on the membrane surface. Researchers are extensively exploring innovative materials with enhanced performance and resistance to fouling. Furthermore, adapting these membranes for industrial applications presents a substantial challenge. Ensuring their durability and long-term stability under harsh operational conditions is crucial, affecting their effectiveness and feasibility in industrial environments [2]. Key challenges associated with membrane-based separation technologies include maintaining membrane stability, fouling, and mitigating degradation.

6.1. Chemical and Thermal Stability

Many studies have been conducted to improve the thermal properties of polymeric membranes and expand their applications. Various types of polymeric membranes have been investigated, including cellulose acetate (CA), PAN, PSf, PES, and poly(vinylidene fluoride) (PVDF), with a focus on advancing membranes for water and wastewater treatment. Table 1 shows some of the high-temperature-resistant polymers. Nevertheless, most polymeric membranes are not resilient to severe chemical and thermal environments. To ensure optimal performance of these membranes, heat exchanger units are often employed to lower the temperature of the inlet stream to below 50 °C, increasing operational costs and energy consumption [152].

Table 1. Some of the high-temperature-resistant polymers [153–155].

Abbreviation	Polymer	Tg (°C)	CUT [a] (°C)	HDT [b] (°C)
PAI	Polyamide-imide	275	250	279
PI	Polyimide	250	288	246
PES	Polyethersulfone	224	177–204	204–238
LCP	Liquid-crystal polymer	_	204–232	321–335
PEI	Polyetherimide	213	177–204	199–216
PSF	Polysulfone	190	149–171	171–182
PEK	Polyetherketone	165	260–288	>316
PEEK	Polyetheretherketone	143	204–232	177–321
PPA	Polyphthalamide	134	204–232	277–285
PPS	Polyphenylene sulfide	92	204–232	149–288
PEKK	Polyetherketoneketone	165	232–260	>316
FEP	Fluorinated ethylene propylene	149		
PFA	Perfluoroalkoxy	_	240–260	_

[a] Continuous use of temperature. [b] Heat deflection temperature.

Polymers with high thermal stability are typically insoluble in most common organic solvents. This resistance is attributed to their melting temperatures (Tm) and high glass transition (Tg) resulting from their rigid molecular structures and robust intermolecular forces. Although these properties contribute to their durability, they also pose difficulties in processing these polymers into various shapes, including films and fibers. Additionally, many thermally stable polymers are not biodegradable, have high environmental resistance, and can release harmful derivatives, posing environmental and human health risks [156]. To address these challenges, biodegradable and environmentally friendly polymers offer a promising alternative [150]. Biopolymers frequently exhibit lower mechanical and thermal stability relative to traditional polymers. Strategies to enhance their performance include blending biopolymers with more robust, stable polymers to improve mechanical strength and durability, cross-linking biopolymer chains to enhance chemical resistance and stability, and fabricating multilayer membranes with strong interlayer adhesion to optimize separation properties.

Additionally, incorporating nanomaterials into biopolymer matrices can significantly improve selectivity and permeability by introducing novel pathways and enhancing overall membrane performance [157,158]. Cellulose is a biopolymer consisting of repeating D-glucose units linked by β-1,4 glycosidic bonds. Its crystalline arrangement and tight chain packing provide superior chemical and thermal stability [159]. Figure 18 illustrates a cellulose-based membrane made by Gohain et al. [137], in which cellulose was taken out of cigarette butts to convert waste into usable products.

Integrating inorganic nanofillers into polymer matrices is a well-established approach for improving thermal stability and enhancing interfacial adhesion between inorganic and organic components, thereby improving the membrane performance [160–162]. Nanomaterials, such as silica (SiO_2), GO, SiO_2@GO, MOFs, salicylate-aluminoxane, zeolitic imidazolate frameworks (ZIFs), and metal oxides, are commonly employed in composite membranes [163–172]. These fillers restrict polymer chain mobility, improving thermal stability [173]. Khorshidi et al. developed high-performance RO nanocomposite membranes via in situ polymerization by integrating surface-modified TiO_2 and oleic acid into a polyamide layer [174]. This approach reduced polymer chain mobility, resulting in a denser, thermally stable membrane. Thermogravimetric analysis showed that adding just 0.0245 wt% TiO_2 increased the onset of an intense degradation temperature between 530 and 550 °C.

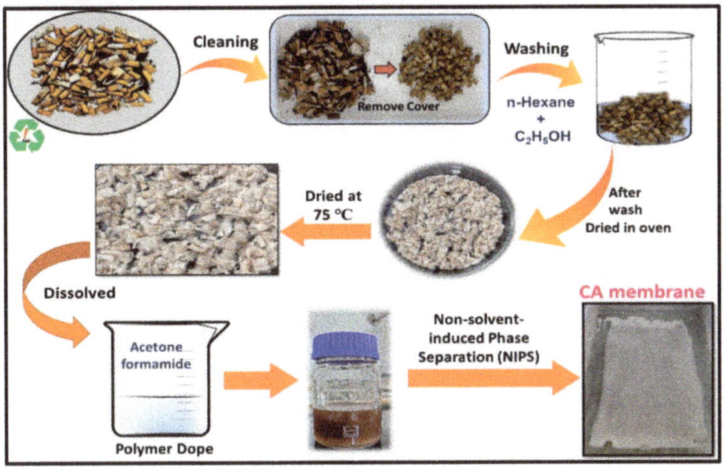

Figure 18. Cellulose acetate membrane fabrication using cigarette butt waste via the phase inversion process [137].

The chemical stability of polymeric membranes can be defined as their ability to withstand different chemical environments, such as bases, acids, oxidizing, or reducing agents, without being chemically degraded. Its chemical stability is an essential determining factor to sustain functioning and preserve its structural integrity for a prolonged period. Resistant to deterioration, chemically stable polymers, such as polyimides, PVDF, PSf, and PTFE, are favored for liquid separation membranes in harsh environments. On the contrary, PVA, PAN, cellulose acetate, and polyethylene are appropriate for not demanding applications, as they can break down in harsh chemical environments, decreasing the membrane's durability and performance [175]. During the liquid separation procedure, swelling and chemical deterioration may result from the interaction between the membranes with solvents, acids, bases, etc. [176,177].

6.2. Fouling and Degradation

During separation processes, particles in the retentate often accumulate on the membrane surface, forming a cake layer that diminishes membrane performance by reducing flux. This issue, known as membrane fouling, is particularly prevalent in liquid separation applications, such as oil–water separation and water desalination. Fouling occurs when suspended impurities, such as iron, aluminum, manganese, clay, silt, organic molecules, and biological materials, including bacteria and biofilms, adhere to the membrane surface, causing blockages [178]. As fouling intensifies, transmembrane pressure rises, necessitating increased energy to sustain the initial flux. Moreover, fouling negatively impacts the membrane's selectivity and rejection efficiency, with common foulants, including macromolecules, salts, colloids, and emulsions, which can cause blockages [179].

Protein adsorption studies are often conducted to assess a membrane's antifouling properties. Hydrophilic surfaces, which attract water and reduce protein adsorption, effectively prevent fouling. Various strategies, including surface modifications using nanoparticles, copolymers, and hydrophilic monomers, have created fouling-resistant polymeric membranes. Techniques such as coating, UV irradiation, and grafting have been reported [2].

To mitigate foulant deposition, one effective strategy is the application of antifouling agents to the membrane surface. As depicted in Figure 19, the antifouling mechanism involves applying a thin coating layer directly onto the membrane surface or using cross-linking agents to improve resistance to fouling. For example, Akbari et al. [180] have shown that coating a polyamide nanofiltration membrane with chitosan markedly enhances its antifouling performance and increases pure water flux. Additionally, the antifouling properties of polymeric membranes can be further improved by using various copolymers or by integrating hydrophilic polymers with the base polymeric materials.

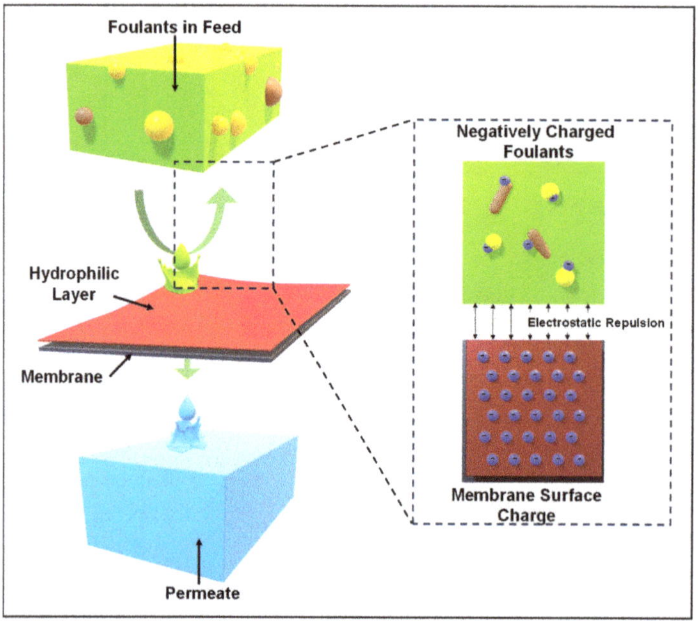

Figure 19. Schematic diagram demonstrating the antifouling mechanism of the membrane.

Polymeric membranes, including those made from materials such as polyurethane, polyacrylonitrile, polyamide-imide, polyacrylonitrile-g-PEO, cellulose acetate, and poly(vinyl pyrrolidone) composites, have shown promising antifouling capabilities. These properties can be enhanced by modifying the membrane surface with various functional groups, such as amines, amides, carboxylates, sulphones, and epoxides. Techniques such as UV irradiation have effectively improved the antifouling performance [178,181].

The employment of advanced polymeric membranes has resulted in improved antifouling properties and membrane durability, as shown by many studies. The membrane's antifouling effectiveness also depends on the design of the membrane module. Cleaning the membrane at definite time intervals is also crucial, as it helps maintain the designed membranes' antifouling behavior [182]. Another important criterion of the membrane that significantly impacts the fouling process is the surface features of the membrane. This suggests that addressing fouling issues in membranes can be achieved by applying antifouling coatings that are carefully engineered to modify the membrane's chemical and physical intrinsic surface. Adopting innovative strategies and methodologies is essential, given that each type of foulant interacts with the membrane differently [183]. Two main approaches are typically employed to fabricate antifouling membranes: passive and active. The passive antifouling strategy aims to prevent the initial foulants from adhesion to the membrane surface without modifying the intrinsic properties of the foulants.

In contrast, active antifouling involves decomposing or removing the foulants' chemical structures. These strategies often fail to prevent bacterial and microbial colonization, forming biofilms. Therefore, a more advanced approach is needed to develop antifouling membranes with enhanced properties that can effectively prevent biofilm growth and control microbial colonization [184]. The application of membranes containing nanomaterials, such as nanoparticles, nanowires, and nanosheets, has significantly broadened the potential for developing new types of antifouling membranes. In the future, research and development should prioritize the development of membranes with zero fouling so that they can compete with other technologies and be scaled up for commercial use.

7. Emerging Technologies and Trends

Membrane-based methods are generally used in desalination, water reuse, wastewater treatment, and water purification, including removing heavy metals [5,140] and micropollutants [10]. These technologies have been applied individually or in combination with others to advance membrane performance. Recent research has focused on synthesizing a new generation of membranes that offer improved antifouling properties, higher recovery flux, biodegradability, and reduced energy requirements, as outlined below [62].

Antifouling membranes: Recent advancements in antifouling membranes encompass the development of hybrid nanocomposite membranes, the incorporation of organic modifiers, techniques involving IP, and the use of acid monomers. The previous study [140] mentioned that thin-film polyamide membranes are fabricated and embedded with COOH-TiO_2 nanoparticles, demonstrating enhanced antifouling properties and effective rejection of heavy metal ions using a novel VP-IP approach.

Membrane surface modification: This process can be accomplished through two primary methods: (a) Physical surface modification, which includes the application of surfactants or polyelectrolytes to the membrane surface or coating the membrane with hydrophilic polymers. These methods enhance electrostatic repulsion and help minimize fouling. (b) Chemical modification, which involves incorporating active functional groups into the membrane through chemical reactions to improve its overall performance [185,186].

Fabrication of membranes utilizing green polymers and solvents: The residual waste from making nanoparticles using green methods has been utilized to synthesize a nano-polymer membrane. This advancement is a component of using membrane technology to minimize pollutants in environmental waste and industrial wastewater before reaching water courses [187]. Green solvents have been employed to create polymeric membranes. To promote environmentally friendly advancements in membrane technology, replacing hazardous solvents with suitable green alternatives is essential [188]. A methodology known as solvent-induced polymerization enhances the yield of uniform, crystalline, porous, and structurally robust polymers. This approach presents valuable prospects for water and wastewater treatment [189]. Greener methods are also employed to fabricate TFN membranes [5,62].

Thin-film nanocomposite membrane: The advancement of polymeric nanofiller-based separation membranes continues to evolve, with several promising trends and developments emerging in the field. Various nanoparticles have been included in polymer matrices, including zeolites, graphene oxide (GO), metal oxide, covalent-organic frameworks (COFs), carbon nanotubes (CNT), and metal-organic frameworks (MOFs), to improve the performance of the separation membranes [4,8,10,18]. Depending on their functional groups or nanoscale holes in the membrane, these nanoparticles can enhance the separation efficiency and provide antibacterial and antifouling properties [141]. Incorporating nanoparticles into membranes can improve solvent recovery and desalination, which are just a few energy-efficient separation processes.

Biodegradable membrane: Biodegradable polymers can be derived from plants, animals, and synthetic materials. Biodegradable polymers are often blended with other polymers to enhance their properties and achieve specific performance characteristics [125–127].

Aside from these, electrospun aramid nanofiber membranes and membrane bioreactors (MBRs) are also gaining prominence for their exceptional performance in various applications. While PAN-based electrospun nanofiber membranes are employed for advanced separation, MBRs have been utilized to produce renewable energy. Polymeric membranes have unlocked numerous opportunities for applications in separation processes. Thus, the potential for new technologies, trends, and applications for membrane utilization continues to expand.

8. Successful Industrial Implementations

Several implementations of membrane-based technology in industries have been reported so far in various applications, including water treatment [190], food and beverages [191], the energy sector [192], gas separation [193], etc. Commercialized membranes are used for water treatment and desalination. Ramdani et al. [190] compared the performance of two commercial TFC NF membranes (NF90 and NF270) regarding fluoride retention, desalination efficiency, and energy consumption. The highest permeate flow rate was found for NF270, while NF90 had the highest retention efficiency. Chandrapala et al. [191] utilized three commercialized NF membranes to separate undissociated lactose from charged lactate anions in an acid whey system. Membranes are also implemented in energy-related applications. A study by Sadrameli et al. [192] utilized membranes having commercialized benefits for the effective purification of biodiesel by removing the impurities present in biodiesel, such as glycerol, diglycerides, and triglycerides. These membranes were utilized on a large scale for the above-mentioned applications [194]. Therefore, due to economic feasibility, easy processing, less wastage, and negligible byproducts, membrane separation technology has emerged as a prominent technology for industrial implementation.

9. Future Directions in Polymeric Membrane Research

The research on polymeric membranes is one of the most evolving and dynamic fields, with many potentials in water treatment, water purification, and biomedical devices. The primary objective of membrane research has continuously been to innovate novel membrane materials with enhanced and modified properties compared to pre-existing membrane materials. Future advancements in membrane materials and fabrication will hinge on the sustainability, innovation, and scalability of novel approaches and the stability of resulting membranes. Additionally, delivering rapid prototype feasibility and engaging in interactive processes with risk-taking and visionary industries is crucial.

The advancement of society and the growing global concern over environmental pollution underscore the critical importance of green and sustainable synthesis. However, only a few studies currently adopt green synthesis approaches for preparing polymeric membranes. Furthermore, most currently used green techniques often fail to achieve complete objectives. However, there is ample potential for progress. Therefore, developing new strategies for polymeric membranes is essential to achieve the ultimate goal of green synthesis. To advance the field, future research should prioritize enhancing the synthesis efficiency of eco-friendly solvents, such as tributyl O-acetyl citrate (ATBC), ethyl lactate, ionic liquids (ILs), supercritical CO_2 ($ScCO_2$), γ-butyrolactone (g-BL), tributyl citrate (TBC), triethyl citrate (TEC), dioctyl sebacate (DOS), triacetate ester of glycerol (triacetin), methyl lactate, and propylene carbonate (PC), aiming to achieve or surpass the performance levels of traditional hazardous solvents. Additional investigations into these green solvents' polarity, viscosity, and cloud point are essential for further development.

Fouling and high energy consumption remain persistent challenges in non-equilibrium pressure-driven processes, highlighting the need for continued research to develop sustainable solutions. These solutions may involve implementing cost-effective yet robust pre-treatment methods or designing membranes that resist fouling. Recent advancements in fabrication technologies, including additive manufacturing (AM), are progressively being investigated for manufacturing membrane modules and desalination cell components with intricate designs [195,196]. These complex structures can mitigate reduced concentration polarization, mass transfer, high fouling, and significant pressure drops. A notable development is 4D printing, an extension of 3D printing, which enables materials to transform their physical properties when exposed to external stimuli. This innovative approach has attracted interest in enhancing the performance of membrane module components. Spiral-wound membrane modules (SWMs) are the most frequently produced using 3D-printing techniques [197].

Emerging modeling techniques, such as molecular dynamics simulation (MDS) and artificial intelligence (AI) tools, including fuzzy logic (FL), genetic algorithms (GA), and artificial neural networks (ANNs), are playing a significant role in advancing sustainability in membrane-based water treatment [198]. Hybridizing AI tools is suggested to achieve more robust global optimization, and expanding their application to include broader water resource management could further enhance their impact [199].

10. Conclusions

Polymeric membranes have emerged as pivotal components in modern separation technologies, offering versatile and efficient solutions for liquid separations across various industries. This review has comprehensively examined the fundamental principles, material innovations, fabrication techniques, and performance enhancement strategies that define the current landscape of polymeric membrane technology. Despite significant advancements, several challenges remain, including membrane stability, fouling, scalability, and environmental impact. Addressing these challenges necessitates ongoing research and development, focusing on advancements in material science and membrane design. Integrating novel materials, such as nanocomposites and advanced hybrid structures, holds promise for overcoming these limitations and achieving superior separation performance. The commercialization of polymeric membranes is advancing steadily, with successful implementations in water treatment and beyond. However, the full potential of these technologies is yet to be realized. Future research should focus on developing more sustainable, cost-effective, and high-performing membranes to meet the growing demands of industrial applications. In conclusion, polymeric membranes are set to play an increasingly important role in addressing global challenges related to resource conservation, environmental protection, and industrial efficiency. By continuing to innovate and refine these technologies, we can unlock new possibilities for liquid separations, driving progress in established and emerging fields.

Author Contributions: L.R.S., methodology, formal analysis, investigation, data curation, writing—original draft; D.Y., methodology, formal analysis, investigation, data curation, writing—original draft; D.B., formal analysis, investigation, data curation, writing—original draft; A.G., formal analysis, investigation, data curation, writing—original draft; S.G., formal analysis, investigation, data curation, writing—original draft; G.H., formal analysis, investigation, data curation, writing—original draft; S.K., formal analysis, investigation, data curation, writing—original draft; M.B.G., formal analysis, investigation, data curation, writing—original draft; S.V.S., formal analysis, investigation, data curation, writing—original draft; S.V.J., formal analysis, investigation, writing—one part; S.C., formal analysis, investigation, writing—one part; P.G.I., conceptualization, methodology, validation, investigation, writing—review and editing, visualization, supervision, project administration, funding acquisition. All authors have read and agreed to the published version of the manuscript.

Funding: We thank the Defence Research and Development Organisation (DRDO) for their funding under Project No. DFTM/07/3603/DIA-CoE, MZU/EWM/P-07 (GPP-0433).

Acknowledgments: We thank the Director of CSIR-NEIST for his enthusiastic support of this Manuscript No: CSIR-NEIST/PUB/2024/088. We thank the Defence Research and Development Organisation (DRDO) for their funding under Project No. DFTM/07/3603/DIA-CoE, MZU/EWM/P-07 (GPP-0433). S.G. and P.G.I. profoundly thank the Science and Engineering Research Board (SERB), New Delhi, India, for their generous funding under the CRG/2022/004008 project (GPP0414). D.B., A.G., S.V.S., and P.G.I. also gratefully acknowledge the Council of Scientific and Industrial Research (CSIR), India, for their crucial financial support under the FBR project (No. FBR080306). D.Y., S.K., and G.H. acknowledge the Department of Science and Technology (DST), India, for Inspire Fellowship Award Nos. IF190678, IF190333, and IF220297, respectively. L.R.S. acknowledges the UGC-NET-JRF fellowship, which has student ID No. 231610156586. M.B.G. acknowledges CSIR for granting financial support through the SRF direct fellowship ID No. 35110319.

Conflicts of Interest: The authors declare no conflicts of interest.

References

1. Cheng, Y.; Ying, Y.; Japip, S.; Jiang, S.D.; Chung, T.S.; Zhang, S.; Zhao, D. Advanced porous materials in mixed matrix membranes. *Adv. Mater.* **2018**, *30*, e1802401. [CrossRef] [PubMed]
2. Karki, S.; Hazarika, G.; Yadav, D.; Ingole, P.G. Polymeric membranes for industrial applications: Recent progress, challenges and perspectives. *Desalination* **2024**, *573*, 117200. [CrossRef]
3. Xu, X.; Yang, Y.; Liu, T.; Chu, B. Cost-effective polymer-based membranes for drinking water purification. *Giant* **2022**, *10*, 100099. [CrossRef]
4. Yadav, D.; Gohain, M.B.; Karki, S.; Ingole, P.G. A novel approach for the development of low-cost polymeric thin-film nanocomposite membranes for the biomacromolecule separation. *ACS Omega* **2022**, *7*, 47967–47985. [CrossRef] [PubMed]
5. Karki, S.; Ingole, P.G. Spray coating vapor-phase interfacial polymerization: A new approach for developing 2D nanosheet enabled thin film nanocomposite membranes to remove heavy metal ions. *Chem. Eng. J.* **2024**, *488*, 150883. [CrossRef]
6. Xu, Q.; Jiang, J. Recent development in machine learning of polymer membranes for liquid separation. *Mol. Syst. Des. Eng.* **2022**, *7*, 856–872. [CrossRef]
7. Bassyouni, M.; Abdel-Aziz, M.H.; Zoromba, M.S.; Abdel-Hamid, S.M.S.; Drioli, E. A review of polymeric nanocomposite membranes for water purification. *J. Ind. Eng. Chem.* **2019**, *73*, 19–46. [CrossRef]
8. Yadav, D.; Karki, S.; Gohain, M.B.; Ingole, P.G. Development of micropollutants removal process using thin-film nanocomposite membranes prepared by green new vapor-phase interfacial polymerization method. *Chem. Eng. J.* **2023**, *472*, 144940. [CrossRef]
9. Shiohara, A.; Prieto-Simon, B.; Voelcker, N.H. Porous polymeric membranes: Fabrication techniques and biomedical applications. *J. Mater. Chem. B* **2021**, *9*, 2129. [CrossRef]
10. Yadav, D.; Gohain, M.B.; Bora, M.; Sarma, S.S.; Karki, S.; Kumar, D.; Ingole, P.G. Greener synthesis of thin-film nanocomposite membranes with varied nanofillers for enhanced organic micropollutant removal. *Sep. Purif. Technol.* **2024**, *335*, 126125. [CrossRef]
11. Zhang, X.R.; Fu, J.; Liu, Y. Resourceful Treatment of Seawater Desalination or High Concentrated Sewage by Renewable Energy. In *Energy Solutions to Combat Global Warming*; Lecture Notes in Energy; Zhang, X., Dincer, I., Eds.; Springer: Cham, Switzerland, 2017; p. 33.
12. Cairone, S.; Hasan, S.W.; Choo, K.H.; Li, C.W.; Zarra, T.; Belgiorno, V.; Naddeo, V. Integrating artificial intelligence modeling and membrane technologies for advanced wastewater treatment: Research progress and future perspectives. *Sci. Total Environ.* **2024**, *944*, 173999. [CrossRef] [PubMed]
13. Ingole, P.G.; Jeon, J.D.; Hazarika, S.; Lee, H.K. Polymeric composite/nanocomposite membranes for diverse applications. In *Handbook of Polymer Nanocomposites for Industrial Applications, Micro and Nano Technologies*; Elsevier: Amsterdam, The Netherlands, 2021; pp. 169–199.
14. Teow, Y.H.; Chiah, Y.H.; Ho, K.C.; Mahmoudi, E. Treatment of semiconductor-industry wastewater with the application of ceramic membrane and polymeric membrane. *J. Clean. Prod.* **2022**, *337*, 130569. [CrossRef]
15. Steven, S.; Mulyono, M.; Yustisia, A.; Soekotjo, E.S.A.; Otivriyanti, G.; Wardani, M.L.D.; Murti, Z.; Sinaga, R.Y.H.; Laili, N.S.; Suantika, G.; et al. Perspectives and research direction on polymeric membrane integration for sustainable aquaculture industries. *J. Environ. Chem. Eng.* **2024**, *12*, 111691. [CrossRef]
16. Hazarika, G.; Jadhav, S.V.; Ingole, P.G. Exploring the potential of polymeric membranes in cutting-edge chemical and biomedical applications: A review. *Mater. Today Commun.* **2024**, *39*, 109022. [CrossRef]
17. Lau, H.S.; Eugenia, A.; Weng, Y.; Yong, W.F. Recent advances in polymers of intrinsic microporosity (PIMs) membranes: Delving into the intrinsic microstructure for carbon capture and arduous industrial applications. *Prog. Mater. Sci.* **2024**, *145*, 101297. [CrossRef]
18. Borban, B.; Gohain, M.B.; Yadav, D.; Karki, S.; Ingole, P.G. Nano-electrospun membranes: Green solutions for diverse industrial needs. *J. Hazard. Mater. Adv.* **2023**, *12*, 100373. [CrossRef]
19. Nour, A.; Iqbal, W.; Navarro-Alapont, J.; Ferrando-Soria, J.; Magarò, P.; Elliani, R.; Tagarelli, A.; Maletta, C.; Mastropietro, T.F.; Pardo, E.; et al. Efficient Nickel and Cobalt Recovery by Metal-Organic Framework-Based Mixed Matrix Membranes (MMM-MOFs). *ACS Sustain. Chem. Eng.* **2024**, *12*, 12014–12028. [CrossRef]
20. Burtch, N.C.; Jasuja, H.; Walton, K.S. Water Stability and Adsorption in Metal-Organic Frameworks. *Chem. Rev.* **2014**, *114*, 10575–10612. [CrossRef]
21. Jaid, G.M.; AbdulRazak, A.A.; Meskher, H.; Al-Saadi, S.; Alsalhy, Q.F. Metal-organic frameworks (MOFs), covalent organic frameworks (COFs), and hydrogen-bonded organic frameworks (HOFs) in mixed matrix membranes. *Mater. Today Sustain.* **2024**, *25*, 100672. [CrossRef]
22. Duan, Y.; Li, L.; Shen, Z.; Cheng, J.; He, K. Engineering Metal-Organic-Framework (MOF)-Based Membranes for Gas and Liquid Separation. *Membranes* **2023**, *13*, 480. [CrossRef]
23. Yadav, D.; Karki, S.; Ingole, P.G. Current advances and opportunities in the development of nanofiltration (NF) membranes in the area of wastewater treatment, water desalination, biotechnological and pharmaceutical applications. *J. Environ. Chem. Eng.* **2022**, *10*, 108109. [CrossRef]

24. Khoo, Y.S.; Seah, M.Q.; Lau, W.J.; Liang, Y.Y.; Karaman, M.; Gürsoy, M.; Meng, J.; Gao, H.; Ismail, A.F. Environmentally friendly approach for the fabrication of polyamide thin film nanocomposite membrane with enhanced antifouling and antibacterial properties. *Sep. Purif. Technol.* **2021**, *260*, 118249. [CrossRef]
25. Albayati, N.; Naser, Z.A.; Ahmed, H.A.B.; Kadhom, M.; Oladoye, P.O. A comprehensive review on the use of $Ti_3C_2T_x$ MXene in membrane-based water treatment. *Sep. Purif. Technol.* **2024**, *345*, 127448. [CrossRef]
26. Mansourpanah, Y.; Ghanbari, A.; Yazdani, H.; Mohammadi, A.G.; Rahimpour, A. Silver-polyamidoamine/graphene oxide thin film nanofiltration membrane with improved antifouling and antibacterial properties for water purification and desalination. *Desalination* **2021**, *511*, 115109. [CrossRef]
27. Elsayed, A.; Li, Z.; Khan, K.; Cormier, R.; de Lannoy, C.F. Predicting membrane cleaning effectiveness in a full-scale water treatment plant using an artificial neural network model. *J. Water Proc. Eng.* **2024**, *66*, 105932. [CrossRef]
28. Fane, A.G.; Wang, R.; Jia, Y. Membrane Technology: Past, Present and Future. In *Membrane and Desalination Technologies. Handbook of Environmental Engineering*; Wang, L.K., Chen, J.P., Hung, Y.T., Shammas, N.K., Eds.; Humana Press: Totowa, NJ, USA, 2011; Volume 13.
29. Zsigmondy, R.; Bachmann, W. Über neue Filter. *Z. Anorg. Allg. Chem.* **1918**, *103*, 119. [CrossRef]
30. Singh, R. Introduction to Membrane Technology. In *Membrane Technology and Engineering for Water Purification*, 2nd ed.; Elsevier: Amsterdam, The Netherlands, 2015; pp. 1–80.
31. Strathmann, H. *Introduction to Membrane Science and Technology*; Wiley-VCH: Weinheim, Germany, 2011.
32. Xia, Q.; Liu, M.; Cao, X.; Wang, Y.; Xing, W.; Sun, S. Structure design and applications of dual-layer polymeric membranes. *J. Membr. Sci.* **2018**, *562*, 85–111. [CrossRef]
33. Rezakazemi, M.; Sadrzadeh, M.; Matsuura, T. Thermally stable polymers for advanced high-performance gas separation membranes. *Prog. Energy Combust. Sci.* **2018**, *66*, 1–41. [CrossRef]
34. Mashhadikhan, S.; Ahmadi, R.; Amooghin, A.E.; Sanaeepur, H.; Aminabhavi, T.M.; Rezakazemi, M. Breaking temperature barrier: Highly thermally heat resistant polymeric membranes for sustainable water and wastewater treatment. *Renew. Sustain. Energy Rev.* **2024**, *189*, 113902. [CrossRef]
35. Cadotte, J.E.; Petersen, R.I. Thin Film Reverse Osmosis Membranes: Origin, Development, and Recent Advances. In *Synthetic Membranes*; ACS Symposium Series; Turbak, A.F., Ed.; American Chemical Society: Washington, DC, USA, 1981; Volume 1, pp. 305–325.
36. Jhaveri, J.H.; Murthy, Z.V.P. A Comprehensive Review on Anti-fouling Nanocomposite Membranes for Pressure Driven Membrane Separation Processes. *Desalination* **2016**, *379*, 137–154. [CrossRef]
37. Satchanska, G.; Davidova, S.; Petrov, P.D. Natural and Synthetic Polymers for Biomedical and Environmental Applications. *Polymers* **2024**, *16*, 1159. [CrossRef] [PubMed]
38. Shrestha, R.; Ban, S.; Devkota, S.; Sharma, S.; Joshi, R.; Tiwari, A.P.; Kim, H.Y.; Joshi, M.K. Technological Trends in Heavy Metals Removal from Industrial Wastewater: A Review. *J. Environ. Chem. Eng.* **2021**, *9*, 105688. [CrossRef]
39. Nqombolo, A.; Mpupa, A.; Moutloali, R.M.; Nomngongo, P.N. Wastewater Treatment Using Membrane Technology. *Wastewater Water Qual.* **2018**, *29*, 30–40.
40. Padaki, M.; Murali, R.S.; Abdullah, M.S.; Misdan, N.; Moslehyani, A.; Kassim, M.A.; Hilal, N.; Ismail, A.F. Membrane Technology Enhancement in Oil-Water Separation. A Review. *Desalination* **2015**, *357*, 197–207. [CrossRef]
41. Gahlot, S.; Kulshrestha, V. Fluoropolymer Nanocomposite Membranes for Fuel Cell Applications. In *Advanced Fluoropolymer Nanocomposites*; Fabrication, Processing, Characterization and Applications; Woodhead Publishing Series in Composites Science and Engineering; Woodhead Publishing: Cambridge, UK, 2023; pp. 597–643.
42. Daramola, M.O.; Aransiola, E.F.; Ojumu, T.V. Potential Applications of Zeolite Membranes in Reaction Coupling Separation Processes. *Materials* **2012**, *5*, 2101–2136. [CrossRef]
43. Valappil, R.S.K.; Ghasem, N.; Al-Marzouqi, M. Current and Future Trends in Polymer Membrane-based Gas Separation Technology: A Comprehensive Review. *J. Ind. Eng. Chem.* **2021**, *98*, 103–129. [CrossRef]
44. Kochkodan, V.; Johnson, D.J.; Hilal, N. Polymeric Membranes: Surface Modification for Minimizing (bio)Colloidal Fouling. *Adv. Coll. Interface Sci.* **2014**, *206*, 116. [CrossRef]
45. Ong, C.S.; Goh, P.S.; Lau, W.J.; Misdan, N.; Ismail, A.F. Nanomaterials for Biofouling and Scaling Mitigation of Thin Film Composite Membrane: A Review. *Desalination* **2016**, *393*, 2–15. [CrossRef]
46. Ingole, P.G.; Ingole, N.P. Methods for separation of organic and pharmaceutical compounds by different polymer materials. *Korean J. Chem. Eng.* **2014**, *31*, 2109–2123. [CrossRef]
47. Jang, J. Classification of membranes: With respect to pore size, material, and module type. In *Current Developments in Biotechnology and Bioengineering*; Membrane Technology for Sustainable Water and Energy Management; Elsevier: Amsterdam, The Netherlands, 2023; pp. 3–17.
48. Mohshim, D.F.; Mukhtar, H.B.; Man, Z.; Nasir, R. Latest Development on Membrane Fabrication for Natural Gas Purification: A Review. *J. Eng.* **2013**, *2013*, 1–7. [CrossRef]
49. Fard, A.K.; McKay, G.; Buekenhoudt, A.; Al Sulaiti, H.; Motmans, F.; Khraisheh, M.; Atieh, M. Inorganic Membranes: Preparation and Application for Water Treatment and Desalination. *Materials* **2018**, *11*, 74. [CrossRef] [PubMed]
50. Baker, R.W. Membranes and Modules. In *Membrane Technology and Applications*; John Wiley & Sons: Hoboken, NJ, USA, 2023; pp. 89–160.

51. Berk, Z. Membrane processes. In *Food Process Engineering and Technology*, 3rd ed.; Elsevier: Amsterdam, The Netherlands, 2018; pp. 261–287.
52. Omidifar, M.; Babaluo, A.A. Fabrication of thin (∼2 μm) pure Ni and Pd–Ni alloy composite membranes by the organic-inorganic activation method for hydrogen separation. *Int. J. Hydrogen Energy* **2024**, *53*, 1025–1036. [CrossRef]
53. Ingole, P.G.; Bajaj, H.C.; Singh, K. Enantiomeric separation of α-amino acids by imprinted terpolymer membrane. *Arab. J. Chem.* **2016**, *9*, S960–S965. [CrossRef]
54. Gu, B.; Kim, D.Y.; Kim, J.H.; Yang, D.R. Mathematical model of flat sheet membrane modules for FO process: Plate-and-frame module and spiral-wound module. *J. Membr. Sci.* **2011**, *79*, 403–415. [CrossRef]
55. Wan, C.F.; Yang, T.; Lipscomb, G.G.; Stookey, D.J.; Chung, T.-S. Design and fabrication of hollow fiber membrane modules. *J. Membr. Sci.* **2017**, *538*, 96–107. [CrossRef]
56. Jiang, S.; Zhang, M.; Zhang, Y.; Chang, H.; Wang, Y.; Wang, H.; Cheng, N.; Hu, L.; Liang, H.; Tang, X. Engineering dual-reinforced antibacterial and anti-fouling ultrafiltration membrane by in-situ synthesising super-dispersion and high-activity silver nanoparticles (AgNPs). *Chem. Eng. J.* **2024**, *500*, 156671. [CrossRef]
57. Khulbe, K.C.; Matsuura, T. Recent progress in polymeric hollow-fibre membrane preparation and applications. *Membr. Technol.* **2016**, *2016*, 7–13. [CrossRef]
58. Zhang, W.; Xue, Z.; Yan, M.; Liu, J.; Xia, Y. Effect of epichlorohydrin on the wet spinning of carrageenan fibers under optimal parameter conditions. *Carbohydr. Polym.* **2016**, *150*, 232–240. [CrossRef]
59. Yan, B.; Zhang, Y.; Li, Z. Electrospun nanofibrous membrane for biomedical application. *SN Appl. Sci.* **2022**, *4*, 172. [CrossRef]
60. Hauru, L.K.J.; Hummel, M.; Michud, A.; Sixta, H. Dry jet-wet spinning of strong cellulose filaments from ionic liquid solution. *Cellulose* **2014**, *21*, 4471–4481. [CrossRef]
61. Ding, X.; Cao, Y.; Zhao, H.; Wang, L. Interfacial morphology between the two layers of the dual-layer asymmetric hollow fiber membranes fabricated by co-extrusion and dry-jet wet-spinning phase-inversion techniques. *J. Membr. Sci.* **2013**, *444*, 482–492. [CrossRef]
62. Shehata, N.; Egirani, D.; Olabi, A.; Inayat, A.; Abdelkareem, M.A.; Chae, K.; Sayed, E.T. Membrane-based water and wastewater treatment technologies: Issues, current trends, challenges, and role in achieving sustainable development goals, and circular economy. *Chemosphere* **2023**, *320*, 137993. [CrossRef] [PubMed]
63. Imbrogno, A.; Calvo, J.I.; Breida, M.; Schwaiger, R.; Schäfer, A.I. Molecular weight cut off (MWCO) determination in ultra- and nanofiltration: Review of methods and implications on organic matter removal. *Sep. Purif. Technol.* **2025**, *354*, 128612. [CrossRef]
64. Lejarazu-Larrañaga, A.; Ortiz, J.M.; Molina, S.; Pawlowski, S.; Galinha, C.F.; Otero, V.; García-Calvo, E.; Velizarov, S.; Crespo, J.G. Nitrate Removal by Donnan Dialysis and Anion-Exchange Membrane Bioreactor Using Upcycled End-of-Life Reverse Osmosis Membranes. *Membranes* **2022**, *12*, 101. [CrossRef] [PubMed]
65. Roy, Y.; Warsinger, D.M.; Lienhard, J.H. Effect of temperature on ion transport in nanofiltration membranes: Diffusion, convection and electromigration. *Desalination* **2017**, *420*, 241–257. [CrossRef]
66. Rijnaarts, T.; Shenkute, N.T.; Wood, J.A.; de Vos, W.M.; Nijmeijer, K. Divalent cation removal by Donnan dialysis for improved reverse electrodialysis. *ACS Sustain. Chem. Eng.* **2018**, *6*, 7035–7041. [CrossRef]
67. Goyal, P.; Sundarrajan, S.; Ramakrishna, S. A Review on Mixed Matrix Membranes for Solvent Dehydration and Recovery Process. *Membranes* **2021**, *11*, 441. [CrossRef]
68. Castro-Muñoz, R.; González-Valdez, J. New Trends in Biopolymer-Based Membranes for Pervaporation. *Molecules* **2019**, *24*, 3584. [CrossRef]
69. Mehanathan, S.; Jaafar, J.; Nasir, A.M.; Rahman, R.A.; Ismail, A.F.; Illias, R.M.; Othman, M.H.D.; Rahman, M.A.; Bilad, M.R.; Naseer, M.N. Adsorptive Membrane for Boron Removal: Challenges and Future Prospects. *Membranes* **2022**, *12*, 798. [CrossRef]
70. Nasir, A.M.; Goh, P.S.; Abdullah, M.S.; Ng, B.C.; Ismail, A.F. Adsorptive nanocomposite membranes for heavy metal remediation: Recent progresses and challenges. *Chemosphere* **2019**, *232*, 96–112. [CrossRef]
71. Goh, P.; Ismail, A. Flat-Sheet Membrane for Power Generation and Desalination Based on Salinity Gradient. In *Membrane-Based Salinity Gradient Processes for Water Treatment and Power Generation*; Sarp, S., Hilal, N., Eds.; Elsevier: Amsterdam, The Netherlands, 2017; pp. 155–174.
72. Loza, S.; Loza, N.; Kovalchuk, N.; Romanyuk, N.; Loza, J. Comparative Study of Different Ion-Exchange Membrane Types in Diffusion Dialysis for the Separation of Sulfuric Acid and Nickel Sulfate. *Membranes* **2023**, *13*, 396. [CrossRef] [PubMed]
73. Khan, M.I.; Shanableh, A.; Khraisheh, M.; AlMomani, F. Synthesis of Porous BPPO-Based Anion Exchange Membranes for Acid Recovery via Diffusion Dialysis. *Membranes* **2021**, *12*, 95. [CrossRef] [PubMed]
74. Tang, Y.; Lin, Y.; Ford, D.M.; Qian, X.; Cervellere, M.R.; Millett, P.C.; Wang, X. A Review on Models and Simulations of Membrane Formation via Phase Inversion Processes. *J. Membr. Sci.* **2021**, *640*, 119810. [CrossRef]
75. Guillen, G.R.; Ramon, G.Z.; Kavehpour, H.P.; Kaner, R.B.; Hoek, E.M. Direct Microscopic Observation of Membrane Formation by Nonsolvent Induced Phase Separation. *J. Membr. Sci.* **2013**, *431*, 212–220. [CrossRef]
76. Graham, P.; McHugh, A. Kinetics of Thermally Induced Phase Separation in a Crystallizable Polymer Solution. *Macromolecules* **1998**, *31*, 2565–2568. [CrossRef]
77. Masaro, L.; Zhu, X. Physical Models of Diffusion for Polymer Solutions, Gels and Solids. *Prog. Polym. Sci.* **1999**, *24*, 731–775. [CrossRef]

78. Phillies, G.D. Universal Scaling Equation for Self-Diffusion by Macromolecules in Solution. *Macromolecules* **1986**, *19*, 2367–2376. [CrossRef]
79. Li, X.; Chen, C.; Li, J. Formation Kinetics of Polyethersulfone with Cardo Membrane via Phase Inversion. *J. Membr. Sci.* **2008**, *314*, 206–211. [CrossRef]
80. Kausar, A. Phase Inversion Technique-Based Polyamide Films and Their Applications: A Comprehensive Review. *Polym.-Plast. Technol. Mater.* **2017**, *56*, 1421–1437. [CrossRef]
81. Wang, D.M.; Lai, J.Y. Recent Advances in Preparation and Morphology Control of Polymeric Membranes formed by Nonsolvent Induced Phase Separation. *Curr. Opin. Chem. Eng.* **2013**, *2*, 229–237. [CrossRef]
82. Liu, M.; Liu, S.H.; Xu, Z.L.; Wei, Y.M.; Yang, H. Formation of Microporous Polymeric Membranes via Thermally Induced Phase Separation: A Review. *Front. Chem. Sci. Eng.* **2016**, *10*, 57–75. [CrossRef]
83. Choi, O.; Ingole, P.G.; Park, P.H. Precision-aiming tuning of membranes prepared by NIPS and its performance enhancement. *J. Clean. Prod.* **2022**, *365*, 132858. [CrossRef]
84. Venault, A.; Chang, Y.; Wang, D.M.; Bouyer, D. A Review on Polymeric Membranes and Hydrogels Prepared by Vapor-Induced Phase Separation Process. *Polym. Rev.* **2013**, *53*, 568–626. [CrossRef]
85. Kim, J.F.; Kim, J.H.; Lee, Y.M.; Drioli, E. Thermally Induced Phase Separation and Electrospinning Methods for Emerging Membrane Applications: A Review. *AIChE J.* **2016**, *62*, 461–490. [CrossRef]
86. Nawi, N.I.M.; Chean, H.M.; Shamsuddin, N.; Bilad, M.R.; Narkkun, T.; Faungnawakij, K.; Khan, A.L. Development of Hydrophilic PVDF Membrane Using Vapour Induced Phase Separation Method for Produced Water Treatment. *Membranes* **2020**, *10*, 121. [CrossRef]
87. Figoli, A.; Marino, T.; Simone, S.; Nicolò, E.D.; Li, X.M.; He, T.; Tornaghi, S.; Drioli, E. Towards Non-Toxic Solvents for Membrane Preparation: A Review. *Green Chem.* **2014**, *16*, 4034–4059. [CrossRef]
88. Fareed, H.; Jang, K.; Lee, W.; Kim, I.S.; Han, S. Tailoring Fully Crosslinked Polyamide layers on Optimized Polyacrylonitrile Supports Via Coactive Delayed Phase Inversion and Alkaline Hydrolysis for Brine Treatment Through Pervaporation. *Sep. Purif. Technol.* **2024**, *337*, 126309. [CrossRef]
89. Ren, S.; Huang, G.; Yao, Y.; Zhang, P.; Zhang, Z.; Wang, Y. Integrating Radical Polymerization and Non-Solvent Induced Phase Inversion Strategy for Functionalized Ultrafiltration Membrane Fabrication. *Desalination* **2024**, *573*, 117220. [CrossRef]
90. Borban, B.; Yadav, D.; Gohain, M.B.; Karki, S.; Mukherjee, A.; Ghosh, P.; Ingole, P.G. Electrospun nanofibrous membranes with functionalized 2D nanofillers for efficient micropollutant removal from water. *Desalination* **2024**, *591*, 118017. [CrossRef]
91. Ahmed, F.E.; Lalia, B.S.; Hashaikeh, R. A Review on Electrospinning for Membrane Fabrication: Challenges and Applications. *Desalination* **2015**, *356*, 15–30. [CrossRef]
92. Munoz, C.R. A Critical Review on Electrospun Membranes Containing 2D Materials for seawater Desalination. *Desalination* **2023**, *555*, 116528. [CrossRef]
93. Behroozi, A.M.; Shaeli, A.M.; Vatanpour, V. Fabrication and Modification of Nanofiltration Membranes by Solution Electrospinning Technique: A Review of Influential Factors and Applications in Water Treatment. *Desalination* **2023**, *558*, 116638. [CrossRef]
94. Mamun, A.; Sabantina, L.; Klocker, M.; Heide, A.; Blachowicz, T.; Ehrmann, A. Electrospinning Nanofiber Mats with Magnetite Nanoparticles Using Various Needle-Based Techniques. *Polymers* **2022**, *14*, 533. [CrossRef] [PubMed]
95. Lee, J.; Moon, S.; Lahann, J.; Lee, K.J. Recent Progress in Preparing Nonwoven Nanofibers via Needleless Electrospinning, Macromol. *Mater. Eng.* **2023**, *308*, 2300057.
96. Raaijmakers, M.J.T.; Benes, N.E. Current Trends in Interfacial Polymerization Chemistry. *Prog. Polym. Sci.* **2016**, *63*, 86–142. [CrossRef]
97. Sarkar, P.; Wu, C.; Yang, Z.; Tang, C.Y. Empowering Ultrathin Polyamide Membranes at the Water–Energy Nexus: Strategies, Limitations, and Future Perspectives. *Chem. Soc. Rev.* **2024**, *53*, 4374–4399. [CrossRef]
98. Khorshidi, B.; Thundat, T.; Fleck, B.A.; Sadrzadeh, M. A Novel Approach Toward Fabrication of High Performance Thin Film Composite Polyamide Membranes. *Sci. Rep.* **2016**, *6*, 22069. [CrossRef]
99. Lin, L.; Feng, C.; Lopez, R.; Coronell, O. Identifying Facile and Accurate Methods to Measure the Thickness of the Active Layers of Thin-Film Composite Membranes—A Comparison of Seven Characterization Techniques. *J. Membr. Sci.* **2016**, *498*, 167–179. [CrossRef]
100. Liu, C.; Zhu, C.Y.; Zhang, C.; Yang, H.C.; Xu, Z.K. Thermodynamic and Kinetic Understanding for Managing the Controllability of Interfacial Polymerization. *Prog. Polym. Sci.* **2024**, *152*, 101815. [CrossRef]
101. Zhao, G.J.; Li, L.L.; Gao, H.Q.; Zhao, Z.J.; Pang, Z.F.; Pei, C.L.; Qu, Z.; Dong, L.L.; Rao, D.W.; Caro, J.; et al. Polyamide Nanofilms through a Non-Isothermal-Controlled Interfacial Polymerization. *Adv. Funct. Mater.* **2024**, *34*, 2313026. [CrossRef]
102. Paul, M.; Jons, S.D. Chemistry and Fabrication of Polymeric Nanofiltration Membranes: A Review. *Polymer* **2016**, *103*, 417–456. [CrossRef]
103. Pan, X.D.; Maydell, E.A.; Milne, R.H.; Fabian, D.J. Diamond-Like Carbon Films Prepared by rf Plasma Deposition. *Vacuum* **1990**, *41*, 1360–1363. [CrossRef]
104. Karan, S.; Samitsu, S.; Peng, X.; Kurashima, K.; Ichinose, I. Ultrafast viscous permeation of organic solvents through diamond-like carbon nanosheets. *Science* **2012**, *335*, 444–447. [CrossRef] [PubMed]

105. Iqbal, A.; Cevik, E.; Mustafa, A.; Qahtan, T.F.; Zeeshan, M.; Bozkurt, A. Emerging developments in polymeric nanocomposite membrane-based filtration for water purification: A concise overview of toxic metal removal. *Chem. Eng. J.* **2024**, *481*, 148760. [CrossRef]
106. Steen, M.; Hymas, L.; Havey, E.D.; Capps, N.E.; Castner, D.G.; Fisher, E.R. Low temperature plasma treatment of asymmetric polysulfone membranes for permanent hydrophilic surface modification. *J. Membr. Sci.* **2001**, *188*, 97–114. [CrossRef]
107. Xu, G.R.; Wang, S.H.; Zhao, H.L.; Wu, S.B.; Xu, J.M.; Li, L.; Liu, X.Y. Layer-by-layer (LBL) assembly technology as promising strategy for tailoring pressure-driven desalination membranes. *J. Membr. Sci.* **2015**, *493*, 428–443. [CrossRef]
108. Ingole, P.G.; Singh, K.; Bajaj, H.C. Enantioselective permeation of α-amino acid isomers through polymer membrane containing chiral metal–Schiff base complexes. *Desalination* **2011**, *281*, 413–421. [CrossRef]
109. Zhao, X.; Yang, M.; Shi, Y.; Sun, L.; Zheng, H.; Wu, M.; Gao, G.; Ma, T.; Li, G. Multifunctional bacterial cellulose-bentonite@polyethylenimine composite membranes for enhanced water treatment: Sustainable dyes and metal ions adsorption and antibacterial properties. *J. Hazard. Mater.* **2024**, *477*, 135267. [CrossRef]
110. Petukhov, D.I.; Johnson, D.J. Membrane Modification with Carbon Nanomaterials for Fouling Mitigation: A Review. *Adv. Colloid Interface Sci.* **2024**, *327*, 103140. [CrossRef]
111. Hosseini, S.S.; Bringas, E.; Tan, N.R.; Ortiz, I.; Ghahramani, M.; Shahmirzadi, M.A.A. Recent Progress in Development of High Performance Polymeric Membranes and Materials for Metal Plating Wastewater Treatment: A Review. *J. Water Process Eng.* **2016**, *9*, 78–110. [CrossRef]
112. Labuto, G.; Sanches, S.; Crespo, J.G.; Pereira, V.J.; Huertas, R.M. Stability of Polymeric Membranes to UV Exposure Before and After Coating with TiO_2 Nanoparticles. *Polymers* **2022**, *14*, 124. [CrossRef] [PubMed]
113. Ji, Y.; Qian, W.; Yu, Y.; An, Q.; Liu, L.; Zhou, Y.; Gao, C. Recent Developments in Nanofiltration Membranes based on Nanomaterials. Chin. *J. Chem. Eng.* **2017**, *25*, 1639–1652.
114. Raja, R.I.; Rashid, K.T.; Toma, M.A.; Abdul Razak, A.A.; Shehab, M.A.; Hernadi, K. A novel Polyethersulfone/Chamomile (PES/Chm) mixed matrix membranes for wastewater treatment applications. *J. Saudi Chem. Soc.* **2024**, *28*, 101805. [CrossRef]
115. Mahajan, R.; Burns, R.; Schaeffer, M.; Koros, W.J. Challenges in forming Successful Mixed Matrix Membranes with Rigid Polymeric Materials. *J. Appl. Polym. Sci.* **2002**, *86*, 881–890. [CrossRef]
116. El-Sawaf, A.K.; Hemdan, M.; Selim, H.; Nassar, A.A.; Mubarak, M.F. Revolutionizing water treatment: Enhanced flux and selectivity in polyethersulfone mixed matrix membrane through magnetic CuO-functionalized Fe_3O_4 nanoparticles for synthetic oily produced water remediation. *Surf. Interfaces* **2024**, *46*, 104142. [CrossRef]
117. Ottino, J.M.; Shah, N. Analysis of Transient Sorption and Permeation of Small Molecules in Multiphase Polymer Systems. *Polym. Eng. Sci.* **1984**, *24*, 153–162. [CrossRef]
118. Karki, S.; Ingole, P.G. Graphene-based thin film nanocomposite membranes for separation and purification. *Compr. Anal. Chem.* **2020**, *91*, 73–97.
119. Bhaskar, V.V.; Kaleekkal, N.J. Next-Generation Thin-Film Composite Nanofiltration Membranes for Water Remediation: A Review. *Emergent Mater.* **2022**, *5*, 1373–1390. [CrossRef]
120. Yadav, D.; Ingole, P.G. Polyurethane Membranes Preparation, Characterization and Their Diverse Applications. In *Polyurethanes: Preparation, Properties, and Applications Volume 1: Fundamentals*; Chapter 4; ACS Symposium Series; American Chemical Society: Washington, DC, USA, 2023; Volume 1452, pp. 59–78.
121. Paseta, L.; Górriz, C.E.; Téllez, C.; Coronas, J. Vapor Phase Interfacial Polymerization: A Method to Synthesize Thin Film Composite Membranes without Using Organic Solvents. *Green Chem.* **2021**, *23*, 2449–2456. [CrossRef]
122. Jose, T.; George, S.C.; Maya, M.G.; Maria, H.J.; Wilson, R.; Thomas, S. Effect of Bentonite Clay on the Mechanical, Thermal, andPervaporation Performance of the Poly(vinyl alcohol)Nanocomposite Membranes. *Ind. Eng. Chem. Res.* **2014**, *53*, 16820–16831. [CrossRef]
123. Dong, H.; Wu, L.; Zhang, L.; Chen, H.; Gao, C. Clay nanosheets as charged filler materials for high-performance and fouling-resistant thin film nanocomposite membranes. *J. Membr. Sci.* **2015**, *494*, 92–103. [CrossRef]
124. Selvan, B.K.; Thiyagarajan, K.; Das, S.; Jaya, N.; Jabasingh, S.A.; Saravanan, P.; Rajasimman, M.; Vasseghian, Y. Synthesis and characterization of nano zerovalent iron-kaolin clay (nZVI-Kaol) composite polyethersulfone (PES) membrane for the efficacious As_2O_3 removal from potable water samples. *Chemosphere* **2022**, *288*, 132405. [CrossRef]
125. Hardian, R.; Alammar, A.; Holtzl, T.; Szekely, G. Fabrication of sustainable organic solvent nanofiltration membranes using cellulose–chitosan biopolymer blends. *J. Membr. Sci.* **2022**, *658*, 120743. [CrossRef]
126. Jiang, Y.H.; Zhang, Y.Q.; Gao, C.; An, Q.D.; Xiao, Z.Y.; Zhai, S.R. Superhydrophobic aerogel membrane with integrated functions of biopolymers for efficient oil/water separation. *Sep. Purif. Technol.* **2022**, *282*, 120138. [CrossRef]
127. Tomietto, P.; Russo, F.; Galiano, F.; Loulergue, P.; Salerno, S.; Paugam, L.; Audic, J.L.; Bartolo, L.D.; Figoli, A. Sustainable fabrication and pervaporation application of bio-based membranes: Combining a polyhydroxyalkanoate (PHA) as biopolymer and Cyrene™ as green solvent. *J. Membr. Sci.* **2022**, *643*, 120061. [CrossRef]
128. Peeva, L.; Marchetti, P.; Livingston, A. Nanofiltration Operations in Nonaqueous Systems. In *Comprehensive Membrane Science and Engineering*, 2nd ed.; Drioli, E., Giorno, L., Fontananova, E., Eds.; Elsevier: Amsterdam, The Netherlands, 2017; Volume 2, pp. 36–78.
129. Ingole, P.G. Application of sustainable nanocomposites in membrane technology. In *Sustainable Polymer Composites and Nanocomposites*; Springer International Publishing: Berlin/Heidelberg, Germany, 2019; pp. 935–960.

130. Gupta, V.K.; Sethi, B. Liquid–Liquid Separation Through Polymeric Membranes. In *Transport Properties of Polymeric Membranes*; Thomas, S., Wilson, R., Kumar, S.A., George, S.C., Eds.; Elsevier: Amsterdam, The Netherlands, 2017; pp. 217–241.
131. SDG 6 Infographics, UN-Water Publications. United Nations (UN). Available online: https://www.unwater.org/publications/sdg-6-infographics (accessed on 8 May 2015).
132. Jiang, Q.; Zhang, K. Optimization of Preparation Conditions of Poly(m-phenylene isophthalamide) PMIA Hollow Fiber Nanofiltration Membranes for Dye/Salt Wastewater Treatment. *Membranes* **2022**, *12*, 1258. [CrossRef]
133. Li, P.P.; Xu, Z.L.; Fang, Y.X.; Pandaya, D.; Gao, L.L.; Ding, R. Novel TFC membrane with crosslinked PMIA active layer fabricated by the T-FLO technique for dye desalination. *J. Membr. Sci.* **2023**, *681*, 121779. [CrossRef]
134. Usha, Z.R.; Babiker, D.M.D.; Yu, R.; Yang, J.; Chen, W.; Chen, X.; Li, L. Super hydrophilic modified biaxially oriented polypropylene microporous membrane for excellent gravity-driven oil/water emulsion separation. *J. Membr. Sci.* **2022**, *660*, 120840. [CrossRef]
135. Le, T.M.H.; Singto, S.; Sajomsang, W.; Mongkolnavin, R.; Nuisin, R.; Painmanakul, P.; Sairiam, S. Hydrophobic PVDF hollow fiber membrane modified with pulse inductively coupling plasma activation and chloroalkylsilanes for efficient dye wastewater treatment by ozonation membrane contactor. *J. Membr. Sci.* **2021**, *635*, 119443.
136. Gohain, M.B.; Karki, S.; Yadav, D.; Yadav, A.; Thakare, N.R.; Hazarika, S.; Lee, H.K.; Ingole, P.G. Development of antifouling thin-film composite/nanocomposite membranes for removal of phosphate and malachite green dye. *Membranes* **2022**, *12*, 768. [CrossRef]
137. Gohain, M.B.; Karki, S.; Ingole, P.G. Cellulose acetate, a source from discarded cigarette butts for the development of mixed matrix loose nanofiltration membranes for selective separation. *Int. J. Biol. Macromol.* **2024**, *271*, 132197. [CrossRef] [PubMed]
138. Yadav, D.; Karki, S.; Gohain, M.B.; Ingole, P.G. Two-Dimensional Nanomaterials-Based Polymer Nanocomposite Membranes for Liquid and Gas Separation. In *Two-Dimensional Nanomaterials-Based Polymer Nanocomposites: Processing, Properties and Applications*; John Wiley & Sons, Inc.: Hoboken, NJ, USA, 2024; pp. 681–712.
139. Otrembska, P.; Gega, J. Separation of nickel(II) and cadmium(II) ions with ion-exchange and membrane processes. *Sep. Sci. Technol.* **2016**, *51*, 2675–2680. [CrossRef]
140. Karki, S.; Ingole, P.G. Development of polymer-based new high performance thin-film nanocomposite nanofiltration membranes by vapor phase interfacial polymerization for the removal of heavy metal ions. *Chem. Eng. J.* **2022**, *446*, 137303. [CrossRef]
141. Karki, S.; Gohain, M.B.; Yadav, D.; Thakare, N.R.; Pawar, R.R.; Hazarika, S.; Ingole, P.G. Building rapid water transport channels within thin-film nanocomposite membranes based on 2D mesoporous nanosheets. *Desalination* **2023**, *547*, 116222. [CrossRef]
142. Li, X.; Mo, Y.; Qing, W.; Shao, S.; Tang, C.Y.; Li, J. Membrane-based technologies for lithium recovery from water lithium resources: A review. *J. Membr. Sci.* **2019**, *591*, 117317. [CrossRef]
143. Wen, X.; Ma, P.; Zhu, C.; He, Q.; Deng, X. Preliminary study on recovering lithium chloride from lithium-containing waters by nanofiltration. *Sep. Purif. Technol.* **2006**, *49*, 230–236. [CrossRef]
144. Zante, G.; Boltoeva, M.; Masmoudi, A.; Barillon, R.; Trébouet, D. Lithium extraction from complex aqueous solutions using supported ionic liquid membranes. *J. Membr. Sci.* **2019**, *580*, 62–76. [CrossRef]
145. Zhu, Y.; Yang, Y.; Fu, L.; Wu, Y. A porous gel-type composite membrane reinforced by nonwoven: Promising polymer electrolyte with high performance for sodium ion batteries. *Electrochim. Acta* **2017**, *224*, 405–411. [CrossRef]
146. Atsbha, M.W.; Farkash, L.; Bartal, G.; Nir, O. Extracting liquid fertilizer from ammonia-laden pharmaceutical wastewater using Hollow fiber membrane contactors. *J. Clean. Prod.* **2023**, *418*, 138183. [CrossRef]
147. Shahgodari, S.; Llorens, J.; Labanda, J. Viability of Total Ammoniacal Nitrogen Recovery Using a Polymeric Thin-Film Composite Forward Osmosis Membrane: Determination of Ammonia Permeability Coefficient. *Polymers* **2024**, *16*, 1834. [CrossRef]
148. Kurama, H.; Poetzschke, J.; Haseneder, R. The application of membrane filtration for the removal of ammonium ions from potable water. *Water Res.* **2002**, *36*, 2905–2909. [CrossRef]
149. Kannan, A.D.; Dillavou, J.; Gamage, K.; Randig, E.; Hettiarachchi, G.M.; Parameswaran, P. Recovery of high-quality calcium phosphate fertilizer products from anaerobic membrane bioreactor treated swine wastewater. *Chem. Eng. J.* **2023**, *53*, 139539. [CrossRef]
150. Santos, C.; Ribeiro, A.; Teixeira, M.R. Phosphorus recovery from waters using nanofiltration. *Desalin. Water Treat.* **2015**, *55*, 1308–1315. [CrossRef]
151. Gerardo, M.L.; Zacharof, M.P.; Lovitt, R.W. Strategies for the recovery of nutrients and metals from anaerobically digested dairy farm sludge using cross-flow microfiltration. *Water Res.* **2013**, *47*, 4833–4842. [CrossRef] [PubMed]
152. Liu, C.; Mao, H.; Zheng, J.; Zhang, S. Tight Ultrafiltration Membrane: Preparation and Characterization of Thermally Resistant Carboxylated Cardo Poly (arylene ether ketone)s (PAEK-COOH) Tight Ultrafiltration Membrane for Dye Removal. *J. Membr. Sci.* **2017**, *530*, 1–10. [CrossRef]
153. Minnesota Rubber and QMR Plastics. *Online Technical Report, Section 5: Plastic & Thermoplastic Elastomer Materials*; Minnesota Rubber and QMR Plastics: Hartland, WI, USA, 2003.
154. Biron, M. Plastics Solutions for Practical. In *Thermoplastics and Thermoplastic Composites*; Elsevier: Amsterdam, The Netherlands, 2018; pp. 883–1038.
155. Sudarsan, V. Materials for Hostile Chemical Environments. In *Materials Under Extreme Conditions*; Elsevier: Amsterdam, The Netherlands, 2017; pp. 129–158.
156. Balart, R.; Montanes, N.; Dominici, F.; Boronat, T.; Torres-Giner, S. Environmentally Friendly Polymers and Polymer Composites. *Material* **2020**, *13*, 4892. [CrossRef]

157. Bandehali, S.; Sanaeepur, H.; Ebadi Amooghin, A.; Shirazian, S.; Ramakrishna, S. Biodegradable Polymers for Membrane Separation. *Sep. Purif. Technol.* **2021**, *269*, 118731. [CrossRef]
158. Mir, I.S.; Riaz, A.; Roy, J.S.; Fréchette, J.; Morency, S.; Gomes, O.P.; Dumée, L.F.; Greener, J.; Messaddeq, Y. Removal of cadmium and chromium heavy metals from aqueous medium using composite bacterial cellulose membrane. *Chem. Eng. J.* **2024**, *490*, 151665. [CrossRef]
159. Wróblewska-Krepsztul, J.; Rydzkowski, T.; Borowski, G.; Szczypiński, M.; Klepka, T.; Thakur, V.K. Recent Progress in Biodegradable Polymers and Nanocomposite-Based Packaging Materials for Sustainable Environment. *Int. J. Polym. Anal. Charact.* **2018**, *23*, 383–395. [CrossRef]
160. Jadav, G.L.; Singh, P.S. Synthesis of Novel Silica-Polyamide Nanocomposite Membrane with Enhanced Properties. *J. Membr. Sci.* **2009**, *328*, 257–267. [CrossRef]
161. Jadav, G.L.; Aswal, V.K.; Singh, P.S. SANS Study to Probe Nanoparticle Dispersion in Nanocomposite Membranes of Aromatic Polyamide and Functionalized Silica Nanoparticles. *J. Colloid Interface Sci.* **2010**, *351*, 304–314. [CrossRef] [PubMed]
162. Xu, G.-R.; Wang, J.-N.; Li, C.-J. Strategies for Improving the Performance of the Polyamide Thin-film Composite (PA-TFC) Reverse Osmosis (RO) Membranes: Surface Modifications and Nanoparticles Incorporations. *Desalination* **2013**, *328*, 83–100. [CrossRef]
163. Nasir, R.; Mukhtar, H.; Man, Z.; Mohshim, D.F. Material Advancements in Fabrication of Mixed-Matrix Membranes. *Chem. Eng. Technol.* **2013**, *36*, 717–727. [CrossRef]
164. He, Y.; Tang, Y.P.; Ma, D.; Chung, T.-S. UiO-66 incorporated thin-film nanocomposite membranes for efficient selenium and arsenic removal. *J. Membr. Sci.* **2017**, *541*, 262–270. [CrossRef]
165. Wei, Z.; Li, J.; Wang, C.; Cao, J.; Yao, Y.; Lu, H.; Li, Y.; He, X. Thermally Stable Hydrophobicity in Electrospun Silica/Polydimethylsiloxane Hybrid Fibers. *Appl. Surf. Sci.* **2017**, *392*, 260–267. [CrossRef]
166. Li, Z.-K.; Lang, W.-Z.; Miao, W.; Yan, X.; Guo, Y.-J. Preparation and Properties of PVDF/ SiO_2@GO Nanohybrid Membranes via Thermally Induced Phase Separation Method. *J. Membr. Sci.* **2016**, *511*, 151–161. [CrossRef]
167. Mokhtari, S.; Rahimpour, A.; Shamsabadi, A.A.; Habibzadeh, S.; Soroush, M. Enhancing Performance and Surface Antifouling Properties of Polysulfone Ultrafiltration Membranes with Salicylate-Alumoxane Nanoparticles. *Appl. Surf. Sci.* **2017**, *393*, 93–102. [CrossRef]
168. Meng, N.; Priestley, R.C.E.; Zhang, Y.; Wang, H.; Zhang, X. The Effect of Reduction Degree of GO Nanosheets on Microstructure and Performance of PVDF/GO Hybrid Membranes. *J. Membr. Sci.* **2016**, *501*, 169–178. [CrossRef]
169. Ding, S.Y.; Wang, W. Covalent organic frameworks (COFs): From design to applications. *Chem. Soc. Rev.* **2013**, *42*, 548–568. [CrossRef] [PubMed]
170. Wan, L.; Zhou, C.; Xu, K.; Feng, B.; Huang, A. Synthesis of Highly Stable UiO-66-NH_2 Membranes with High Ions Rejection for Seawater Desalination. *Microporous Mesoporous Mater.* **2017**, *252*, 207–213. [CrossRef]
171. Liu, X.; Demir, N.K.; Wu, Z.; Li, K. Highly Water-Stable Zirconium Metal-Organic Framework UiO-66 Membranes Supported on Alumina Hollow Fibers for Desalination. *J. Am. Chem. Soc.* **2015**, *137*, 6999–7002. [CrossRef] [PubMed]
172. Zhu, Y.; Gupta, K.M.; Liu, Q.; Jiang, J.; Caro, J.; Huang, A. Synthesis and Seawater Desalination of Molecular Sieving Zeolitic Imidazolate Framework Membranes. *Desalination* **2016**, *385*, 75–82. [CrossRef]
173. Dasgupta, J.; Chakraborty, S.; Sikder, J.; Kumar, R.; Pal, D.; Curcio, S.; Drioli, E. The Effects of Thermally Stable Titanium Silicon Oxide Nanoparticles on Structure and Performance of Cellulose Acetate Ultrafiltration Membranes. *Sep. Purif. Technol.* **2014**, *133*, 55–68. [CrossRef]
174. Khorshidi, B.; Biswas, I.; Ghosh, T.; Thundat, T.; Sadrzadeh, M. Robust Fabrication of Thin-film Polyamide-TiO_2 Nanocomposite Membranes With Enhanced Thermal Stability and Anti-biofouling Propensity. *Sci. Rep.* **2018**, *8*, 784. [CrossRef] [PubMed]
175. Aristizábal, L.; Lively, R.; Nunes, P. Solvent and Thermally Stable Polymeric Membranes for Liquid Molecular Separations: Recent Advances, Challenges, and Perspectives. *J. Membr. Sci.* **2023**, *685*, 121972. [CrossRef]
176. Izák, P.; Hovorka, Š.; Bartovský, T.; Bartovská, L.; Crespo, J.G. Swelling of Polymeric Membranes in Room Temperature Ionic Liquids. *J. Membr. Sci.* **2007**, *296*, 131–138. [CrossRef]
177. Fan, H.; Elimelech, M.; Biesheuvel, P.M. Theory of Expansion and Compression of Polymeric Materials: Implications for Membrane Solvent Flow Under Compaction. *J. Membr. Sci.* **2024**, *697*, 122576. [CrossRef]
178. Zhang, R.; Liu, Y.; He, M.; Su, Y.; Zhao, X.; Elimelech, M.; Jiang, Z. Antifouling Membranes for Sustainable Water Purification: Strategies and Mechanisms. *Chem. Soc. Rev.* **2016**, *45*, 5888–5924. [CrossRef]
179. Ayyavoo, J.; Nguyen, T.P.N.; Jun, B.M.; Kim, I.C.; Kwon, Y.N. Protection of Polymeric Membranes with Antifouling Surfacing via Surface Modifications. *Colloids Surf. A Physicochem. Eng. Asp.* **2016**, *506*, 190–201. [CrossRef]
180. Akbari, A.; Derikvandi, Z.; Rostami, S.M.M. Influence of Chitosan Coating on the Separation Performance, Morphology and Anti-fouling Properties of the Polyamide Nanofiltration Membranes. *J. Ind. Eng. Chem.* **2015**, *28*, 268–276. [CrossRef]
181. Wang, S.Y.; Fang, L.F.; Zhu, B.K.; Matsuyama, H. Enhancing the antifouling property of polymeric membrane via surface charge regulation. *J. Colloid Interface Sci.* **2021**, *593*, 315–322. [CrossRef] [PubMed]
182. Li, Q.; Imbrogno, J.; Belfort, G.; Wang, X.L. Making Polymeric Membranes Antifouling via "Grafting from" Polymerization of Zwitterions. *J. Appl. Polym. Sci.* **2015**, *132*, 41781. [CrossRef]
183. Rana, D.; Matsuura, T. Surface Modifications for Antifouling Membranes. *Chem. Rev.* **2010**, *110*, 2448–2471. [CrossRef] [PubMed]
184. Ahmad, N.N.R.; Mohammad, A.W.; Mahmoudi, E.; Ang, W.L.; Leo, C.P.; Teow, Y.H. An overview of the Modification Strategies in Developing Antifouling Nanofiltration Membranes. *J. Membr. Sci.* **2022**, *12*, 1276. [CrossRef] [PubMed]

185. Zhang, X.; Jin, P.; Xu, D.; Zheng, J.; Zhan, Z.M.; Gao, Q.; Yuan, S.; Xu, Z.L.; Bruggen, B.V. Triethanolamine modification produces ultra-permeable nanofiltration membrane with enhanced removal efficiency of heavy metal ions. *J. Membr. Sci.* **2022**, *644*, 120127. [CrossRef]
186. Landaburu-Aguirre, J.; García-Pacheco, R.; Molina, S.; Rodríguez-Sáez, L.; Rabadán, J.; García-Calvo, E. Fouling prevention, preparing for re-use and membrane recycling. Towards circular economy in RO desalination. *Desalination* **2016**, *393*, 16–30. [CrossRef]
187. Ravi, B.; Duraisamy, P.; Marimuthu, T. A novel integrated circular economy approach in green synthesis of copper oxide nanoparticles from waste printed circuit boards and utilization of its residue for preparation of carbon engulfed nano polymer membrane. *J. Clean. Prod.* **2023**, *383*, 135457. [CrossRef]
188. Mehrabani, S.A.N.; Vatanpour, V.; Koyuncu, I. Green solvents in polymeric membrane fabrication: A review. *Separ. Purif. Technol.* **2022**, *298*, 121691. [CrossRef]
189. Wu, Y.; Wang, Y.; Xu, F.; Qu, K.; Dai, L.; Cao, H.; Xia, Y.; Lei, L.; Huang, K.; Xu, Z. Solvent-induced interfacial polymerization enables highly crystalline covalent organic framework membranes. *J. Membr. Sci.* **2022**, *659*, 120799. [CrossRef]
190. Ramdani, A.; Deratani, A.; Taleb, S.; Drouiche, N.; Lounici, H. Performance of NF90 and NF270 commercial nanofiltration membranes in the defluoridation of Algerian brackish water. *Desalination Water Treat.* **2021**, *212*, 286–296. [CrossRef]
191. Chandrapala, J.; Duke, M.C.; Gray, S.R.; Weeks, M.; Palmer, M.; Vasiljevic, T. Nanofiltration and nanodiafiltration of acid whey as a function of pH and temperature. *Sep. Purif. Technol.* **2016**, *160*, 18–27. [CrossRef]
192. Tajziehchi, K.; Sadrameli, S.M. Optimization for free glycerol, diglyceride, and triglyceride reduction in biodiesel using ultrafiltration polymeric membrane: Effect of process parameters. *Process. Saf. Environ. Prot.* **2020**, *148*, 34–46. [CrossRef]
193. Maab, H.; Touheed, A. Polyazole polymers membranes for high pressure gas separation technology. *J. Membr. Sci.* **2022**, *642*, 119980. [CrossRef]
194. Pardo, F.; Gutiérrez-Hernández, S.V.; Rodríguez-San Miguel, P.; Zarca, G.; Urtiaga, A. Polymer/ionic liquid pilot scale membrane prototype for the recovery of difluoromethane (R-32) from refrigerant mixtures. *Sep. Purif. Technol.* **2023**, *320*, 124115. [CrossRef]
195. Armillotta, A. Direct texturing for additive manufacturing: Software support and build tests. *Rapid Prototyp. J.* **2020**, *26*, 881–894. [CrossRef]
196. Chowdhury, M.R.; Steffes, J.; Huey, B.D.; McCutcheon, J.R. 3D printed polyamide membranes for desalination. *Science* **2018**, *361*, 682–685. [CrossRef]
197. Roy Barman, S.; Pratik, G.; Chowdhury, S.; Chatterjee, K.; Nain, A. 3D-printed materials for wastewater treatment. *JACS Au* **2023**, *11*, 2930–2947. [CrossRef]
198. Icaga, Y. Fuzzy evaluation of water quality classification. *Ecol. Indic.* **2007**, *7*, 710–718. [CrossRef]
199. Yuan, S.; Ajam, H.; Sinnah, Z.A.B.; Altalbawy, F.M.; Ameer, S.A.A.; Husain, A.; Cao, Y. The roles of artificial intelligence techniques for increasing the prediction performance of important parameters and their optimization in membrane processes: A systematic review. *Ecotoxicol. Environ. Saf.* **2023**, *260*, 115066. [CrossRef]

polymers

MDPI

Article

Facile Fabrication of Hierarchical Structured Anodic Aluminum Oxide Molds for Large-Scale Production of Superhydrophobic Polymer Films

Athinarayanan Balasankar [1,2,*], Raja Venkatesan [3,*], Dae-Yeong Jeong [2,4], Tae Hwan Oh [3], Seong-Cheol Kim [3], Alexandre A. Vetcher [5] and Subramaniyan Ramasundaram [3,*]

1 Department of Physics, Gobi Arts and Science College, Gobichettipalayam 638453, India
2 Nano-Hybrid Technology Research Center, Korea Electrotechnology Research Institute, 9 Beon-gil, 12 Bulmosan-gil, Seongsan-gu, Changwon 51543, Republic of Korea; dyjeong@keri.re.kr
3 School of Chemical Engineering, Yeungnam University, 280 Daehak-Ro, Gyeongsan 38541, Republic of Korea; taehwanoh@ynu.ac.kr (T.H.O.); sckim07@ynu.ac.kr (S.-C.K.)
4 Nanoeco. Co., Ltd., Technology Start-up Center, Seongju-dong, 10 Jeongiyigil, Seongsan-ku, Changwon 50062, Republic of Korea
5 Institute of Biochemical Technology and Nanotechnology, Peoples' Friendship University of Russia n.a. P. Lumumba (RUDN), 6 Miklukho-Maklaya St., 117198 Moscow, Russia; avetcher@gmail.com
* Correspondence: balphd1987@gmail.com (A.B.); rajavenki101@gmail.com (R.V.); ramasundaram79@hotmail.com (S.R.)

Abstract: Anodized aluminum oxide (AAO) molds were used for the production of large-area and inexpensive superhydrophobic polymer films. A controlled anodization methodology was developed for the fabrication of hierarchical micro–nanoporous (HMN) AAO imprint molds (HMN-AAO), where phosphoric acid was used as both an electrolyte and a widening agent. Heat generated upon repetitive high-voltage (195 V) anodization steps is effectively dissipated by establishing a cooling channel. On the HMN-AAO, within the hemispherical micropores, arrays of hexagonal nanopores are formed. The diameter and depth of the micro- and nanopores are 18/8 and 0.3/1.25 μm, respectively. The gradual removal of micropatterns during etching in both the vertical and horizontal directions is crucial for fabricating HMN-AAO with a high aspect ratio. HMN-AAO rendered polycarbonate (PC) and polymethyl methacrylate (PMMA) films with respective water contact angles (WCAs) of 153° and 151°, respectively. The increase in the WCA is 80% for PC (85°) and 89% for PMMA (80°). On the PC and PMMA films, mechanically robust arrays of nanopillars are observed within the hemispherical micropillars. The micro–nanopillars on these polymer films are mechanically robust and durable. Regular nanoporous AAO molds resulted in only a hydrophobic polymer film (WCA = 113–118°). Collectively, the phosphoric acid-based controlled anodization strategy can be effectively utilized for the manufacturing of HMN-AAO molds and roll-to-roll production of durable superhydrophobic surfaces.

Keywords: superhydrophobic; polymeric films; anodic aluminum oxide; phosphoric acid anodizing; hierarchically porous; UV lithography

1. Introduction

Superhydrophobic surfaces in which the water contact angle is >150° have been intensively studied over a decade due to the importance of their industrial applications [1–10]. Superhydrophobic surfaces can be prepared either by tailoring the surface morphology [11,12] or by coating the appropriate hydrophobic materials [13]. A typical example of superhydrophobic surfaces is a lotus leaf, where an array of nanopillars is organized in nipple-like protruded microstructures [14,15]. In order to mimic the surface morphology of a superhydrophobic lotus leaf, surfaces with micro- and nanopatterns are formed using various fabrication techniques such as soft lithography [16], chemical etching [17], solvent-assisted UV-lasering [18], 3D printing [19], and electro-brush plating [20]. One of the ways to mass-produce such superhydrophobic surfaces is replication of the surface morphology of molds by rapid imprinting methods such as thermal imprinting [21]. Specifically, the most important prerequisite for mass production of superhydrophobic polymeric films is the large-scale and cost-effective fabrication of imprint molds with tailored micro- and nanoporous patterns.

Metallic molds possess higher wear resistance and a longer life than molds based on rubber, wood, polymers, or plastics. Metallic molds have been fabricated by various methods, including thermal nanoimprint lithography [22], focused ion beam, femto-second laser, eximer laser, and dicing techniques [23], advanced edge lithography [24], and anodization [25]. Anodization is known as one of the most simple, cheap, and reproducible methods to fabricate imprinting molds. Due to valve metal characteristics, the formed AAO works as a strong protective layer against natural corrosion and enhances the life of the mold surface [26–28]. The surface of AAO imprinting molds is usually designed with a variety of organized hexagonal arrays of columnar nanopore channels. The large number of air pockets trapped between water droplets and gaps between pillars on the surface is believed to be responsible for the superhydrophobicity of the surface [29]. The self-ordering is mainly influenced by the applied anodization voltage as well as the purity of the aluminum used [30,31]. The inter-pore distance varies between tens and hundreds of nm in size, depending on the anodization regime [31–33]. Besides that, as the pores grow in a direction perpendicular to the anodized surface, a homogenous and smooth surface is required to form a self-ordered array of well-defined nanopores. Also, self-ordering is significantly promoted by performing a two-step anodization [31]. In order to form a relatively thick AAO layer, first the aluminum plate was anodized for a relatively long duration, and then the AAO layer was etched out to leave an array of concave nanoseeds on the aluminum surface [34]. Anodization is repeated multiple times so that the nanopores grow in a well-ordered way at the center of the nanoseeds.

Anodization of aluminum has been widely carried out in three different acidic electrolytes with various voltages: 0.3 M sulfuric at 25 V, 0.3 M oxalic at 40 V, and 0.1 M phosphoric at 195 V [35–37]. Self-ordering regimes were limited at higher voltages for oxalic acid and sulfuric acid anodizing. Thus, producing AAO molds with arrays of sub-micron pores requires tedious process control. As a self-ordering regime is achieved at a high voltage in phosphoric acid, it is feasible to produce a sub-micron porous AAO mold with an inter-pore distance of 500 nm. Therefore, a phosphoric acid electrolyte in a 195 V anodizing regime was found to be an appropriate option. Moreover, aluminum is more stable against phosphoric acid than sulfuric acid. Phosphoric acid can also be used as both an electrolyte and a widening agent. However, because of the application of high voltage (195 V), the control of heat dissipation in the phosphoric acid-based anodization process is considered a highly crucial step. Poor heat dissipation causes a catastrophic increase in induced current, which leads to the burning of oxide films and the collapse of the mold surface. The other important issue pertaining to AAO-based imprint molds is the requirement for strong and highly expensive etchants, such as aluminum etchant type A. These kinds of strong etchants are known to be damaging to the interface between the photoresist (PR) used in the patterning process and the aluminum and cause both horizontal and vertical etching. So, it is very difficult to obtain microporous patterns with the required depth. The aforementioned issues must be addressed to obtain low-cost imprint molds as well as the mass production of polymeric superhydrophobic films.

From Table 1, existing methods are not adequate for mass production by industries due to cost, complex procedures, time consumption, and long life. These drawbacks are overcome by the present work. In the present investigation, efforts were made to prepare an AAO imprint mold with a hierarchical micro–nanoporous structure. To address the heat dissipation issue, a phosphoric acid-based anodizing system installed with separate sample- and electrolyte-cooling channels was proposed. A controlled, four-step anodizing–

etching procedure was used. Phosphoric acid was used as an electrolyte for anodization and as a widening agent for etching. The fabricated hierarchical micro–nanoporous AAO imprint mold (HMN-AAO) was used to produce superhydrophobic poly(carbonate) (PC) and poly(methyl methacrylate) (PMMA) films using thermal nanoimprint lithography (NIL). For comparison, a nanoporous AAO imprint mold (N-AAO) was also prepared. The surface morphologies of the AAO imprint mold and superhydrophobic polymer films were studied in detail by a field emission scanning electron microscope (FE-SEM). The wetting behavior of the polymeric films was studied using contact angle goniometry.

Table 1. Comparison table for fabrication of large-scale superhydrophobic films.

Methodology	Advantages	Challenges	References
Electrochemical deposition method	Even on a large scale, it is very easy to control the thickness of the deposition layer	This will not work in non-conductive materials	[38]
Spin coating methods	Minimum curing time	Poor binding energy between the substrates and polymers	[39]
Chemical vapor deposition method	Portable, precise thickness control	Very expensive and not adequate for mass production	[40]
Chemical etching method	Stable due to single-material work as a substrate	Expensive, time-consuming, and poor mechanical strength	[41]
Spray coating method	Simple operation procedure	Consumed too many coating materials and did not have strong binding	[42]
Combined methodology of UV lithography and anodization for master mold fabrication	Very cheap, simple, high throughput, mechanically stable	Stamping only possible for flexible polymers	Current work

2. Materials and Methods

2.1. Fabrication of Nanoporous AAO Imprinting Molds

Figure 1 shows a schematic of the whole process of fabricating an AAO imprinting mold with an array of conical nanopores. The nanoporous AAO imprinting molds were fabricated using a four-step anodization and etching technique: First, the electro-polished aluminum plate was anodized at −1.2 °C by applying 195 V for 10 h in a 0.1 M phosphoric acid electrolyte. The anodization was performed in a 2 L double-walled glass beaker under vigorous stirring. A constant power supply (OPS 603, ODA Technologies, Incheon, Republic of Korea) was used as a DC power source. During the anodizing process, the current induced was measured and monitored using a Keithley−2100 digital multi-meter interfaced with LabVIEW software. The diameter of the anodized surface was 4 cm. The AAO surface was etched in a mixed solution of 0.19 M chromic acid and 0.1 M phosphoric acid solutions for 60 min at 60 °C. The resultant aluminum surface, which had a highly ordered array of nano-sized concave seeds, was re-anodized for 15 min. The AAO surface formed by the second anodization was etched for 60 min in 0.1 M phosphoric acid at 30 °C. The etched AAO surface was subjected to a third anodization for 15 min and etched for 60 min. Finally, the etched AAO surface was anodized for a fourth time in a series for 15 min and etched for 60 min. Then, the AAO specimen was cleaned with ethanol and dried at 60 °C for 15 min. Except for time duration, in the first to fourth anodization processes, all conditions were the same. The conditions followed for the third and fourth etchings were identical to those of the second etching.

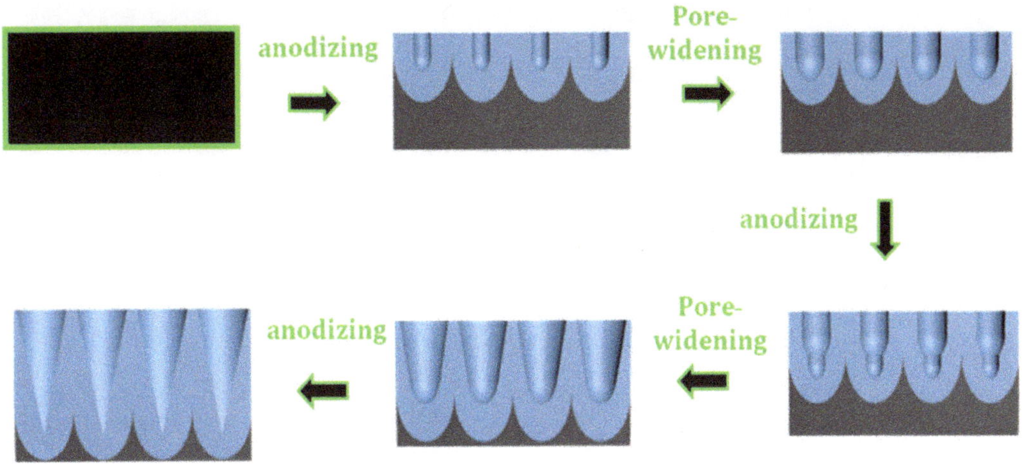

Figure 1. Schematic of the process used for fabricating an AAO imprinting mold with an array of conical nanopores.

2.2. Fabrication of Hierarchical Micro–Nanoporous Imprinting Molds

The HMN-AAO imprinting molds were fabricated on an electro-polished aluminum plate via anodizing, followed by UV lithography. First, the photo resist (PR) (AZ5214E, AZ Electronic Materials, Charlotte, NC 28269, USA) was spin coated on the electro-polished aluminum plate at a spin speed of 4000 rpm for 30 s using a SPIN−1200D (Midas system, Daejeon, Republic of Korea) spin coater. The average thickness of the spin-coated PR film was 2.5 μm. The thickness was measured by a cross-section analysis using a field emission scanning electron microscope (FE-SEM, Hitachi FE-SEM S4800, Tokyo, Japan). The PR-coated specimen was soft-baked at 110 °C for 120 s, covered with a photomask, and exposed to UV light for 100 s using a UV light lithography machine (LABSYS LIT-2000, NEXTRON, Busan, Korea). A negative-type photomask used in the experiment had hexagonally arranged circles (8 μm in diameter) with an inter-circle distance of 24 μm. After UV lithography, the specimen was dried in an oven at 110 °C for 2 min, cooled down to room temperature for 15 min, and developed for 120 s in a developer solution (AMF300, AZ Electronic Materials, Charlotte, NC 28269, USA). The specimen covered with a micropatterned PR layer was hard-baked at 110 °C for 20 min and then anodized at 195 V for 10 h in a 0.1 M phosphoric acid electrolyte. The anodizing temperature was −1.2 °C. Then, the anodized surface was etched with a mixture of 0.19 M chromic acid and 0.1 M phosphoric acid aqueous solutions at 60 °C for 60 min. The etched aluminum plate was subjected to a second anodizing process in a 0.1 M phosphoric acid electrolyte (195 V for 10 min) at −1.2 °C. After the second anodizing, the widening process was performed by immersing the anodized surface in 0.1 M phosphoric acid. Then, the second anodizing and widening processes were identically repeated two times. After electropolishing, successive anodizing, and etching processes, the surface was cleaned by immersing in DI water, ethanol, and acetone in a series for a min and dried at 30 °C for 5 min.

2.3. Fabrication of Polymeric Replica Films

First, to remove the oxygen adsorbed on the pore wall surfaces, the fabricated HMN-AAO molds were cleaned using a 1:1 weight/weight mixture of hydrogen peroxide and sulfuric acid for an hour at room temperature (RT). Then, the mold was rinsed with water and ethanol and dried at 60 °C for 20 min. The dried mold was lubricated by immersing in toluene mixed with trichloro (octadecyl)-silane (2 drops/20 g toluene) for 15 h at RT. Then, the mold was rinsed with ethanol and dried at 120 °C for 2 h. The polymeric substrates to be imprinted (PC and PMMA) were also cleaned by sonicating with methanol. The above-cleaned mold and polymeric substrates were mounted on the nanoimprinting lithography (NIL) machine (model no. EITRE® 6, OBDUCAT CO., Nytänkargatan 4, SE-223 63 Lund, Sweden). The mold mounted on the NIL machine was heated up to 130 to 170 °C (glass transition temperature range of polymeric substrates). Then, the pattern on the mold was imprinted on the polymeric substrates by applying a pressure of 30 bars for 5 min. Then, the replica films were detached from the mold, cooled down to room temperature for 15 min, cleaned by using DI water, and dried at 50 °C.

2.4. Characterization

The surface morphologies of the N-AAO and HMN-AAO molds and the PC and PMMA replica films were examined using a field emission scanning electron microscope (FE-SEM, Hitachi FE-SEM S4800) operated at 10 kV. The specimens were coated with a layer of osmium by using an osmium plasma coating machine before the FE-SEM observations. The water contact angles (WCAs) of the replica film surfaces were measured at RT using a Phoenix300 contact angle/interface system (Surface Electro Optics Co., Ltd., Suwon, Republic of Korea). They were measured at three different locations on each replica film, and the average values were reported. The volume of the water droplet placed on the surfaces of the replica films was 5 μL.

3. Results and Discussion

A conical-shaped nanoporous structure of the AAO imprint mold is essential to increasing the air interface between the surface and water and also to ensuring smooth surface finishing. An increase in the air interface increases the WCA [43]. A smooth surface resulted from an array of strong pillars formed out of a conical-shaped nonporous structure, which facilitates the easy recovery of the replica film from the imprint mold. So, to check the feasibility of our strategy of controlled anodization using a phosphoric acid electrolyte, the first N-AAO was prepared. Figure 2a–f depicts the FE-SEM images of a template obtained as a function of the increase in the number of anodization and subsequent wet etching (widening) steps during the N-AAO process. The latter was performed to increase the pore diameter as well as induce a change in pore shape [43]. In Figure 2, the surface morphology images are organized on the left (Figure 2a,c,e), and the respective cross-sectional images (Figure 2b,d,f) are provided on the right. After the second anodization−widening step, the circular and random pores (206 nm in diameter) were formed on the surface. Notably, these pores were surrounded by a hexagonal array of seeds, formed at the bottom of the barrier layer just after the first anodization−etching step.

Figure 3 shows the FE-SEM images of PC (a and b) and PMMA (e and f) replica films prepared by thermal imprinting of N-AAO. The WCA images of pristine and replica films are shown in Figure 3c,g and Figure 3d,h, respectively. In the FE-SEM images with 10 μm resolution, the surface of the replica film was filled with uniformly arranged pillars (top diameter = 154 nm). The conical shape of the pillars was clearly visible in the respective images with 1 μm resolution. The inter-pillar distance was 550 nm, the same as the inter-pore distance noticed in the AAO imprint mold. The pillars appeared intact, and there was no sign of damage. The presence of structurally intact pillars confirmed that the conically shaped nanopores in the AAO mold facilitated the easy separation of the replica film after imprinting without any damage. The WCAs of the PC (Figure 3d) and PMMA (Figure 2h) replica films were 118° and 113°, respectively. After imprinting with N-AAO, the surface of the polymer films was turned from hydrophilic to hydrophobic. The WCA of the pristine PC is 85° (Figure 3c), and it is 80° (Figure 3g) for pristine PMMA. The surface morphology and WCA of the PC and PMMA replica films proved the usability of N-AAO prepared by the proposed 0.1 M phosphoric acid electrolyte-based controlled anodization method for the large-scale production of hydrophobic polymer films.

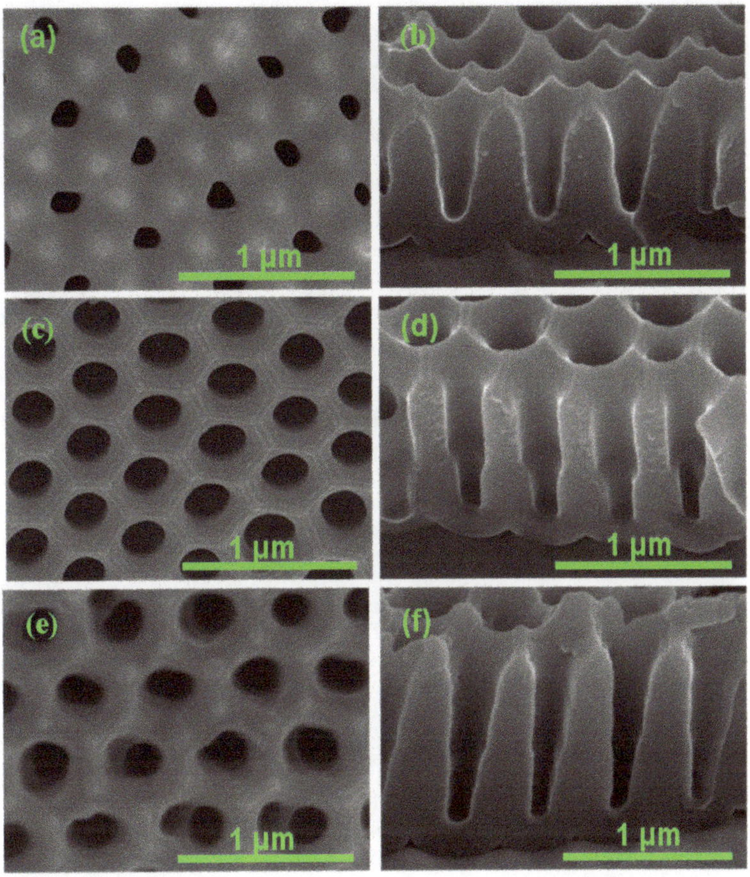

Figure 2. FE-SEM images showing the changes in pore morphology on the top surface (**a**,**c**,**e**) and the cross-section (**b**,**d**,**f**) of AAO templates during the N-AAO process involving multiple anodization and wet-etching steps. The images (**a**,**b**), (**c**,**d**), and (**e**,**f**) were taken after the 2nd, 3rd, and 4th repetitions, respectively.

The cross-sectional view confirmed the formation of conical pores with a depth of 536 nm. Well-ordered hexagonal pores with a diameter of 220 nm were seen after the third anodization−widening, and the depth of the conical pores was 721 nm. After the fourth anodization, the diameter of the pore was further increased to 227 nm, and the pore depth became 1030 nm. With respect to the increase in the number of anodization steps, the pore entrance and the pore diameter proved to be gradually increased. The depth-to-diameter ratio (aspect ratio) of the pores was also increased with the increase in the number of anodization steps; it was 2.77, 3.27, and 4.53 after the second, third, and fourth anodizations, respectively. It is important to emphasize that the increase in pore size, depth, and depth-to-diameter ratio with respect to the increase in the number of anodization steps did not collapse the pore walls. The inter-pore distance (550 nm) also remained unaltered after three repetitive anodization and widening processes. The above results proved that the proposed strategy of controlled anodization with a 0.1 M phosphoric acid electrolyte was useful for preparing an AAO mold with a uniform surface and conical pores.

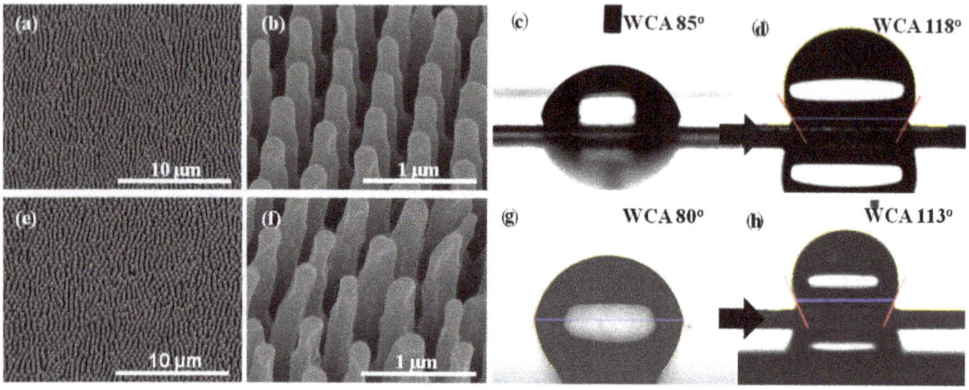

Figure 3. FE-SEM images showing the surface morphology of PC (**a**,**b**) and PMMA (**e**,**f**) films replicated using an N-AAO mold using the thermal imprinting method. (**b**) and (**f**) are the enlarged images of (**a**) and (**e**), respectively. The images (**c**,**d**) and (**g**,**h**) are photographs of water droplets taken on the surfaces of PC and PMMA films, respectively. The images (**c**,**g**) and (**d**,**h**) are WCAs of pristine and nanopillared PC and PMMA replica films.

Figure 4 depicts the schematic of the procedure followed for preparing HMN-AAO by the combination of UV lithography and a controlled anodization strategy optimized for preparing N-AAO. UV lithography is one of the conventional methods used for micropatterning. The shortcomings, such as the need for expensive chemical etching and difficulties in forming high-depth microporous patterns, restrict the use of UV lithography for preparing cost-effective AAO imprint molds. Therefore, to make high-depth microporous arrays, the fabrication process was modified as follows: The aluminum surface covered with the PR layer, micropatterned using UV lithography, was anodized. In order to form the arrays of hexagonal nanoseeds and micropores, the formed AAO was etched by a mixture of phosphoric acid and chromic acid. Then, the surface was re-anodized and etched repeatedly to make conical nanopores on the nanoseeds. As a result, AAO molds with hierarchical structures consisting of arrays of nanopores and higher-depth micropores could be obtained.

The changes in the morphology of the aluminum template during the various steps of the HMN-AAO fabrication process (Figure 4) are shown in Figure 5. Uniform micropatterns of PR (MPPR) can be seen on the micropatterned aluminum surface subjected to UV lithography (Figure 5a). The inter-distance between neighboring circles was 24 μm. The average diameter of the single circle was estimated to be 6.7 μm (inset of Figure 5a). The aluminum surface with the MPPR arrays was anodized at 195 V in a phosphoric acid electrolyte for 10 h. In conventional anodization, the nanopore channels used to be straight and uniform. In contrast, after anodizing, both declined, and straight nanopore channels were formed (Figure 5b). For better understanding, in Figure 5c, the straight pore channels, declined nanopore channels, and the boundary are marked as regions 1, 2, and 3, respectively. The interface area between regions 2 and 3 was covered by a PR layer along with nanoporous AAO. This interfacial PR layer acted as a sealant and prevented the diffusion of the electrolyte to the aluminum. The 45° titled cross-sectional view of the declined pore channel (Figure 5d) and boundary (Figure 5e) discerns that the growth rate of the straight pore channel is higher than the declined pore channel. The formation of the straight and declined pore channels during anodization can be explained using the stability of the binding interface between the PR and aluminum and the electrolyte diffusion. In conventional anodization, numerous closely packed pores are formed, and there is no free space in the horizontal direction. Thus, due to the constraints posed by inter-pore repulsive forces, the formed pores grow vertically at the center of the circles as straight channels. In the controlled anodization used here, only in the region of the straight pore channel do AAO pores start to grow up in directions perpendicular to the aluminum surface. Growth in other regions is restricted due to strong binding between the PR layer and aluminum surfaces. However, AAO pores in the vicinity of the circle edges start to grow gradually in all possible directions into aluminum under the PR layer (region 2) because there are no constraints in the radial (horizontal) direction. As a result, the pore channels start to bend to the region under the PR layer. Due to bending and a 40% volume expansion caused by AAO formation [44] at the edges of the circle, the crevices occur at the interface of the PR and aluminum. The electrolyte starts to penetrate through these crevices, and the pores under MPPR grow in both vertical and horizontal directions simultaneously. The growth of both the horizontal and vertical pores continued until the point at which they deviated from the MPPR circles. These points of deviation appeared like barrier layers between the pores grown in the vertical and horizontal directions (Figure 5d,e).

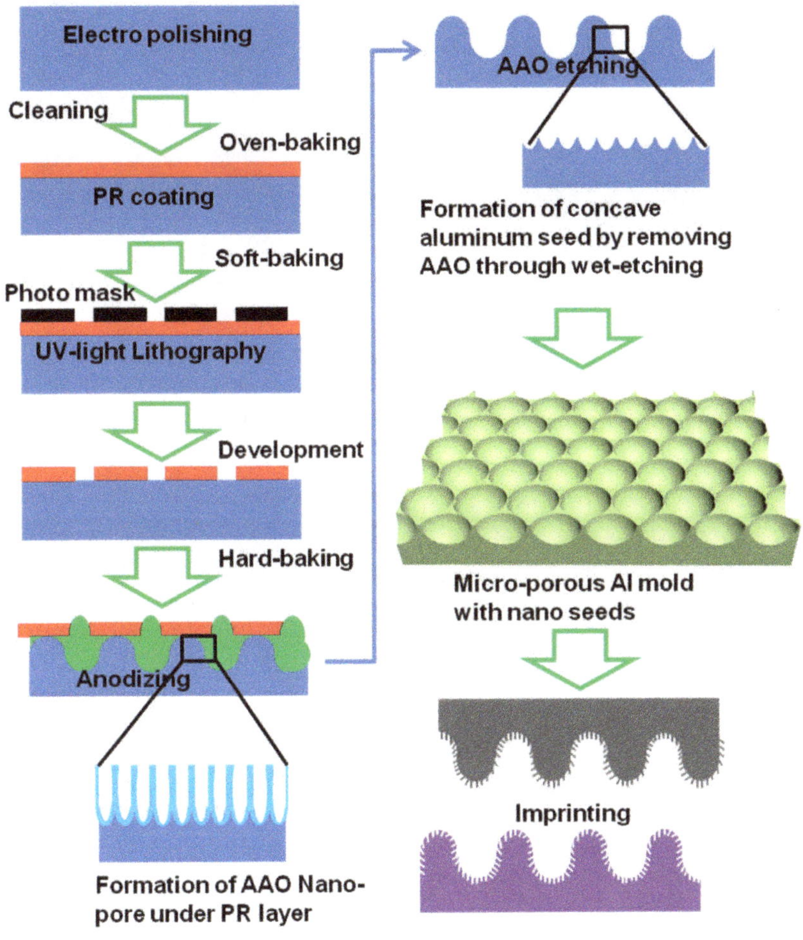

Figure 4. Schematic of the process used for fabricating an HMN-AAO imprint mold using the controlled anodization method.

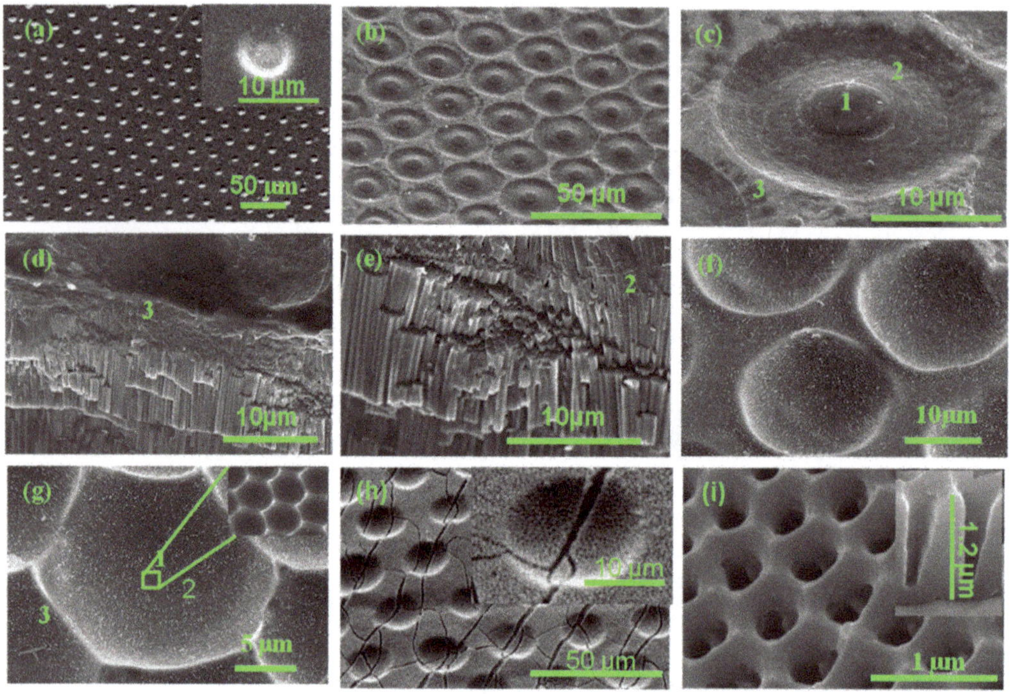

Figure 5. FE-SEM images of a template at various stages of the HMN-AAO fabrication process. (**a**) Micropatterned template after the UV lithography; (**b**,**c**) after anodization in a phosphoric acid electrolyte for 10 h; (**d**) a 45° tilted image of an accidentally cleaved portion near zone 3 in (**c**); (**e**) a 45° tilted image of an intentionally cracked portion near zone 2 in (**c**); (**f**) after etching of the AAO layer formed during the anodization; (**g**) enlarged images of (**f**) showing an array of aluminum seeds in sub-micron size; (**h**,**i**) surface of the HMN-AAO imprint mold after bending.

To further obtain insights about the controlled anodization process, the mechanism behind the formation of deep micropores (Figure 5b–e) must be explained. The anodized layer thickness in region 1 gradually decreases when approaching region 3. As a result of a 40% volume expansion during the anodization [44] and a constraint posed by the PR layer, the edges were suppressed and region 1 was protruded (Figure 5c). The height of region 1 appeared to be 4 µm larger than that of region 3. Therefore, the lengths of the straight (vertical) pore channel were deeper (longer) than the bent or declined (horizontal) pore channel. When the entire anodized layer etched out, an array of pores with a depth almost equal to their radius was formed. The PR layer was predicted to remain up to half of the (4–5.3 h) total anodization process (10 h). In the typical aluminum etching, within an hour, the PR layer was usually removed by the aluminum etchant [43]. Due to that, the pores formed therein were not deeper. The average diameter of the micropores in Figure 5g was estimated to be ~18.0 µm. The depth of the micropores in Figure 5g was estimated to be 8.0 µm. The smaller depth of the micropores in region 2 (Figure 5f,g) than their diameter is attributed to the detachment of the PR layer during the anodization. Thus, the shape of the microporous appears concave. Also, it can be inferred that the initial anodization period is crucial for determining the pore depth. The inset in Figure 5g exhibited the formation of a uniform array of hexagonal nanoseeds inside the micropores.

Figure 5h,i illustrates the formation of the hierarchical micro-/nanoporous structure upon completion of fourth anodization and accompanying the widening process. The cracks seen on these images were formed due to folding of the sample prepared for effective observation of the cross-section and top surface. The array of the AAO micropores can be seen in Figure 5h. Due to the growth of the nanopores in the radial direction (inset of Figure 5h), the diameter (17.5 μm) of these AAO micropores was found to be less than the diameter (18.4 μm) of the microporous aluminum with the nanoseed (Figure 5f,g). Due to the cracks, the inter-pore distance appears to be larger than the microporous aluminum with the nanoseed. As seen in the morphology of a single AAO micropore, shown in the inset of Figure 5h, the arrays of nanopores were distributed on both the boundary and micropore region. As seen in Figure 5i and inset therein, these nanopores were hexagonal in shape with a diameter and depth of 300 nm and 1.2 μm, respectively. The obtained HMN-AAO was used for preparing the polymeric replica film.

Figure 6 shows the FE-SEM images of PC (a–c) and PMMA (e–g) films replicated using the HMN-AAO. The WCA photographs of the resultant PC and PMMA replica films are shown in Figure 6d,h. FE-SEM images with 10 and 5 μm resolution were provided to effectively explain the micro- and nanostructure on the replica films. Upon replication, the well-organized micropillars comprising the array nanopillars were created on both the PC and PMMA surfaces. The diameters of the micro and nanopillars were 21.42 and 0.306 μm, respectively. The absence of breakages discerns that the formed micro- and nanopillars possess sufficient mechanical stability to withstand forces exerted on the replica films while imprinting, especially during the separation of the mold after imprinting. When compared with replica films prepared using N-AAO (Figure 2e,f), the HMN-AAO led to an increase in the WCAs of PC (153°) and PMMA (151°), respectively. In other words, the N-AAO mold enhanced the hydrophobicity, and HMN-AAO imparted the superhydrophobicity. The significant increase in the WCA was attributed to the presence of an array of nanopillars within the micropillar structures. Specifically, this hierarchal micro–nanopore structure increases the effective number of air pockets trapped between water droplets and gaps between micropillars. The increase in the number of air pockets increased the WCA. Overall, the morphology and WCA proved that HMN-AAO fabricated by the controlled anodization process is suitable for the mass production of superhydrophobic polymeric films using the thermal nanoimprinting method. This strategy is more advantageous than the previous methods reported for the fabrication of micro-/nanopillared superhydrophobic PDMSs [45], where silicon nanocrystals were pasted on a micropatterned PDMS obtained by a two-step soft-lithography process. Also, due to the weaker binding of pasted nanocrystals on the micropatterned surface, the superhydrophobic property may decay over a short span of time and use. The mechanically robust micro–nanopillars created on the replica films by imprinting HMN-AAO are considered permanent and durable over a longer time.

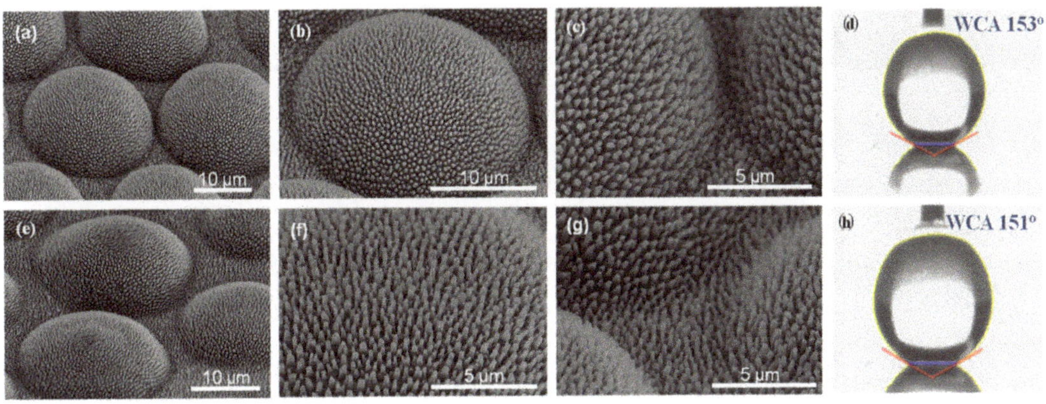

Figure 6. FE-SEM images of PC (**a–c**) and PMMA (**e–g**) replica films prepared using an HMN-AAO imprinting mold. The magnification was increased in the order of (**a–c**), and (**e–g**), respectively. The images (**d**) and (**h**) are WCA photographs of PC and PMMA replica films, respectively.

4. Conclusions

The low-cost and considerably safer phosphoric acid electrolyte-based controlled anodization method was established for the fabrication of HMN-AAO imprint molds. An HMN-AAO mold with the dimension of 100 × 100 mm^2 was fabricated by phosphoric acid anodizing at 195 V. The relevant thermodynamic parameters were controlled and optimized by establishing a special cooling system. The total production time and cost were minimized by using phosphoric acid as both the electrolyte and widening agent. The morphology of the HNM-AAO mold observed by FE-SEM confirmed the formation of arrays of hexagonal nanopores within the hemispherical micropore structures with a high aspect ratio. Both the micro- and nanopore structures were uniform and had distinct boundaries. The superhydrophobic PC and PMMA films exhibiting WCAs of 153° and 151°, respectively, were prepared by thermally imprinting the HMN-AAO mold. Imprinting of the HMN-AAO mold resulted in the formation of mechanically robust hierarchical micro–nanopillars on the surface of the polymeric films, whereas thermal imprinting of the N-AAO mold only increased the WCA of the hydrophilic PC (85°) and PMMA (80°) to hydrophobic PC (118°) and PMMA (113°). The above-established phosphoric acid electrolyte-based controlled anodization strategy was proved to have the potential for the large-scale commercial production of HMN-AAO imprint molds as well as durable superhydrophobic polymeric films.

Author Contributions: Conceptualization, A.B. and R.V.; methodology, D.-Y.J. and R.V.; software, A.B.; validation, A.A.V. and D.-Y.J.; formal analysis, A.B. and R.V.; investigation, R.V. and S.R.; resources, T.H.O. and A.A.V.; data curation, T.H.O. and S.R.; writing—original draft preparation, A.B. and S.R.; writing—review and editing, R.V. and S.-C.K.; visualization, A.A.V. and S.R.; supervision, S.-C.K. and T.H.O.; project administration, T.H.O. and S.-C.K.; funding acquisition, T.H.O. All authors have read and agreed to the published version of the manuscript.

Funding: The present study was financially supported by the Korean Ministry of Science, ICT, and Future Planning (15-12-N0101-90).

Data Availability Statement: Upon reasonable request, the data supporting this investigation are available from the corresponding authors.

Conflicts of Interest: The author Dae-Yeong Jeong is employed by the company of Nanoeco. Co., Ltd., Republic of Korea. The remaining authors declare that the research was conducted in the absence of any commercial or financial relationships that could be construed as a potential conflict of interest.

References

1. Shibuichi, S.; Yamamoto, T.; Onda, T.; Tsujii, K. Super Water and Oil-Repellent Surfaces Resulting from Fractal Structure. *J. Colloid Interface Sci.* **1998**, *208*, 287–294. [CrossRef] [PubMed]
2. Nakajima, A.; Hashimoto, K.; Watanabe, T.; Takai, K.; Yamauchi, G.; Fujishima, A. Transparent Superhydrophobic Thin Films with Self-Cleaning Properties. *Langmuir* **2000**, *16*, 7044–7047. [CrossRef]
3. Wang, B.; Liang, W.X.; Guo, Z.G.; Liu, W.M. Biomimetic super-lyophobic and super-lyophilic materials applied for oil/water separation: A new strategy beyond nature. *Chem. Soc. Rev.* **2015**, *44*, 336–361. [CrossRef] [PubMed]
4. Valipour, N.M.; Birjandi, F.C.; Sargolzaei, J. Super-non-wettable surfaces: A review. *Colloids Surface A* **2014**, *448*, 93–106. [CrossRef]
5. Guo, Z.G.; Liu, W.M.; Su, B.L. Superhydrophobic Surfaces: From Natural to Biomimetic to Functional. *J. Colloid Interface Sci.* **2011**, *353*, 335–355. [CrossRef] [PubMed]
6. Lv, J.Y.; Song, Y.L.; Jiang, L.; Wang, J.J. Bio-Inspired Strategies for Anti-Icing. *ACS Nano* **2014**, *8*, 3152–3169. [CrossRef] [PubMed]
7. Jeevahan, J.; Chandrasekaran, M.; Joseph, G.B.; Durairaj, R.; Mageshwaran, G. Superhydrophobic surfaces: A review on fundamentals, applications, and challenges. *J. Coat. Technol. Res.* **2018**, *15*, 231–250. [CrossRef]
8. Das, S.; Kumar, S.; Samal, S.K.; Mohanty, S.; Nayak, S.K. A review on superhydrophobic polymer nanocoatings: Recent development, application. *Ind. Eng. Chem. Res.* **2018**, *57*, 2727–2745. [CrossRef]
9. Shome, A.; Das, A.; Borbora, A.; Dhar, M.; Manna, U. Role of chemistry in bio-inspired liquid wettability. *Chem. Soc. Rev.* **2022**, *51*, 5452–5497. [CrossRef] [PubMed]
10. Hu, Z.; Chu, F.; Shan, H.; Wu, X.; Dong, Z.; Wang, R. Understanding and Utilizing Droplet Impact on Superhydrophobic Surfaces: Phenomena, Mechanisms, Regulations, Applications, and Beyond. *Adv. Mater.* **2023**, *36*, 2310177. [CrossRef] [PubMed]
11. Gou, X.; Guo, Z. Surface topographies of biomimetic superamphiphobic materials: Design criteria, fabrication and performance. *Adv. Colloid Interface Sci.* **2019**, *269*, 87–121. [CrossRef] [PubMed]

12. Ma, M.L.; Hill, R.M.; Lowery, J.L.; Fridrikh, S.V.; Rutledge, G.C. Electrospun Poly(Styrene-Block-Dimethylsiloxane) Block Copolymer Fibers Exhibiting Superhydrophobicity. *Langmuir* **2005**, *21*, 5549–5554. [CrossRef] [PubMed]
13. Wang, H.; Dai, D.; Wu, X. Fabrication of superhydrophobic surfaces on aluminum. *Appl. Surf. Sci.* **2008**, *254*, 5599–5601. [CrossRef]
14. Wang, H.X.; Fang, J.; Cheng, T.; Ding, J.; Qu, L.T.; Dai, L.M.; Wang, X.G.; Lin, T. One-Step Coating of Fluoro- Containing Silica Nanoparticles for Universal Generation of Surface Superhydrophobicity. *Chem. Commun.* **2008**, *7*, 877–879. [CrossRef] [PubMed]
15. Yu, S.; Guo, Z.G.; Liu, W.M. Biomimetic transparent and superhydrophobic coatings: From nature and beyond nature. *Chem. Commun.* **2015**, *51*, 1775–1794. [CrossRef] [PubMed]
16. Wang, F.; Li, S.; Wang, L. Fabrication of artificial super-hydrophobic lotus-leaf-like bamboo surfaces through soft lithography. *Colloid Surface A* **2017**, *513*, 389–395. [CrossRef]
17. Shi, T.; Xue, S.; Ma, X.; Peng, H.; Du, J.; Zheng, B.; Xia, Z. Fabrication of superhydrophobic micro-nanostructured aluminum alloy surface via a cost-effective processing using an ultra-low concentration of fluoroalkylsilane. *Appl. Phys. A* **2021**, *127*, 399. [CrossRef]
18. Zhang, S.; Jiang, Q.; Xu, Y.; Guo, C.F.; Wu, Z. Facile Fabrication of Self-Similar Hierarchical Micro-Nano Structures for Multifunctional Surfaces via Solvent-Assisted UV-Lasering. *Micromachines* **2020**, *11*, 682. [CrossRef] [PubMed]
19. Wei, Y.; Hu, Y.; Li, M.; Li, D. Fabrication of Sr-functionalized micro/nano-hierarchical structure ceramic coatings on 3D printing titanium. *Surf. Eng.* **2021**, *37*, 373–380. [CrossRef]
20. Liu, H.; Wang, X.; Ji, H. Fabrication of lotus-leaf-like superhydrophobic surfaces via Ni-based nano-composite electro-brush plating. *Appl. Surf. Sci.* **2014**, *288*, 341–348. [CrossRef]
21. Brock, L.; Sheng, J. Robust fabrication of polymeric nanowire with anodic aluminum oxide templates. *Micromachines* **2020**, *11*, 46. [CrossRef] [PubMed]
22. Kwon, S.; Kim, Y.; Lim, H.; Kim, J.; Choi, K.; Lee, J.; Kim, G. Fabrication of a Metal Roller Mold with Nanoimprinted Pattern Using Thermal Nanoimprint Lithography. *Sci. Adv. Mater.* **2020**, *12*, 481–485. [CrossRef]
23. Youn, W.; Takahashi, M.; Goto, H.; Maeda, R. Fabrication of micro-mold for glass embossing using focused ion beam, femto-second laser, eximer laser and dicing techniques. *J. Mater. Process. Technol.* **2007**, *187*, 326–330. [CrossRef]
24. Sakamoto, J.; Nishino, T.; Kawata, H.; Yasuda, M.; Hirai, Y. High aspect ratio nano mold fabrication by advanced edge lithography without CVD. *Microelectron. Eng.* **2011**, *88*, 1992–1996. [CrossRef]
25. Lee, W.; Park, S.-J. Porous anodic aluminum oxide: Anodization and templated synthesis of functional nanostructures. *Chem. Rev.* **2014**, *114*, 7487–7556. [CrossRef]
26. Parvate, S.; Dixit, P.; Chattopadhyay, S. Superhydrophobic Surfaces: Insights from Theory and Experiment. *J. Phys. Chem. B* **2020**, *124*, 1323–1360. [CrossRef]
27. Darmanin, T.; Guittard, F. Superhydrophobic and superoleophobic properties in nature. *Mater. Today* **2015**, *18*, 273–285. [CrossRef]
28. Jeong, C.; Choi, C.-H. Single-step direct fabrication of pillar-on-pore hybrid nanostructures in anodizing aluminum for superior superhydrophobic sufficiency. *ACS Appl. Mater. Interfaces* **2012**, *4*, 842–848. [CrossRef] [PubMed]
29. Bravo, J.; Zhai, L.; Wu, Z.; Cohen, R.-E.; Rubner, M.-F. Transparent superhydrophobic films based on silica nanoparticles. *Langmuir* **2007**, *23*, 7293–7298. [CrossRef]
30. Michalska-Domańska, M.; Norek, M.; Stępniowski, W.J.; Budner, B. Fabrication of high quality anodic aluminum oxide (AAO) on low purity aluminum—A comparative study with the AAO produced on high purity aluminum. *Electrochim. Acta* **2013**, *105*, 424–432. [CrossRef]
31. Lee, W.; Ji, R.; Gösele, U.; Nielsch, K. Fast fabrication of long-range ordered porous alumina membranes by hard anodization. *Nat. Mater.* **2006**, *5*, 741–747. [CrossRef] [PubMed]
32. Surawathanawises, K.; Cheng, X. Nanoporous anodic aluminum oxide with a long-range order and tunable cell sizes by phosphoric acid anodization on pre-patterned substrates. *Electrochim. Acta* **2014**, *117*, 498–503. [CrossRef] [PubMed]
33. Masuda, H.; Fukuda, K. Ordered Metal Nanohole Arrays Made by a Two-Step Replication of Honeycomb Structures of Anodic Alumina. *Science* **1995**, *268*, 1466–1468. [CrossRef] [PubMed]
34. Kim, D.; Kim, J.; Park, H.-C.; Lee, K.-H.; Hwang, W. A superhydrophobic dual-scale engineered lotus leaf. *J. Micromech. Microeng.* **2008**, *18*, 5019. [CrossRef]
35. Nielsch, K.; Choi, J.; Schwirn, K.; Wehrspohn, R.B.; Gosele, U. Self-ordering regimes of porous alumina: The 10% porosity rule. *Nano Lett.* **2002**, *2*, 677–680. [CrossRef]
36. Zhao, N.Q.; Jiang, X.X.; Shi, C.S.; Li, J.J.; Zhao, Z.G.; Du, X.W. Effects of anodizing conditions on anodic alumina structure. *J. Mater. Sci.* **2007**, *42*, 3878–3882. [CrossRef]
37. Masuda, H.; Yada, K.; Osaka, A. A Self-Ordering of Cell Configuration of Anodic Porous Alumina with Large-Size Pores in Phosphoric Acid Solution. *Jpn. J. Appl. Phys.* **1998**, *37*, L1340–L1342. [CrossRef]
38. Tian, J.; Bao, J.; Li, L.; Sha, J.; Duan, W.; Qiao, M.; Cui, J.; Zhang, Z. Facile fabrication of superhydrophobic coatings with superior corrosion resistance on LA103Z alloy by one-step electrochemical synthesis. *Surf. Coat. Technol.* **2023**, *452*, 129090. [CrossRef]
39. Pawar, P.G.; Xing, R.; Kambale, R.C.; Kumar, A.M.; Liu, S.; Latthe, S.S. Polystyrene assisted superhydrophobic silica coatings with surface protection and selfcleaning approach. *Prog. Org. Coat.* **2017**, *105*, 235–244. [CrossRef]
40. Fu, J.; Sun, Y.; Ji, Y.; Zhang, J. Fabrication of robust ceramic based superhydrophobic coating on aluminum substrate via plasma electrolytic oxidation and chemical vapor deposition methods. *J. Mater. Process. Technol.* **2022**, *306*, 117641. [CrossRef]

41. Sinha, A.; Gupta, M.C. Microscale patterning of semiconductor c-Si by selective laser-heating induced KOH etching. *Semicond. Sci. Technol.* **2021**, *36*, 085002. [CrossRef]
42. Shen, Y.; Li, K.; Chen, H.; Wu, Z.; Wang, Z. Superhydrophobic F-SiO2@PDMS composite coatings prepared by a two-step spraying method for the interface erosion mechanism and anti-corrosive applications. *Chem. Eng. J.* **2021**, *413*, 127455. [CrossRef]
43. Athinarayanan, B.; Jeong, D.Y.; Kang, J.H.; Koo, B.H. Fabrication of hydrophobic and anti-reflective polymeric films using anodic aluminum-oxide imprints. *J. Korean Phys. Soc.* **2015**, *67*, 1977–1985. [CrossRef]
44. Pu, Y.; Hu, J.J.; Yao, T.; Li, L.; Zhao, J.; Guo, Y. Influence of anodization parameters on film thickness and volume expansion of thick- and large-sized anodic aluminum oxide film. *J. Mater. Sci. Mater. Electron.* **2021**, *32*, 13708–13718. [CrossRef]
45. Gao, X.; Yan, X.; Yao, X.; Xu, L.; Zhang, K.; Zhang, J.; Yang, B.; Jiang, L. The Dry-Style Antifogging Properties of Mosquito Compound Eyes and Artificial Analogues Prepared by Soft Lithography. *Adv. Mater.* **2007**, *19*, 2213–2217. [CrossRef]

polymers

MDPI

Article

Correlation Between Conditions of Polyaniline Interlayer Formation and the Structure and Performance of Thin-Film Composite Membranes for Nanofiltration Prepared via Interfacial Polymerization

Katsiaryna S. Burts [1], Tatiana V. Plisko [1,*], Anastasia V. Penkova [1], Bingbing Yuan [2], Sergey S. Ermakov [1] and Alexandr V. Bildyukevich [3]

[1] Department of Analytical Chemistry, Institute of Chemistry, St. Petersburg State University, 7/9 Universitetskaya nab., 199034 St. Petersburg, Russia; katyaburt@gmail.com or e.burt@spbu.ru (K.S.B.); a.penkova@spbu.ru (A.V.P.); s.ermakov@spbu.ru (S.S.E.)

[2] Key Laboratory of Green Chemical Media and Reactions Ministry of Education, School of Chemistry and Chemical Engineering, Henan Normal University, Xinxiang 453007, China; yuanbingbing@htu.edu.cn

[3] Institute of Physical Organic Chemistry, National Academy of Sciences of Belarus, Surganov str., 13, 220072 Minsk, Belarus; uf@ifoch.bas-net.by

* Correspondence: t.plisko@spbu.ru or plisko.v.tatiana@gmail.com

Abstract: Correlations between conditions of the polyaniline (PANI) interlayer formation on the surface of a polysulfone (PSF) porous membrane substrate and the structure and performance of thin-film composite (TFC) membranes for nanofiltration with a polyamide (PA) selective layer prepared via interfacial polymerization (IP) were studied. It was shown that application of the PANI layer significantly enhanced hydrophilicity (the water contact angle decreased from 55 ± 2° down to 26–49 ± 2°), decreased pore size and porosity, and increased the surface roughness of the selective layer surface of porous PSF/PANI membrane substrates due to the formation of bigger PANI globules, which affect the formation of the PA layer of TFC membranes via IP. It was shown that the application of the PANI intermediate layer yielded the formation of a thinner PA selective layer, a decline in surface roughness, and an increase in hydrophilicity (the water contact angle declined from 28 to <10°) and crosslinking degree of the selective layer of TFC NF membranes. The developed approach allows us to enhance the water permeation up to 45–64 $L{\cdot}m^{-2}{\cdot}h^{-1}$ at ΔP = 0.5 MPa and improve membrane selectivity (rejection coefficient of $MgSO_4$—>99.99%; LiCl—5–25%; sulfadimetoxine—80–95%) and also ensure enhanced long-term operational stability of TFC nanofiltration membranes with a PANI interlayer. Moreover, Mg^{2+}/Li^{+} separation factor values were found to increase to 37 and 58 for PANI-modified membranes compared to 9 and 8 for the reference NF-PSF membranes.

Keywords: nanofiltration; thin-film composite membrane; intermediate layer; polyaniline; interfacial polymerization

1. Introduction

Thin-film composite (TFC) membranes with a polyamide (PA) selective layer have become a key focus within the field of nanofiltration membrane preparation [1]. Nanofiltration (NF) is a pressure-driven process that is usually applied to separate divalent/multivalent ions, salts, and antibiotics in water treatment [2,3]. Nanofiltration is characterized by

 https://doi.org/10.3390/polym17091199

a number of advantages, such as low energy consumption, high purification efficiency, flexibility in application, and lack of secondary pollution [2]. The PA selective layer is usually formed on the surface of a porous ultrafiltration (UF) or microfiltration (MF) membrane substrate via the interfacial polymerization (IP) reaction between multifunctional amine monomers in the aqueous solution and multifunctional acyl chloride in the organic phase [4]. The commonly used multifunctional amines are m-phenylenediamine, p-phenylenediamine, and piperazine (PIP) [5–7]. In addition to the mentioned amine monomers, polyethyleneimine (PEI) is also used as an amine component to obtain a PA selective layer [8]. The commonly used multifunctional acyl chloride monomers are trimesoyl chloride (1,3,5-benzenetricarbonyl trichloride, TMC) and isophthaloyl chloride (IPC) [9,10]. The uncontrolled diffusion of amine monomers to an interphase boundary poses a significant challenge in achieving precise control over the morphology and structure of the resulting PA layer. This lack of control often results in the underperformance of nanofiltration membranes compared to their expected efficiency [11].

Due to the trade-off between the permeability and selectivity of nanofiltration membranes, optimization of the PA layer is still relevant (varying the nature and concentrations of the reagents in IP reaction, the duration of soaking in solutions, and the drying temperature) [12–15]. PIP and TMC are more typically used for the preparation of PA selective layers for NF membranes [1,16–18].

Moreover, there are some other techniques to break the permeability–selectivity trade-off: hydrophilization of the membrane substrate, formation of the intermediate layer, introduction of different fillers (nanoparticles, metal–organic frameworks, and surfactants) to the PA selective layer [19–24]. The formation of the intermediate layer was found to be the prospective technique for improving the nanofiltration membrane's performance. The interlayer can enhance hydrophilicity, reduce the thickness of the PA layer, and create nanostructures by controlling the IP reaction [25]. It was reported that the intermediate layers were formed from TiO_2 [26], MoS_2 [27,28], nanocellulose [29,30], carbon nanotubes [31–33], metal–organic frameworks (MOFs) and covalent organic frameworks (COFs) [34–38], polymer/MOF (COF) composites [39–42], polydopamine (PDA) [20,43,44], poly (sodium 4-styrenesulfonate)/PEI [45], PEI [46], sodium alginate [47], polyphenols [48–50], etc.

Unlike inorganic nanomaterials, polymeric coatings can establish robust adhesion forces of a PA selective layer with the surface of a membrane substrate through electrostatic interactions, hydrogen bonds, or Van der Waals forces. These interactions contribute to enhancing the structural stability of the TFC membrane [51].

Polydopamine (PDA) is widely used for preparation of the intermediate layer of nanofiltration membranes [20,43,52–55]. Al-Nahari et al. [20] obtained nanofiltration membranes by deposition of a PDA intermediate layer on the surface of a PSF ultrafiltration membrane with a molecular weight cut-off (MWCO) of 70 kDa, followed by the formation of a PA selective layer from PIP and TMC via IP. It was found that the deposition of the PDA intermediate layer resulted in a remarkably thin PA layer of 8 nm and a low average surface roughness parameter (R_a) of 3.2 nm. The resulting PDA-TFC nanofiltration membranes exhibited high performance, achieving a Na_2SO_4 rejection coefficient of 99.98%, with a pure water permeance of 34 $L \cdot m^{-2} \cdot h^{-1} \cdot bar^{-1}$ [20].

Yang et al. [43] prepared nanofiltration PANI/PDA/PA membranes similar to membranes developed by Al-Nahari et al. [20] but with a denser selective layer. It was revealed [43] that PANI/PDA/PA nanofiltration membranes were characterized by lower water permeance (15 $L \cdot m^{-2} \cdot h^{-1} \cdot bar^{-1}$) and a Na_2SO_4 rejection coefficient of 90–95%, depending on the deposition time of the intermediate layer compared to PDA/PA membranes [20].

A literature analysis revealed that there are a lot of studies dedicated to the development of nanofiltration membranes with a modified PDA intermediate layer. Thus, the PDA intermediate layer was formed with the addition of PEI [52,53], graphene oxide (GO) [54], UiO-66 carbon nanotubes [42], and MXene [55]. It was shown that these additives significantly increased the surface hydrophilicity and enhanced the water permeance without a significant decline in selectivity.

According to the literature analysis, it was revealed that different polymers were used for the formation of the intermediate layer: poly (sodium 4-styrenesulfonate) (PSS)/polyethyleneimine (PEI) with the polyelectrolyte complex formation [45], sodium alginate [47], polyvinyl alcohol (PVA) [25], PVA/tannic acid [56], hyaluronic acid [57], PEI crosslinked with glutaraldehyde [46], and glutaraldehyde crosslinked chitosan [58].

It was found that nanofiltration membranes with an intermediate layer based on the above-mentioned polymers except the PEI/GA interlayer [46] resulted in the hydrophilization of the selective layer. Regarding the membrane performance, the formation of a polymer intermediate layer resulted in enhanced water permeance, excluding membranes with interlayers based on hyaluronic acid [57] and chitosan crosslinked by glutaraldehyde [58]. However, it was revealed that the mentioned membranes with a polymer intermediate layer possessed an advanced selectivity compared to the TFC membranes for nanofiltration without an interlayer.

Zha et al. applied a PANI interlayer by an in situ polymerization method on the surface of a hydrolyzed polyacrylonitrile (HPAN) ultrafiltration membrane (MWCO—50 kDa), followed by PA selective layer formation via IP between PEI and TMC [51]. The performance of the prepared membranes was investigated in organic-solvent nanofiltration. It was shown that the PA/PANI/HPAN membrane was characterized by an ethanol permeance of 3.1 $L{\cdot}m^{-2}{\cdot}h^{-1}{\cdot}bar^{-1}$ and a rejection coefficient for Eriochrome black T, Acid fuchsin, Congo red, Brilliant blue R, and Rose Bengal dyes of over 98%.

Guo et al. obtained sulfonated polyaniline (SPANI) nanofibers in order to apply them for intermediate layer formation on the surface of polyethersulfone (PES) microfiltration (MF) membranes with an effective pore size of 0.45 μm [59,60]. The selective PA layer was prepared via the traditional IP process using PIP and TMC solutions. It was found that the NF membrane with a SPANI intermediate layer was characterized by a water permeance of 29 $L{\cdot}m^{-2}{\cdot}h^{-1}{\cdot}bar^{-1}$ and a Na_2SO_4 rejection of 99% compared to the reference membrane with a water permeance of 12 $L{\cdot}m^{-2}{\cdot}h^{-1}{\cdot}bar^{-1}$ and a Na_2SO_4 rejection of 94% [59,60].

However, according to the literature analysis, no works devoted to the study of the effect of the application modes of the PANI layer via oxidative polymerization (concentration of aniline and duration of PANI intermediate layer deposition) on the formation of a PA selective layer via IP and the structure and performance of the resulting TFC nanofiltration membranes were reported. The aim of this research was to investigate the patterns of the formation of the PA selective layer via IP depending on the conditions of intermediate PANI layer formation. Moreover, the influence of the concentrations of reagents during IP on the structure and performance of TFC membranes for nanofiltration was studied. The developed membranes were characterized by scanning electron microscopy (SEM), atomic force microscopy (AFM), contact angle measurements, and X-ray photoelectron spectroscopy (XPS). Membrane performance was studied in ultra- and nanofiltration using polyvinylpyrrolidone (PVP K30, M_n = 40,000 $g{\cdot}mol^{-1}$), salts and antibiotic sulfadimetoxine, and mixtures of magnesium sulfate and lithium chloride as test substances. The novelty of the present work is that for the first time, the correlation between the conditions of the PANI interlayer formation on the surface of a polysulfone (PSF) ultrafiltration (UF) membrane substrate and the structure and performance of thin-film composite (TFC) membranes for nanofiltration with a polyamide (PA) selective layer prepared via interfacial polymerization (IP) were revealed.

2. Materials and Methods

2.1. Materials

Polysulfone Ultrason S 6010 (PSF, M_n = 55,000 g·mol^{-1}) and N,N-dimethylacetamide (DMAc) were acquired from BASF (Ludwigshafen, Germany) and used for membrane substrate preparation. Polyvinylpyrrolidone (PVP K30, M_n = 40,000 g·mol^{-1}), used as a pore-forming agent, was provided by Fluka (Munich, Germany). Aniline (ANI, 99.5% purity) was sourced from Sisco Research Laboratories Pvt. Ltd. (Mumbai, India). Concentrated hydrochloric acid (HCl) was purchased from AnalytComplect (Minsk, Belarus). Ammonium persulfate (APS, $(NH_4)_2S_2O_8$) was obtained from Vecton (St. Petersburg, Russia). Piperazine (PIP, 98% purity) and trimesoyl chloride (TMC, benzene-1,3,5-tricarbonyl chloride, 98% purity) were supplied by Sisco Research Laboratories Pvt. Ltd. (Mumbai, India) and Tokyo Chemical Industry Co., Ltd. (Tokyo, Japan), respectively.

Magnesium sulfate ($MgSO_4 \cdot 7H_2O$, AnalytComplect, Minsk, Belarus), sodium chloride (NaCl, Belleshimkomplekt, Minsk, Belarus), calcium chloride ($CaCl_2$, anhydrous, Belleshimkomplekt, Minsk, Belarus), magnesium chloride ($MgCl_2 \cdot 6H_2O$, Serva, Heidelberg, Germany), sodium sulfate ($Na_2SO_4 \cdot 10H_2O$, AnalytComplect, Minsk, Belarus), lithium chloride (LiCl, Lenreactiv, St. Petersburg, Russia), and sulfadimetoxine (SDM, M = 310.33 g·mol^{-1}, BLDpharm, Shanghai, China) were applied as test substances to investigate membrane performance in nanofiltration.

2.2. Methods

2.2.1. Membrane Preparation

The polymer solutions were prepared in a glass flask under constant stirring using an IKA RW20 (IKA Werke GmbH & Co. KG, Staufen im Breisgau, Germany) overhead stirrer at 120 °C for 4–5 h. The resulting homogeneous solution was then left to stand overnight to remove air bubbles.

UF membranes were prepared via a non-solvent induced phase separation technique using a water coagulation bath at room temperature. Polymer solution was cast on the polyester nonwoven fabric by the casting knife with the gap of 250 µm. After polymer solution casting, the film was immersed in the water coagulation bath. Prepared UF membranes were left for one day to remove residual solvent and wash out the pore-forming agent.

The polyaniline (PANI) interlayer was formed by oxidative polymerization of ANI on the surface of the UF PSF membrane substrate at the temperature of 4–10 °C by physical adsorption from the aniline solution in 1 M HCl with ammonium persulfate as an initiator under constant stirring (Figure S1). The scheme of oxidative polymerization of aniline is presented in the Supporting Information (Figure S1). The PSF membrane was fixed upside down in such a way that only the selective layer surface contacted the solution, where ANI oxidative polymerization occurred. After ANI polymerization, the membrane was kept in water for 24 h to remove unreacted monomer from the membrane surface. The duration of PANI layer formation was 0.5 h or 1 h. The concentrations of ANI in the aqueous solution were 0.05, 0.1, and 0.3 wt.%. The codes of the prepared UF membranes with the PANI layer are presented in Table 1.

The polyamide (PA) selective layer was formed by interfacial polymerization (IP) via the reaction between PIP aqueous solutions and TMC solution in Nefras C2 (80/120, "Vershina" LLC, Vsevolozhsk, Russia). The membrane was kept in each solution for 10 s in such a way that only a selective layer contacted with the solutions of monomers. The excess

PIP aqueous solution was gently removed by a rubber roller. After applying the PA layer, the membrane was dried for 10 min at 80 °C. The PIP concentration was varied from 1 to 4 wt.%, while the mass ratio PIP/TMC remained constant. (±)-Camphor-10-sulfonic acid (CSA, Maclin, Shanghai, China) was used as an additive to the PIP aqueous solution at a concentration similar to the PIP concentration. The codes of prepared TFC NF membranes are listed in Table 1.

Table 1. The codes of UF and NF membranes.

<table>
<tr><th rowspan="2">Membrane Code</th><th colspan="2">PANI Interlayer</th><th colspan="2">PA Selective Layer</th></tr>
<tr><th>ANI Concentration, wt.%</th><th>Duration of PANI Layer Application, h</th><th>PIP Concentration, wt.%</th><th>TMC Concentration, wt.%</th></tr>
<tr><td>UF-PSF</td><td colspan="4">-</td></tr>
<tr><td>UF-PSF/PANI 0.05</td><td>0.05</td><td rowspan="3">0.5</td><td colspan="2">-</td></tr>
<tr><td>UF-PSF/PANI 0.1</td><td>0.1</td><td colspan="2">-</td></tr>
<tr><td>UF-PSF/PANI 0.3</td><td>0.3</td><td colspan="2">-</td></tr>
<tr><td>UF-PSF/PANI 0.1–1</td><td>0.1</td><td>1.0</td><td colspan="2">-</td></tr>
<tr><td>NF-PSF–2</td><td colspan="2" rowspan="2">-</td><td>2.0</td><td>0.12</td></tr>
<tr><td>NF-PSF–4</td><td>4.0</td><td>0.24</td></tr>
<tr><td>NF-PSF/PANI 0.05–0.5–2</td><td rowspan="2">0.5</td><td rowspan="5">0.5</td><td>2.0</td><td>0.12</td></tr>
<tr><td>NF-PSF/PANI 0.05–0.5–4</td><td>4.0</td><td>0.24</td></tr>
<tr><td>NF-PSF/PANI 0.1–0.5–1</td><td rowspan="4">0.1</td><td>1.0</td><td>0.06</td></tr>
<tr><td>NF-PSF/PANI 0.1–0.5–2</td><td>2.0</td><td>0.12</td></tr>
<tr><td>NF-PSF/PANI 0.1–0.5–4</td><td rowspan="2">4.0</td><td rowspan="2">0.24</td></tr>
<tr><td>NF-PSF/PANI 0.1–1–4</td><td>1.0</td></tr>
<tr><td>NF-PSF/PANI 0.3–0.5–2</td><td rowspan="2">0.3</td><td rowspan="3">0.5</td><td>2.0</td><td>0.12</td></tr>
<tr><td>NF-PSF/PANI 0.3–0.5–4</td><td rowspan="2">4.0</td><td rowspan="2">0.24</td></tr>
<tr><td>NF-PSF/PANI 0.5–0.5–4</td><td>0.5</td></tr>
</table>

2.2.2. Membrane Characterization

Study of Membrane Structure by Scanning Electron Microscopy (SEM)

To investigate the membrane morphology, a scanning electron microscope Zeiss AURIGA Laser (Carl Zeiss AG, Oberkochen, Germany) was used. Membranes were immersed in 20 wt.% glycerol aqueous solution and dried in air before SEM studies. Impregnation of membranes in glycerol aqueous solution allows the membrane pore structure to be preserved upon drying. Membrane cleavages were prepared by submerging membrane samples in liquid nitrogen and fracturing perpendicular to the surface.

Study of Membrane Surface Structure by Atomic Force Microscopy (AFM)

The surface topography of the developed UF and NF membranes was studied by an NTMDT nTegra Maximus atomic force microscope with standard silicon cantilevers with a rigidity of 15 N·m^{-1} (NT-MDT Spectrum Instruments, Zelenograd, Russia) in tapping mode.

Fourier-Transform Infrared (FTIR) Spectroscopy

The composition of selective layers of PSF membrane substrates was analyzed by Fourier-transform infrared (FTIR) spectroscopy using a Nicolet Is50 spectrometer

(Thermofisher Scientific, Waltham, MA, USA). Spectra were acquired in the range of 400–4000 cm^{-1} at 25 °C. Prior to analysis, all membrane samples were air-dried at ambient temperature for 5 days to ensure complete water evaporation.

X-Ray Photoelectron Spectroscopy (XPS)

X-ray photoelectron spectroscopy (XPS) measurements were conducted using an ESCALAB™ 250Xi spectrometer (Thermo Fisher Scientific, Waltham, MA, USA) equipped with a monochromatic Al-Kα X-ray source (650 μm spot size). Analyses were performed at a 90° take-off angle relative to the surface plane. Elemental composition was quantified from survey spectra (0–1350 eV) acquired with a 100 eV pass energy, a 0.5 eV step size, a 50 ms dwell time, and 5 scan accumulations. Atomic concentrations were derived by integrating peak intensities with Smart-type background subtraction. Five different surface locations were analyzed on each sample surface, and average values were calculated.

High-resolution spectra of C1s, O1s, and N1s core levels were recorded with a 20 eV pass energy, a 0.05 eV step resolution, a 50 m·s dwell time, and 50 accumulated scans. Spectral deconvolution was performed using Thermo Scientific™ Advantage software, applying a Smart-type background subtraction with constrained parameters: 1.5 eV full width at half maximum (FWHM) and 35% Gaussian–Lorentzian peak shape mixing.

The polyamide network architecture comprises two distinct structural domains:

- A crosslinked domain (m) formed when all three acyl chloride functionalities of TMC react with PIP amine groups;
- A linear domain (n) resulting from partial reaction where only two acyl chlorides participate in amide bond formation [32].

These structural motifs exhibit characteristic oxygen-to-nitrogen stoichiometric ratios:
A 3:3 O:N ratio for fully crosslinked segments (m);
A 4:2 O:N ratio for linear segments (n).

The degree of crosslinking (D) was quantitatively determined through systematic analysis using Equations (1) and (2), which account for these structural variations and stoichiometric relationships:

$$\frac{O}{N} = \frac{3m + 4n}{3m + 2n}, \tag{1}$$

$$D = \frac{m}{m + n} \times 100\% = \frac{4N - 2O}{N + O} \times 100\%. \tag{2}$$

In Equations (1) and (2), O (%) and N (%) indicate the respective ratios of O and N.

Zeta Potential of the TFC NF Membrane Delective Layer

Zeta potentials of the selective layers of TFC NF membranes were determined using an Anton Paar SurPass 3 (Austria, Graz) solid surface analyzer. The membrane samples were kept in MilliQ water for 24 h before tests. The sample holder for discs was used; the gap height between two discs of the membrane sample was set at 100–106 μm. Aqueous potassium chloride solution in MilliQ water with a concentration of 1 mmol·L^{-1} was used as an electrolyte. The zeta potential was measured in the pH range of 3.0 to 10.0, with steps of 0.5–0.7 pH units.

Contact Angle

The contact angles of UF and NF membranes with a PANI layer were analyzed by the attached bubble method using an LK-1 goniometer (RPC «Open Science Ltd.»). Contact angle measurements were conducted in a three-phase system comprising water, a membrane surface, and an air bubble. For each batch, measurements were performed on five membranes. The measurement error was within ±2°.

Preparation of the Solutions of Test Substances

To evaluate the performance of TFC NF membranes, aqueous solutions of salts $MgSO_4$ (2 $g{\cdot}L^{-1}$), Na_2SO_4 (2 $g{\cdot}L^{-1}$), $MgCl_2$ (2 $g{\cdot}L^{-1}$), $CaCl_2$ (2 $g{\cdot}L^{-1}$), LiCl (1 $g{\cdot}L^{-1}$), NaCl (0.5 $g{\cdot}L^{-1}$), and SDM (20 $mg{\cdot}L^{-1}$) were prepared at room temperature by dissolving a predetermined amount of the test substance in distilled water with continuous stirring.

Membrane Performance

Membrane performance was assessed using a custom-designed filtration cell with a volume of 400 mL and a membrane area of 24.6 cm^2. The separation efficiency of the reference UF PSF membrane substrate was evaluated by measuring the pure water flux (J_{water}, $L{\cdot}m^{-2}{\cdot}h^{-1}$) and the rejection coefficients for polyvinylpyrrolidone (PVP K30, M_n = 40 kDa) (R, %). Specifically, the UF membrane was conditioned by filtering distilled water at 0.1 MPa for 30 min at room temperature while stirring at 250–300 rpm, after which the pure water flux was recorded. The rejection coefficients for PVP K-30 were measured 20 min after the filtration of a 0.3 wt% PVP K-30 aqueous solution.

To determine the membrane rejection coefficients, the concentration of PVP K-30 in the feed and permeate was determined using an ITR-2 interferometer (Zagorsk Optical-Mechanical Plant, Sergiyev Posad, Russia) with a cell length of 40 mm, 150-fold lens magnification, and an incandescent lamp as a light source. The interferometer LIR-2 determines the concentration of the solution via the difference between the refractive indices of the studied solution and a standard solution or liquid with a known refractive index. The principle of operation of the interferometer LIR-2 is based on double-slit diffraction phenomenon.

The membrane performance of TFC NF membranes was systematically evaluated through filtration experiments in stirring mode conducted at ambient temperature (25 ± 1 °C) under an applied pressure of 0.5 MPa. The investigation encompassed three feed solutions: distilled water, saline solutions, and antibiotic-containing aqueous systems. A stirring rate of 400 rpm was maintained throughout the experiments to minimize concentration polarization effects.

Prior to performance characterization, membranes underwent a stabilization protocol consisting of 1 h distilled water filtration to establish steady-state operation. Initial water permeability measurements were subsequently recorded. The feed solution was then replaced with the test solutions for rejection studies.

Comprehensive performance assessment included:

- Water permeation flux determination;
- Quantitative evaluation of salt rejection efficiency;
- Antibiotic removal capacity analysis.

All performance characteristics were measured following a 2 h equilibration period to ensure system stability and reproducible results.

The membrane rejection coefficient was determined by measuring the concentrations of salts in both the permeate and feed solution using an ITR-2 interferometer. The rejection coefficients for the antibiotic SDM were assessed with a UV/Visible Spectrophotometer SP-8001 (Metertech Inc., Taipei, Taiwan) at a wavelength of 280 nm.

The membrane's ion rejection was evaluated using a mixed aqueous solution containing LiCl and $MgSO_4$ with the following characteristics:

- Total dissolved salt concentration: 5 $g{\cdot}L^{-1}$;
- Mg^{2+}/Li^{+} mass ratio: 18:1.

Feed and permeate compositions were quantitatively analyzed using inductively coupled plasma atomic emission spectroscopy (ICP-AES, Vista PRO, Varian, Palo Alto,

CA, USA). The separation factor ($S_{Mg^{2+},Li^{+}}$), representing the membrane's preferential selectivity between divalent and monovalent cations, was calculated using Equation (3):

$$SMg^{2+}, Li^{+} = \frac{(C_{Mg2+}/C_{Li+})f}{(C_{Mg2+}/C_{Li+})p} = \frac{100 - R_{Li+}}{100 - R_{Mg2+}}, \quad (3)$$

where $C_{Mg}{}^{2+}$ and C_{Li+} represent concentrations of Mg^{2+} and Li^{+} in feed solution (f) and permeate (p), respectively, while $R_{Li}{}^{+}$ and $R_{Mg}{}^{2+}$ denote the rejection coefficients of Li^{+} and Mg^{2+}, respectively.

Pore Size Distribution of TFC NF Membranes

The mean effective pore size of the fabricated nanofiltration membranes was characterized using a neutral solute rejection method. Aqueous solutions containing polyethylene glycol (PEG) with molecular weights of 200, 300, 400, and 600 Da (each at 0.3 wt.% concentration) were employed as probe molecules for filtration experiments. The hydrodynamic Stokes radius (r_s, nm) of each PEG molecule was calculated using Equation (4), enabling quantitative assessment of the membrane's pore size distribution based on size-exclusion characteristics.

$$r_s = 16.74 \times 10^{-3} \times M_W^{0.557} \quad (4)$$

The hydrodynamic radius of polyethylene glycol (PEG) molecules was determined using Equation (4). Membrane pore structure parameters, including mean effective pore diameter (μ_p) and pore size distribution, were quantified following established protocols [61]. The analytical procedure involved the following steps:

- Construction of a semi-logarithmic probability plot of solute rejection versus Stokes radius;
- Determination of μ_p as the intercept at 50% rejection from the fitted curve;
- Calculation of the geometric standard deviation (σ_g) from the ratio of solute radii at 84.13% and 50% rejection thresholds.

Under the assumption of negligible steric and hydrodynamic interactions between solute molecules and pore walls, the obtained μ_p and σ_g values correspond to the actual mean pore size (μ_s) and size distribution (σ_g), respectively. The pore size distribution was mathematically described by the probability density function (PDF) given in Equation (5):

$$\frac{dR_T(r_p)}{dr_p} = \frac{1}{r_p \ln\sigma_p \sqrt{2\pi}} \exp\left(-\frac{\left(\ln r_p - \ln\mu_p\right)^2}{2(\ln\sigma_p)^2}\right). \quad (5)$$

Determination of the Molecular Weight Cut-Off (MWCO) of TFC NF Membranes

The investigation of the molecular weight cut-off (MWCO) was conducted during the filtration of 0.3 wt. % aqueous solutions of PEG with different molecular weights (200, 300, 400, and 600 Da). To obtain MWCO values of TFC NF membranes, the rejection coefficients of tested PEGs were calculated, and the relationship between the rejection and PEG molecular weight was plotted. The MWCO was considered as a molecular weight at 90% rejection.

Study of TFC NF Membrane Chemical Stability

Chlorine resistance evaluation: The membrane was immersed in 2000 ppm NaClO solution for oxidation treatment, and the pH of NaClO was controlled at pH = 8 with hydrogen chloride and sodium hydroxide solutions. The NaClO solution was replaced every hour for 12 h to prevent any concentration change in the solution. Thereafter, the water permeation flux at $\Delta P = 0.5$ MPa and $MgSO_4$ rejection were measured.

Acid resistance evaluation: The membrane was immersed in 10 wt % H_2SO_4 acid solution and heated to 80 °C for 3 h. Thereafter, the water permeation flux at $\Delta P = 0.5$ MPa and $MgSO_4$ rejection were measured.

3. Results

3.1. Effect of PANI Layer Deposition at Different Aniline Concentrations During Oxidative Polymerization on the UF PSF Membrane Structure and Performance

The structure of the reference PSF membrane substrate and PSF membranes with a deposited PANI layer was studied by SEM (Figure 1). It was revealed that the formation of a PANI layer on the surface of the UF-PSF reference membrane resulted in partial blocking of the pores of the selective layer of the membrane substrate and a reduction in the porosity of the selective layer (Figure 1). Moreover, PANI formations of irregular shape were observed on the selective layer surface of UF-PSF/PANI membranes. The PANI layer was undetectable via SEM of the membrane cross-section; therefore, it was impossible to determine the PANI layer thickness from SEM micrographs.

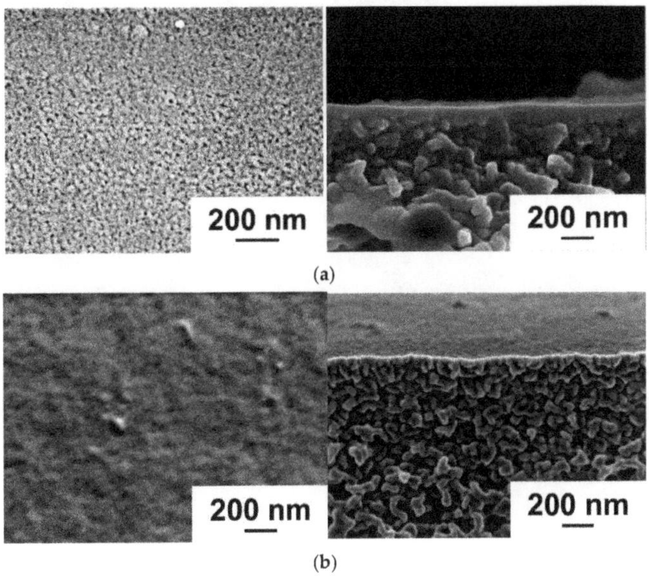

Figure 1. SEM microphotographs of the surface and cross-section of the selective layer of ultrafiltration PSF membrane substrate: (**a**) reference UF-PSF; (**b**) UF-PSF/PANI 0.1.

The topography of the surface of UF membranes with a deposited PANI layer was investigated by atomic force microscopy (AFM) (Figure 2, Table 2). According to AFM investigations, the surface of the selective layer consisted of polymer nodules, similar to other membranes prepared by the NIPS method. It was demonstrated that the formation of a PANI layer on the selective layer surface of the PSF membrane led to an increase in the surface roughness parameters (average surface roughness (R_a) and root-mean-squared surface roughness (R_q)) (Figure 2, Table 2). This is due to the formation of PANI globules of irregular shape on the selective layer surface proved by SEM studies (Figure 1). Moreover, an increase in the concentration of monomers (ANI) in the oxidative polymerization process resulted in a significant rise in the roughness of the UF membrane surface due to the formation of bigger PANI globules (Table 2). It was found that increasing the time of

PANI layer deposition from 0.5 h up to 1 h at an ANI concentration of 0.1 yields an increase in surface roughness parameters from R_a = 2.19 nm and R_q = 2.74 nm for the UF-PSF/PANI 0.1 membrane to R_a = 3.94 nm and R_q = 5.24 nm for UF-PSF/PANI 0.1–1. Visually, a darker membrane surface was formed when the PANI layer was applied for 1 h compared to 0.5 h.

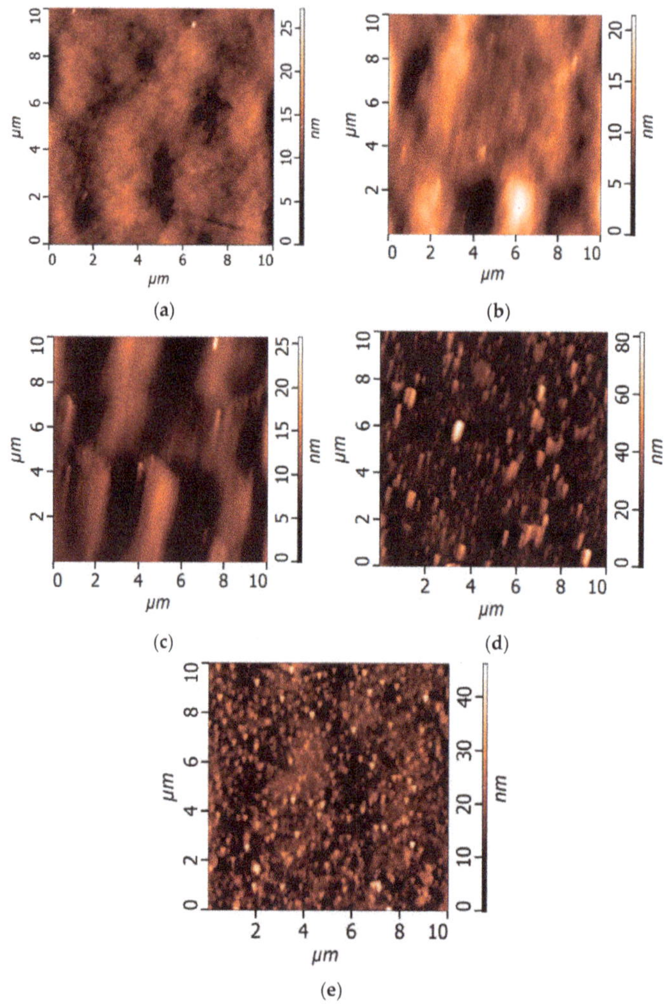

Figure 2. AFM microphotographs of UF PSF membranes with PANI layer: (**a**) reference UF-PSF; (**b**) UF- PSF/PANI 0.05; (**c**) UF- PSF/PANI 0.1; (**d**) UF-PSF/PANI 0.3, (**e**) UF- PSF/PANI 0.1–1.

The hydrophilic–hydrophobic balance of the selective layer surface of UF PSF/PANI membranes was studied by the determination of the water contact angle (Table 2). It was found that an increase in the concentration of ANI during the formation of the PANI layer on the surface caused a significant enhancement of surface hydrophilicity, confirmed by the decrease in water contact angle (Table 2). The water contact angle of UF PSF/PANI membranes was shown to decline down to 26–49°, compared to 55° for the initial UF PSF membrane, which indicated significant surface hydrophilization (Table 2). It should be mentioned that the initial membrane substrate UF-PSF had a lower water contact angle than pristine PSF (77–80°), attributed to the addition of very hydrophilic PVP K30 to the casting PSF solution.

Table 2. The surface roughness parameters and water contact angles of PSF UF membranes with a PANI layer obtained by physical adsorption in the process of oxidative polymerization of ANI for 0.5 h and 1 h.

Membrane Code	Roughness Parameters		Water Contact Angle, °
	R_a, nm	R_q, nm	
UF-PSF	1.87	2.32	65 ± 2
UF-PSF/PANI 0.05	2.25	2.95	49 ± 2
UF-PSF/PANI 0.1	2.19	2.74	38 ± 2
UF-PSF/PANI 0.3	5.46	7.88	26 ± 2
UF-PSF/PANI 0.1–1	3.94	5.24	30 ± 2

The formation of a PANI layer on the surface of UF PSF membranes was proved by FTIR studies (Figure S2). It was found that vibrations of the main membrane material PSF (1150, 1586, 1487, 1150, 1243, and 2960 cm^{-1}) predominate in all spectra (Figure S2). The peak at 1650 cm^{-1} is due to the vibrations of the carbonyl group of PVP K-30. The presence of polyaniline on the membrane surface is proved by the appearance of vibrations of the secondary amine group at 3293 cm^{-1} (Figure S2).

Moreover, it was shown that rise in the duration of PANI layer deposition from 0.5 h up to 1 h at an ANI concentration of 0.1 results in a decline in water contact angle down to 30 ± 2°, which demonstrated the formation of a more hydrophilic selective layer surface.

UF PSF/PANI membranes were examined during water and PVP K30 aqueous solution ultrafiltration (Figure 3).

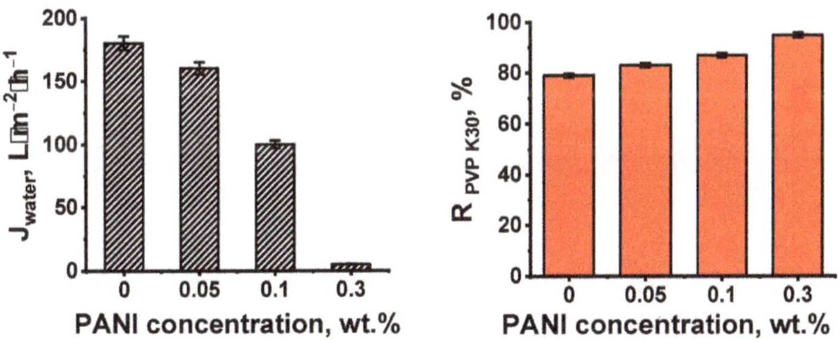

(a) (b)

Figure 3. Performance of UF membranes with a PANI layer obtained during 0.5 h of application depending on ANI concentration during oxidative polymerization: (**a**) pure water flux; (**b**) rejection coefficient of PVP K30.

It was shown that the formation of a PANI layer on the surface of a PSF porous membrane resulted in a decrease in pure water flux (Figure 3). An increase in ANI concentration during oxidative polymerization led to a significant decline in membrane permeability due to the decrease in porosity and pore size of the selective layer, since pores are blocked by the PANI layer. It was also found that an increase in ANI concentration up to 0.3 wt.% during oxidative polymerization on the surface of the UF PSF membrane support yielded a crucial decline in water permeability down to 5 $L{\cdot}m^{-2}{\cdot}h^{-1}$ at ΔP = 0.1 MPa (Figure 3a). This is likely due to the increase in the PANI layer thickness and decrease in its pore size and porosity. It was revealed that membranes with a layer applied from 0.05 and 0.1 wt.% ANI solution are characterized by a pure water flux of 160 and 100 $L \cdot m^{-2} \cdot h^{-1}$ at $\Delta P = 0.1$ MPa, respectively. An increase in the concentration of ANI in the aqueous solution resulted in a decline in the pore size of the selective layer, which caused a rise in membrane rejection—the PVP K30 rejection coefficient rose from 79 to 83–95% (Figure 3b). It was shown that the increase in the deposition time of the PANI layer from 0.5 h up to 1 h at an ANI concentration of 0.1 wt.% yields a decrease in pure water flux from 100 $L \cdot m^{-2} \cdot h^{-1}$ at $\Delta P = 0.1$ MPa down to 84 $L \cdot m^{-2} \cdot h^{-1}$ at $\Delta P = 0.1$ MPa. The rejection of PVP K30 remains at the same level of 87%. This is due to the formation of a thicker PANI layer with lower porosity. The obtained results showed that UF-PSF/PANI 0.05 and UF-PSF/PANI 0.10 membranes were characterized by a lower molecular weight cut-off (MWCO)—approximately 20 kDa, compared to the reference UF PSF membrane with an MWCO of approximately 50 kDa.

Thus, it was found that deposition of a PANI layer via oxidative polymerization on the surface of a PSF ultrafiltration membrane yields partial pore blocking, an increase in surface roughness, and hydrophilization of the membrane selective layer. When a higher ANI concentration and longer time of PANI layer formation are applied, a rougher and more hydrophilic selective layer with smaller pores is formed. It is expected that these substantial changes in structure, porosity, and hydrophilic–hydrophobic properties of the membrane selective layer will affect the PA layer formation via IP during TFC membrane preparation.

3.2. *Effect of PANI Interlayer Formation on the Structure and Performance of TFC Membranes for Nanofiltration Prepared via IP*

TFC nanofiltration membranes with a PANI interlayer were obtained via IP. The PANI interlayer was applied for 0.5 h or 1 h using concentrations of 0.05 wt.%, 0.1 wt.%, 0.3 wt.%, and 0.5 wt.% of the monomer ANI in the aqueous solution during oxidative polymerization. The effect of the modes of PANI interlayer formation on the structure and performance of TFC membranes for nanofiltration was revealed. Moreover, the influence of PIP and TMC concentrations during IP on the structure and performance of TFC NF membranes was studied.

3.2.1. The Effect of Time of PANI Interlayer Deposition on the Structure and Performance of TFC NF Membranes

At the first stage, the influence of the duration of the deposition of the PANI interlayer from 0.1 wt.% ANI solution on the structure and performance of TFC NF membranes was studied (Figures 4 and 5).

It was found that the application of the PANI interlayer on the surface of the PSF membrane substrate from 0.1 wt.% ANI solution for 0.5 h and 1 h resulted in a substantial decrease in the thickness of the PA layer from 104 nm down to 50 and 31 nm, respectively (Figure 4). When the PANI interlayer was applied for 0.5 h and 1 h from 0.1 wt.% ANI solution, the formation of smaller PA globules on the selective layer surface of TFC NF membranes compared to the reference TFC NF membrane occurred (Figure 4). It was found that the application of the PANI interlayer and the increase in the duration of the interlayer deposition led to a substantial decrease in the surface roughness parameters (Table 3). This is due to the hydrophilization of the membrane substrate selective layer when PANI was deposited, which led to its better impregnation by PIP solution. It is supposed that when the hydrophilic PANI layer is adsorbed on the pore walls, the capillary effect of the substrate is increased, and more PIP molecules are entrapped inside the pores. Moreover, the rate of PIP release from the PSF/PANI membrane substrate is decreased, yielding a lower rate of IP reaction. This leads to the formation of a PA layer with higher uniformity, lower surface roughness, and lower thickness in comparison with the initial TFC membrane (Figure 4, Table 3).

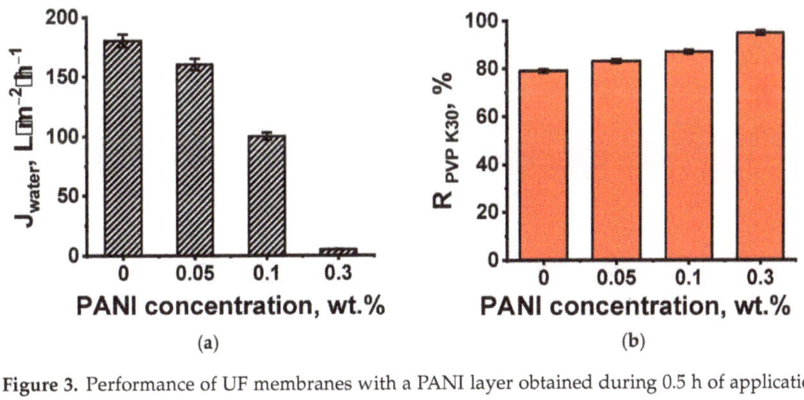

Figure 4. SEM and AFM microphotographs of the selective layer surface and cross-sections of TFC NF membranes depending on the duration of PANI interlayer formation: (**a**) NF-PSF–4; (**b**) NF-PSF/PANI 0.1–0.5–4; (**c**) NF-PSF/PANI 0.1–1–4. Figure 4a was adapted with permission from Ref. [62]. Copyright 2025, copyright Elsevier B.V.

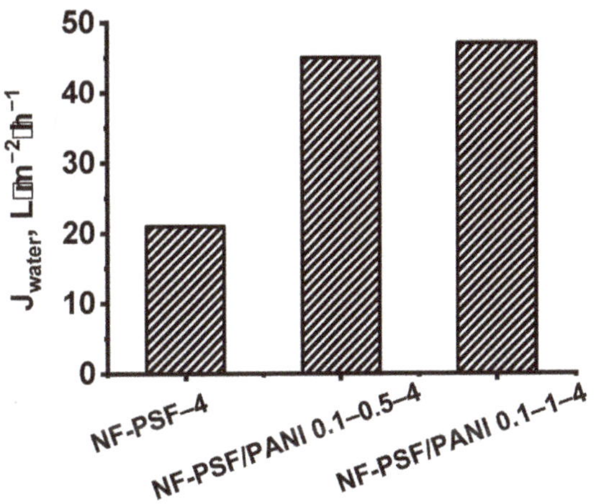

Figure 5. Dependence of water permeation (J_{water}) of TFC NF membranes on the time of application of the PANI intermediate layer (PA layer—4 wt.% PIP, 4 wt.% CSA, 0.24 wt.% TMC) at ΔP = 0.5 MPa.

Table 3. The surface roughness parameters of TFC NF membranes with a PANI interlayer obtained from 0.1 wt.% ANI solution.

Membrane Code	Roughness Parameters		Water Contact Angle, °
	R_a, nm	R_q, nm	
NF-PSF-4	18.95	25.10	28 ± 2
NF-PSF/PANI 0.1–0.5–4	10.98	15.48	<10
NF-PSF/PANI 0.1–1–4	5.54	8.20	<10

Moreover, it was shown that the application of the PANI interlayer led to the significant hydrophilization of the selective layer of TFC NF membranes due to the incorporation of highly hydrophilic PANI groups into the PA layer (Table 3). It was found that the water contact angle of the selective layer surface of TFC NF membranes decreased from 28 ± 2° down to $\theta < 10°$ when the PANI interlayer was deposited, regardless the time of deposition.

The nanofiltration performance of TFC membranes depending on the duration of the PANI layer application was studied using aqueous solutions of salts (Figure 5, Table 4). It was found that the formation of an intermediate layer based on PANI led to a significant increase in water permeation (J) of TFC NF membranes by more than two times: up to 45 and 47 $L{\cdot}m^{-2}{\cdot}h^{-1}$ (at ΔP = 0.5MPa) for NF-PSF/PANI 0.1–0.5–4 and NF-PSF/PANI 0.1–1–4 membranes, respectively, compared to 21 $L{\cdot}m^{-2}{\cdot}h^{-1}$ for the reference NF-PSF–4 membrane. This was due to the decrease in the selective layer thickness as well as substantial hydrophilization of the selective layer when the PANI interlayer was applied and the time of PANI deposition increased (Figure 4, Table 3). However, it was revealed that the permeability of the TFC NF membranes with a PANI interlayer did not change much with the rise in the duration of PANI layer application from 0.5 to 1 h (Figure 5).

Table 4. Salt and SDM rejection coefficients of TFC NF membranes with different times of application of the PANI intermediate layer (PA layer—4 wt.% PIP, 4 wt.% CSA, 0.24 wt.% TMC) at ΔP = 0.5 MPa.

Salt	Rejection		
	NF-PSF–4	**NF-PSF/PANI 0.1–0.5–4**	**NF-PSF/PANI 0.1–1–4**
$MgSO_4$	92	99.99	93
Na_2SO_4	90	90	90
$MgCl_2$	91	99	92
$CaCl_2$	91	92	91
NaCl	28	25	22
LiCl	28	25	22
SDM	95	96	83

It was shown that according to the rejection coefficients, salts can be arranged as follows for both the reference membrane and TFC membranes with a PANI interlayer:

$MgSO_4 > MgCl_2 > CaCl_2 > Na_2SO_4 > NaCl \approx LiCl$

It was found that the depositions of the PANI intermediate layer for 0.5 h resulted in a slight increase in the rejection coefficients of divalent salts ($MgSO_4$, $MgCl_2$, and $CaCl_2$) compared to the reference NF-PSF–4 membrane without intermediate layer (Table 4). However, the selectivity of monovalent salts NaCl and LiCl was shown to decrease, and selectivity of Na_2SO_4 was not changed compared to the reference membrane (Table 4). The reason for the increased selectivity may be the formation of a more uniform and more crosslinked selective layer when the PANI interlayer was applied for 0.5 h. Thus, when the PANI interlayer was deposited for 0.5 h, both the permeation and selectivity of the TFC membrane increased compared to the reference membrane.

With an increase in PANI interlayer application time from 0.5 h to 1 h, a looser PA layer was probably formed that caused a reduction in the rejection coefficient (R) for $MgSO_4$ from 99.99% to 93%, for $MgCl_2$ from 99 to 92%, for $CaCl_2$ from 92 to 91%, and for NaCl and LiCl from 25% to 22% for membranes with PANI interlayers applied for 0.5 and 1 h, respectively (Table 4). However, $MgSO_4$ rejection of the NF-PSF/PANI 0.1–1–4 TFC membrane is slightly higher compared to the reference TFC membrane (Table 4). It is worth noting that rejection of the NF-PSF/PANI 0.1–1–4 membrane is practically at the same level as for the reference membrane; however, the water permeation for NF-PSF/PANI 0.1–1–4 is significantly higher compared to the reference membranes.

It was found that SDM rejection slightly increases when the PANI interlayer was deposited for 0.5 h compared to the reference TFC NF membrane. Application of the PANI interlayer for 1 h led to a substantial decrease in SDM rejection compared to both NF-PSF-4 and NF-PSF/PANI 0.1–0.5–4 membranes.

According to the obtained results, 0.5 h was selected as the optimal duration for PANI layer application for further experiments.

3.2.2. The Effect of the Concentration of ANI, PIP, and TMC on the Structure and Properties of TFC NF Membranes

The effect of the ANI concentration (0.05 wt.%, 0.1 wt.%, 0.3 wt.%, and 0.5 wt.%) in the aqueous solution applied to form the intermediate layer on the performance of TFC NF membranes was studied. The duration of PANI interlayer deposition was 0.5 h. The effect of the concentration of monomers in IP reaction during the formation of the selective PA layer on the structure and performance of TFC NF membranes with a PANI interlayer obtained by oxidative polymerization was investigated (Figures 6–9, Tables 5 and 6). The PA layer was formed using three different concentrations of PIP, namely 1%, 2%, and 4%, and three different concentrations of TMC, namely 0.06%, 0.12%, and 0.24%, respectively.

SEM microphotographs of the selective layer surface and cross-sections of TFC NF membranes with different ANI, PIP, and TMC concentrations used for PANI interlayer and PA selective layer formation are shown in Figures 6 and 7. AFM images and surface roughness parameters of the developed TFC NF membranes are presented in Figure 8 and Table 5.

As discussed above, when the ANI concentration in oxidative polymerization increased, the surface roughness and hydrophilicity of the selective layer surface of the membrane substrate were enhanced and the pore size declined (Section 3.1, Figures 1 and 2, Table 2).

It was found that the surface of the selective layer of all studied TFC membranes is characterized by the presence of PA globular formations, which are the typical structure of TFC NF membranes based on piperazine amide (Figure 6). Application of a PANI interlayer and an increase in ANI concentration in oxidative polymerization lead to a significant decrease in the size of PA globules. Moreover, the shape of PA globules becomes more regular and the size distribution narrows when the ANI concentration in oxidative polymerization increases both for membranes prepared using 2% PIP and 4% PIP (Figure 6). It was revealed that an increase in the concentrations of PIP and TMC yields an increase in the size of PA globules on the selective layer surface due to the increase in the number of molecules that participate in the IP reaction (Figure 6e,f,i).

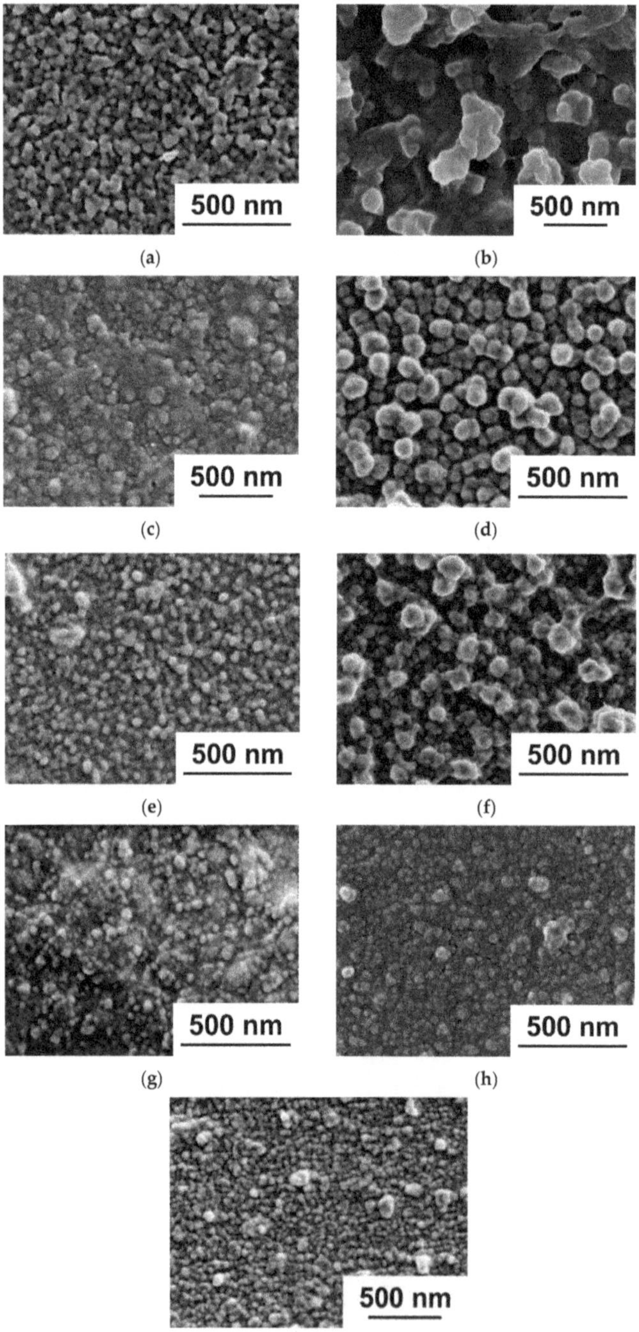

Figure 6. SEM microphotographs of the selective layer surface of TFC NF membranes with different ANI, PIP, and TMC concentrations upon PANI interlayer and PA selective layer formation: (**a**) NF-PSF–2; (**b**) NF-PSF–4; (**c**) NF-PSF/PANI 0.05–0.5–2; (**d**) NF-PSF/PANI 0.05–0.5–4; (**e**) NF-PSF/PANI 0.1–0.5–2; (**f**) NF-PSF/PANI 0.1–0.5–4; (**g**) NF-PSF/PANI 0.3–0.5–2; (**h**) NF-PSF/PANI 0.3–0.5–4; (**i**) NF-PSF/PANI 0.1–0.5–1.

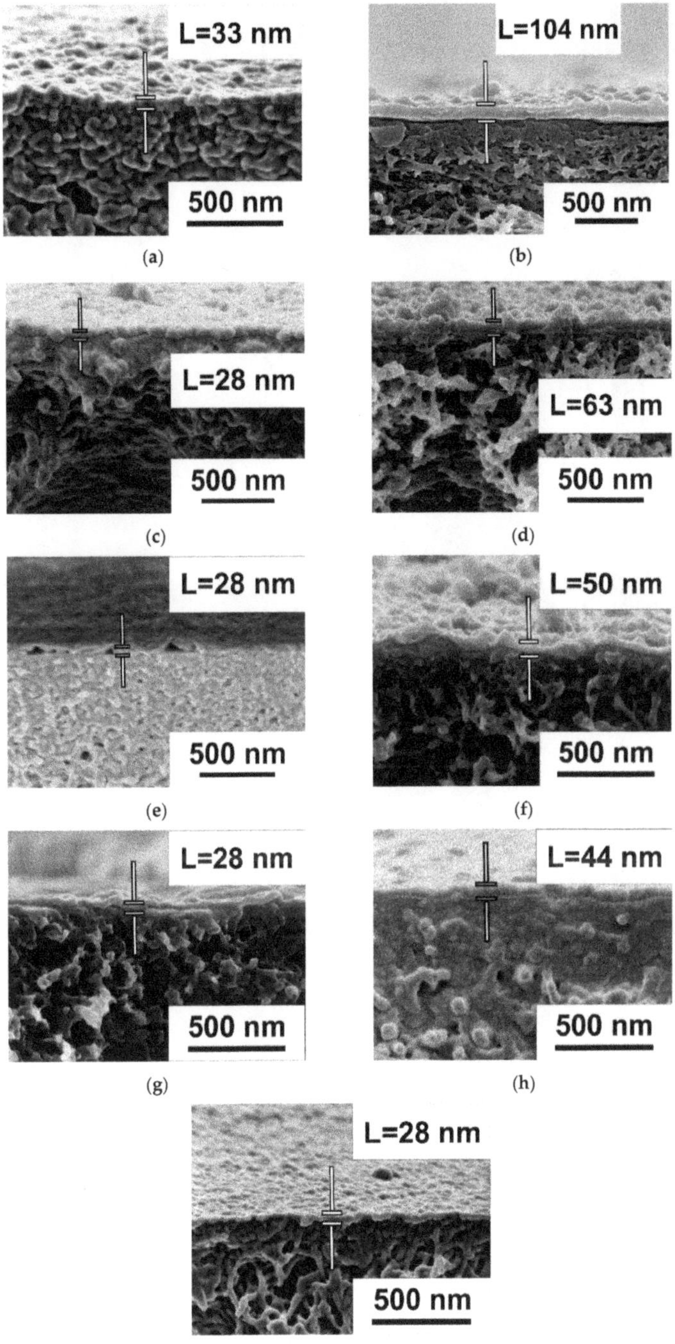

Figure 7. SEM microphotographs of the cross-sections of TFC NF membranes depending on the ANI concentration under different compositions of the PA selective layer: (**a**) NF-PSF–2; (**b**) NF-PSF–4; (**c**) NF-PSF/PANI 0.05–0.5–2; (**d**) NF-PSF/PANI 0.05–0.5–4; (**e**) NF-PSF/PANI 0.1–0.5–2; (**f**) NF-PSF/PANI 0.1–0.5–4; (**g**) NF-PSF/PANI 0.3–0.5–2; (**h**) NF-PSF/PANI 0.3–0.5–4; (**i**) NF-PSF/PANI 0.1–0.5–1.

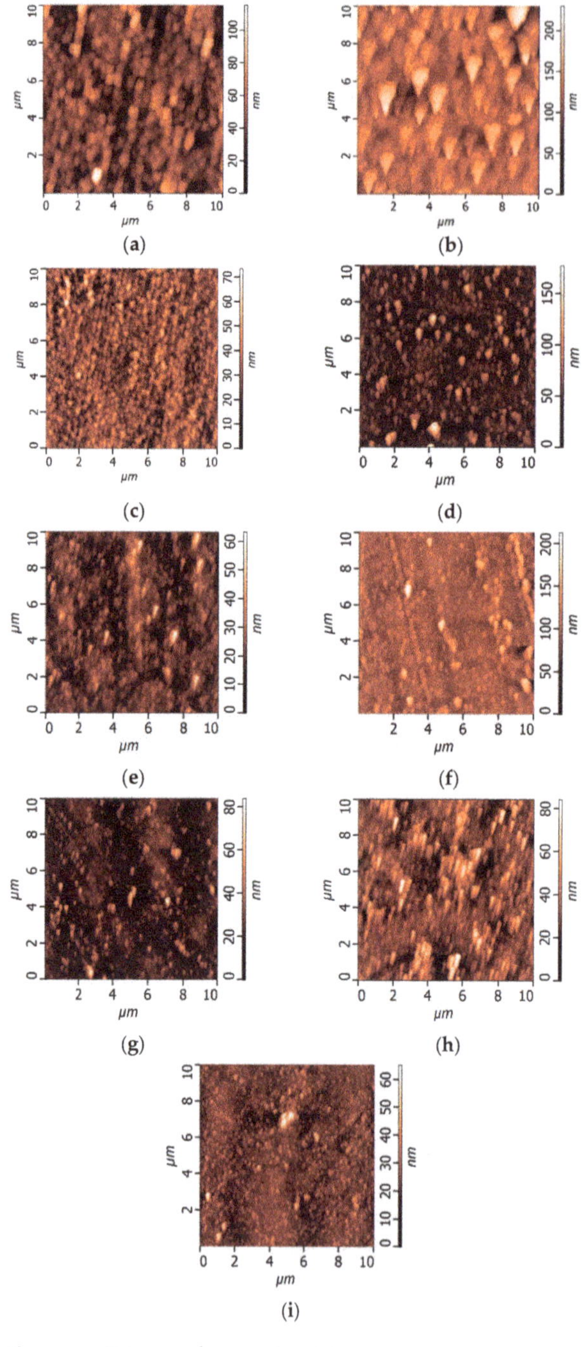

Figure 8. AFM microphotographs of the selective layer surface NF/PANI 0.5 h/PA membranes depending on the ANI concentration under different compositions of the PA selective layer: (**a**) NF-PSF–2; (**b**) NF-PSF–4; (**c**) NF-PSF/PANI 0.05–0.5–2; (**d**) NF-PSF/PANI 0.05–0.5–4; (**e**) NF-PSF/PANI 0.1–0.5–2; (**f**) NF-PSF/PANI 0.1–0.5–4; (**g**) NF-PSF/PANI 0.3–0.5–2; (**h**) NF-PSF/PANI 0.3–0.5–4; (**i**) NF-PSF/PANI 0.1–0.5–1.

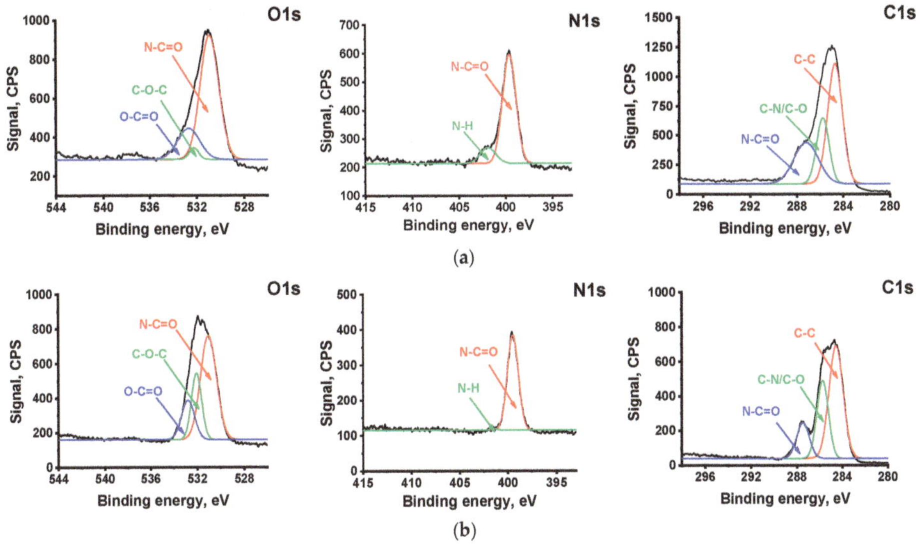

Figure 9. The high-resolution XPS spectra of the selective layer surface of TFC NF membranes: (**a**) NF-PSF–4; (**b**) NF-PSF/PANI 0.1–0.5–4.

Table 5. The surface roughness parameters of TFC NF membranes with a PANI interlayer obtained for 0.5 h and PA layer.

Membrane Code	Roughness Parameters		Water Contact Angle, °
	Ra, nm	Rq, nm	
NF-PSF–2	10.51	13.54	28 ± 2
NF-PSF–4	18.95	25.10	30 ± 2
NF-PSF/PANI 0.05–0.5–2	6.87	8.58	<10
NF-PSF/PANI 0.05–0.5–4	12.35	17.56	<10
NF-PSF/PANI 0.1–0.5–2	6.53	9.07	<10
NF-PSF/PANI 0.1–0.5–4	10.98	15.48	<10
NF-PSF/PANI 0.3–0.5–2	5.68	7.85	<10
NF-PSF/PANI 0.3–0.5–4	8.39	11.05	<10
NF-PSF/PANI 0.1–0.5–1	4.07	5.60	<10

Table 6. Composition of PA selective layer and crosslinking degree of the developed TFC NF membranes studied by XPS.

Membrane	Atom Content in the Selective Layer of TFC NF Membranes, %			Ratio of Peak Intensities		Crosslinking Degree, %
	O	N	C	N–C=O/ O–C=O	N–C=O/ N–H	
NF-PSF–4	19	12	69	3.67	5.82	31.33
NF-PSF/PANI 0.1–0.5–4	20	14	66	3.93	97.9	47.06

It was shown that the application of a PANI interlayer yields a decrease in the thickness of the PA selective layer both for TFC membranes prepared using 2% PIP and 4% PIP (Figure 7). It was found that an increase in ANI concentration from 0.05 wt.% up to 0.3 wt.% during PANI interlayer formation results in a decrease in the thickness of the

PA selective layer for TFC NF membranes prepared using 4 wt.% PIP from 63 down to 44 nm (Figure 7b,d,h,f). However, no changes in PA layer thickness occurred for TFC NF membranes prepared using 2 wt.% PIP when the ANI concentration rose (Figure 7c,e,g). As expected, an increase in monomer concentrations in IP leads to an increase in the PA selective layer thickness regardless of ANI concentration and the presence or absence of a PANI interlayer (Figure 7). For instance, the thickness of the PA selective layer was shown to increase slightly from 28 up to 50 nm with the rise in PIP concentration from 1 to 4 wt.% and TMC concentration from 0.06 wt.% up to 0.24 wt.% correspondingly (Figure 7e,f,i).

It was found that surface roughness parameters of TFC NF membranes significantly declined with the increase in the concentration of ANI during oxidative polymerization applied for PANI interlayer formation (Figure 8, Table 5).

As expected, surface roughness parameters increased when PIP and TMC concentrations used for PA layer formation increased for all studied TFC membranes (Figure 8, Table 5). This is due to the increase in the size of PA globules when the PIP concentration increases from 1 wt.% up to 4 wt.%, which is clearly observed by SEM (Figure 6e,f,i). Thus, a thicker and rougher selective layer was formed with the increase in PIP and TMC concentrations, which is in good agreement with the literature data [63,64].

Overall, with the increase in PIP and TMC concentrations, interface instability increases due to intensified IP reaction, causing a concurrent increase in the thickness and roughness of the PA selective layer. It was reported that the crosslinking degree of the PA selective layer also depends on amine and acyl chloride concentrations during IP. It was shown that the degree of crosslinking of the PA selective layer increased from 78.6% to 98.8% when the concentration of m-phenylenediamine increased from 0.1 wt.% to 10 wt.%. This is due to the sufficient supply of monomers at high concentration, which promoted the migration of amine from the aqueous phase to the organic phase. As a result, more corrugated and thicker PA selective layers were formed via IP [64]. It was found that independent of the monomer concentrations, a highly hydrophilic PA selective layer was formed when the PANI interlayer was applied, which is due to the incorporation of hydrophilic groups of PANI in the selective layer (Table 5).

The changes in TFC membrane structure when the ANI concentration used for PANI interlayer formation increases are attributed to the enhancement of hydrophilicity and decline in the pore size of the membrane substrate selective layer (Figures 1 and 2, Table 2). Moreover, the formation of hydrogen bonds and Van der Waals bonds between PIP and PANI facilitates the storage of more PIP molecules in the pores of the membrane substrate. However, the rate of PIP release to the IP reaction interface decreases, which leads to the formation of a more uniform, thinner, and smoother PA selective layer when the ANI concentration increases during PANI interlayer formation. On the other hand, the constant thickness (28 nm) of the PA selective layer regardless of ANI concentration when a lower concentration (2 wt.%) of PIP is used may be due to the insufficient number of PIP molecules at the IP reaction aqueous–organic interface attributed to their retarded release from the PANI interlayer.

The compositions of the PA selective layer of the reference NF-PSF–4 membrane and NF-PSF/PANI 0.1–0.5–4 TFC membrane with a PANI interlayer were studied using XPS analysis (Figure 9, Table 6).

Figure 9 shows the high-resolution XPS spectra C1s, O1s, and N1s of the selective layer surface of the initial TFC membrane and TFC membranes with a PANI interlayer.

The O1s spectrum was resolved into three typical peaks at 533.2, 532.2, and 530.9 eV, which were attributed to O–C=O, C–O–C, and N–C=O groups, respectively (Figure 9). It is generally accepted that O=C–O groups are generated due to the hydrolysis of unreacted acyl chloride groups. It should be noted that the C–O–C peak can be assigned to the ether groups of PSF, which is a main membrane substrate material.

It was demonstrated that the N1s spectrum can be deconvoluted into two characteristic peaks at around 399.5 eV and 400.9 eV, which were attributed to N–C=O and N–H groups, respectively (Figure 9).

The C1s spectra have N–C=O, C–N/C–O, and C–H/C–C groups at 287.6, 286.0, and 284.8 eV, respectively (Figure 9).

The clear peaks for C, O, and N elements occurred in XPS spectra (Figure 9). There are two sources of N in the XPS spectra: PA and PANI. It was found that the concentration of N and O atoms rose when a PANI interlayer was applied (Table 7). The overall nitrogen concentration enhanced since PANI macromolecules were incorporated into the PA layer, which is indirectly proved by the significant rise in the PA layer's hydrophilicity (Tables 5 and 6). It was found that the total content of C atoms decreased since the concentration of N and O atoms rose (Table 6).

Table 7. Zeta potential of the selective layer surface of TFC NF membranes.

Membrane Code	Zeta-Potential of Membrane Surface, mV			Isoelectric Point, pH
	pH 3	pH 7	pH 10	
NF-PSF–2	8.71	−32.55	−34.68	4.4
NF-PSF/PANI 0.05–0.5–2	15.06	−45.38	−52.22	4.5
NF-PSF–4	5.68	−32.49	−34.32	4.0
NF-PSF/PANI 0.1–0.5–4	14.94	−40.53	−46.67	4.4

The ratios of the peak areas N–C=O/O–C=O (in O 1s spectra) and N–C=O/NH (in N 1s spectra) were calculated. It was demonstrated that the amount of the amide groups rose when the PANI interlayer was deposited (Table 6), which indicated that there was a smaller amount of free unreacted carboxylic and amine groups in the PA layer. This confirms that the formation of the PANI interlayer leads to the formation of a PA layer with a higher crosslinking degree (Table 6).

It was shown that water permeation substantially increased when 0.05 and 0.1 wt.% ANI in the aqueous solution was applied for PANI interlayer deposition both for 2 wt.% and 4 wt.% PIP solutions used for PA layer formation compared to the reference NF-PSF–2 and NF-PSF–4 membranes. The best water permeation was achieved for NF-PSF/PANI 0.05–0.5–2 and NF-PSF/PANI 0.1–0.5–2 membranes and reached 64 $L \cdot m^{-2} \cdot h^{-1}$ at $\Delta P = 0.5$ MPa (Figure 10). The rise in water permeation with the increase in ANI concentration up to 0.05 and 0.1 wt.% is due to the decrease in PA selective layer thickness, which was clearly observed by SEM (Figure 7). The TFC NF membrane series with 2 wt.% PIP demonstrated higher water permeation compared to the TFC NF membrane series with 4 wt.% PIP due to the smaller selective layer thickness (Figure 10). It was revealed that an increase in the concentration of ANI up to 0.3 wt.% yielded a decline in water permeability down to 31 $L \cdot m^{-2} \cdot h^{-1}$ for the NF-PSF/PANI 0.3–0.5–2 membrane in spite of the thin PA layer (28 nm) (Figures 7 and 10). Moreover, it was found that the formation of the PANI intermediate layer from a solution with a monomer concentration of 0.5 wt.% with subsequent application of a PA layer resulted in the preparation of an impermeable TFC NF membrane at $\Delta P = 0.5$ MPa. However, when 0.3 wt.% ANI was used for PANI interlayer formation, NF-PSF/PANI 0.3–0.5–4 was also impermeable. This may be due to the increased mass transfer resistance due to the blockage of pores of the selective layer of the membrane substrate and formation of a thick PANI interlayer, which counterbalances the thin PA layer. This is proved by the substantial decrease in the pure water flux of the membrane substrate UF-PSF/PANI 0.3 down to 5 $L \cdot m^{-2} \cdot h^{-1}$ at $\Delta P = 0.1$ MPa compared to 180 $L \cdot m^{-2} \cdot h^{-1}$ for the reference membrane (Figure 3).

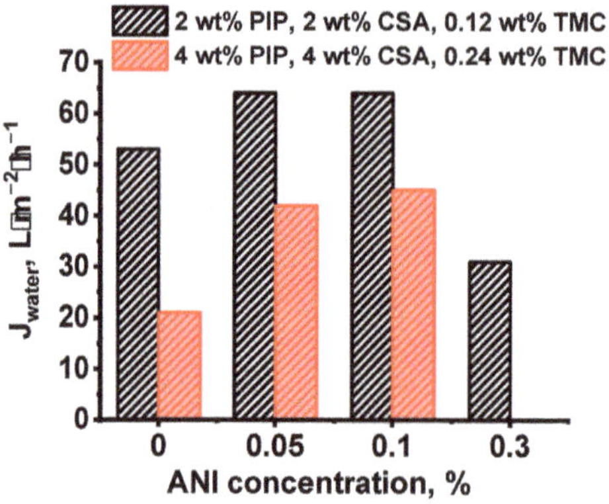

Figure 10. Water permeation (at $\Delta P = 0.5$ MPa) of TFC NF membranes versus the concentration of ANI in aqueous solution.

Thus, the increase in ANI concentration during PANI interlayer formation up to 0.05 and 0.1 wt.% leads to a rise in water permeation due to the increase in PA layer thickness. However, a further increase in ANI concentration up to 0.3 and 0.5 wt.% decreases water permeation, which may be attributed to the blockage of the pores of the membrane substrate, which increases mass transfer resistance.

It was found that an increase in the concentration of monomers in the IP led to a decrease in water permeation from 64 $L \cdot m^{-2} \cdot h^{-1}$ for the NF-PSF/PANI 0.1–0.5–1 and NF-PSF/PANI 0.1–0.5–2 membranes down to 45 $L \cdot m^{-2} \cdot h^{-1}$ for the NF-PSF/PANI 0.1–0.5–4 membrane (at $\Delta P = 0.5$ MPa) (Figure S3). The decrease in water permeation was due to the formation of a thicker and denser selective PA layer (Figure 7e,f,i). However, the effect of the increased selective layer thickness is counterbalanced by the significant increase in surface roughness parameters when the concentration of PIP and TMC rises in the selective layer (Table 5).

It was found that rejection of $MgSO_4$ significantly increases up to 99.99% when 0.05 and 0.1 wt.% ANI is applied for PANI interlayer formation both for 2 wt.% PIP and 4 wt.% PIP TFC NF membrane series (Table 8).

Table 8. Rejection coefficients of NF membranes with different PA layer compositions.

Membrane Code	Rejection, %						
	$MgSO_4$	Na_2SO_4	$MgCl_2$	$CaCl_2$	NaCl	LiCl	SDM
NF-PSF–2	82	70	90	86	16	6	83
NF-PSF/PANI 0.05–0.5–2	>99.99	84	96	79	10	5	80
NF-PSF/PANI 0.1–0.5–2	>99.99	76	84	59	24	10	78
NF-PSF/PANI 0.3–0.5–2	71	54	73	74	22	22	80
NF-PSF–4	92	90	91	91	28	28	95
NF-PSF/PANI 0.05–0.5–4	>99.99	90	99	92	25	13	96
NF-PSF/PANI 0.1–0.5–4	>99.99	90	99	92	25	25	96
NF-PSF/PANI 0.1–0.5–1	>99.99	69	73	45	23	4	69

However, the rejection of $MgSO_4$ for NF-PSF/PANI 0.3–0.5–2 was found to be 71% compared to the 82% for the reference NF-PSF–2 membrane. So, an increase in ANI concentration up to 0.3 wt.% yields a decrease in $MgSO_4$ rejection for the TFC NF membrane series with 2 wt.% PIP. Na_2SO_4 and $MgCl_2$ rejection increased with the maximum for NF-PSF/PANI 0.05–0.5–2 and NF-PSF/PANI 0.1–0.5–4 TFC NF membranes (at 0.05 wt.% ANI for membranes with 2 wt.% PIP and 0.1 wt.% for membranes with 4 wt.% PIP). The rejection coefficients for $CaCl_2$ and SDM decreased with increasing ANI concentration for all developed TFC NF membranes with 2 wt.% PIP used for PA layer formation (Table 8). Rejection coefficients for monovalent salts NaCl and LiCl were revealed to slightly decrease for the NF-PSF/PANI 0.05–0.5–2 membrane and increase for the NF-PSF/PANI 0.1–0.5–2 membrane and NF-PSF/PANI 0.3–0.5–2 membrane compared to the reference NF-PSF–2 membrane (Table 8). Thus, the better rejection demonstrates the NF-PSF/PANI 0.05–0.5–2 membrane compared to the NF-PSF–2 membrane.

It was shown that the rejection coefficient for $MgSO_4$ did not change with the variation in the concentrations of PIP and TMC during PA layer formation via IP and was 99.99% (compare the rejection of NF-PSF/PANI 0.1–0.5–1, NF-PSF/PANI 0.1–0.5–2, and NF-PSF/PANI 0.1–0.5–4 membranes, Table 8). Overall, the rise in the concentration of monomers in the IP reaction led to a significant increase in rejections for all studied salts and antibiotic SDM.

According to the rejections for NF-PSF/PANI 0.1–0.5–1 and NF-PSF/PANI 0.1–0.5–2, salts can be arranged as follows:

$MgSO_4 > MgCl_2 > Na_2SO_4 > CaCl_2 > NaCl > LiCl$

However, with the increase in PIP and TMC concentration, this line slightly changes for the NF-PSF/PANI 0.1–0.5–4 membrane:

$MgSO_4 > MgCl_2 > CaCl_2 > Na_2SO_4 > NaCl > LiCl$

When 4 wt.% PIP is used for PA layer formation, an increase in ANI concentration results in an increase in rejection of all studied salts with the maximum values at 0.1 ANI concentration. This is due to the formation of a PA layer with a higher crosslinking degree (with a higher number of amide groups) when the PANI intermediate layer (0.3 wt.% ANI concentration) was deposited, which was confirmed by XPS (Figure 9, Table 6).

The pattern of change in rejection with the increase in PIP and TMC concentration suggests that at lower monomer concentrations, a PA selective layer with a lower crosslinking degree is formed. This may be attributed to the insufficient supply of monomers at lower concentrations (1 wt.% PIP and 2 wt. % PIP) to the interface of the aqueous and organic phases in the IP reaction. Moreover, as discussed above, a hydrophilic PANI interlayer retards the release of PIP to the reaction interface, leading to an insufficient number of amine molecules in the reaction zone.

The zeta potential of the selective layer surface of NF membranes was determined using an electrokinetic analyzer, and the results are presented in Table 7. It was found that the formation of a PANI interlayer on the surface of PSF membranes via oxidation polymerization yielded an increase in the positive zeta potential at pH 3 and negative zeta potential at pH 7 and 10 (Table 7). Moreover, it was shown that the NF-PSF–2 membrane is characterized by a higher positive zeta potential value at pH 3 compared to the NF-PSF–4 membrane. It was revealed that application of a PANI interlayer yields a slight rise in the isoelectric point of the membrane surface from 4.0 for the reference NF-PSF–4 membrane up to 4.4 for the NF-PSF/PANI 0.1–0.5–4 membrane and from 4.4 for the NF-PSF–2 membrane up to 4.5 for the NF-PSF/PANI 0.05–0.5–2 membrane. The increase in positive zeta potential values at pH 3 and the rise in isoelectric point are due to the protonation of PANI in acidic media, since PANI groups are incorporated in the PA selective layer. The increase in negative zeta potential values at pH 7 and 10 can be attributed to the increase in binding sides for non-specific adsorption of hydroxide and chloride ions during zeta potential measurements.

The results of the investigation of Mg^{2+}/Li^{+} separation are presented in Table 9. It was found that the formation of a PANI interlayer yielded a significant enhancement of the Mg^{2+}/Li^{+} separation factor by more than four and seven times for the NF-PSF/PANI 0.05–0.5–2 and NF-PSF/PANI 0.1–0.5–4 membranes, respectively, compared to the reference NF-PSF–2 and NF-PSF–4 membranes.

Table 9. The Mg^{2+}/Li^{+} separation factor of the developed TFC NF membranes.

Membrane Code	$S_{Mg/Li}$
NF-PSF–2	9
NF-PSF/PANI 0.05–0.5–2	37
NF-PSF–4	8
NF-PSF/PANI 0.1–0.5–4	58

The rejection coefficients of PEGs of the developed TFC NF membranes are presented in Figure 11 It was shown that rejection coefficients for PEGs of different molecular weights were higher for PANI-modified TFC NF membranes in both cases, prepared using 2 wt.% PIP and 4 wt.% PIP aqueous solutions (Figure 11). The rejection coefficients of TFC NF membranes for PEG 200 Da were found to be 72–85 wt.%. The rejection coefficients of PEG 300, 400, and 600 Da were over 90% (Figure 11).

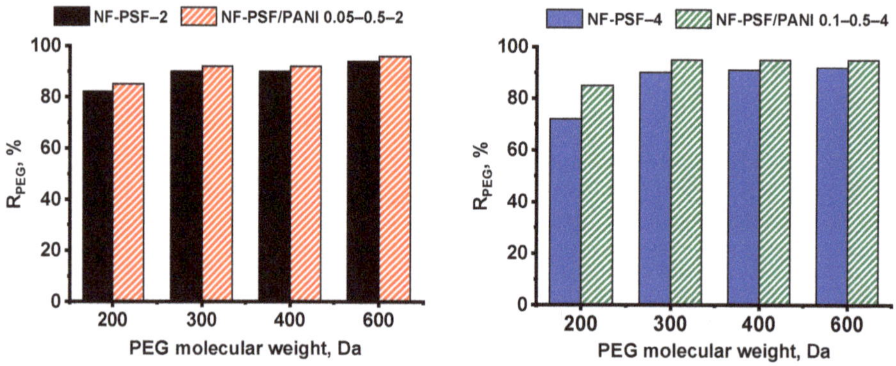

(a) (b)

Figure 11. Rejection coefficients of PEGs of different molecular weights of the developed TFC NF membranes. PIP aqueous solution concentration, wt.%: (**a**) 2, (**b**) 4.

According to the obtained results, the MWCO and average pore size as well as pore size distribution of TFC NF membranes were measured and calculated (Table 10, Figure 12). It was revealed that application of a PANI interlayer caused a decrease in the average pore size together with a decline in the MWCO of the developed membranes compared to the corresponding reference NF-PSF–2 and NF-PSF–4 TFC membranes (Table 10).

Table 10. MWCO and average pore size of the developed TFC NF membranes.

Membrane Code	MWCO, Da	Average Pore Size, nm
NF-PSF–2	300	310
NF-PSF/PANI 0.05–0.5–2	260	307
NF-PSF–4	300	326
NF-PSF/PANI 0.1–0.5–4	240	305

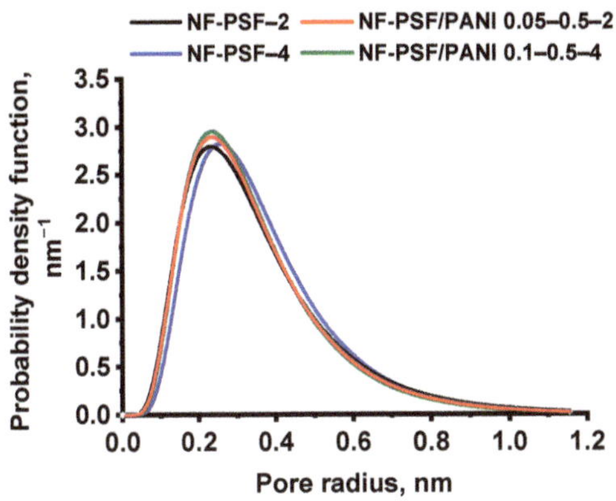

Figure 12. Pore size distribution of TFC NF membranes.

The mean effective pore size of the obtained TFC NF membranes was evaluated using PEG 200–600. The pore size distribution of NF membranes is shown in Figure 12. It was found that the formation of a PANI interlayer led to a slight narrowing the pore size distribution compared to the reference NF-PSF–2 and NF-PSF–4 membranes (Figure 12). This can be explained by better impregnation of the PSF/PANI membrane substrate with PIP, caused by the increase in its hydrophilicity, and as a result the formation of a PA selective layer with a higher crosslinking degree and smaller pores (Figure 12, Tables 6 and 10).

The results on average pore size and pore size distribution studies are in good accordance with the MWCO and rejection coefficients studies (Tables 7, 9 and 10, Figures 11 and 12). The increased selectivity of the developed NF-PSF/PANI 0.05–0.5–2 and NF-PSF/PANI 0.1–0.5–4 membranes (especially for $MgCl_2$) is due to the decrease in pore size and narrowing of the pore size distribution of the selective layer. Moreover, the increase in the negative zeta potential at pH 7 and 10 decreases the transport of chloride ions of $MgCl_2$ providing higher $MgCl_2$ rejection. The transport of chloride ions may decrease the rejection of $MgCl_2$ since Mg^{2+} may be transported through the membrane to provide the electroneutrality of the solution. It often leads to lower $MgCl_2$ rejection compared to $MgSO_4$ and Na_2SO_4 (Table 7).

According to the optimal combination of water permeation and selectivity, NF-PSF/PANI 0.1–0.5–4 was selected for further studies of long-term operation performance. The long-term operation performance of the NF-PSF–4 and NF-PSF/PANI 0.1–0.5–4 membrane during the filtration of water and aqueous solution of magnesium sulphate (2 $g \cdot L^{-1}$) at ΔP = 0.5 MPa is presented in Figure 13. It was revealed that the water permeance of the TFC NF membrane as well as the $MgSO_4$ rejection coefficient were constant for 120 h. Based on the obtained data, it could be concluded that the TFC NF membrane modified with application of a PANI intermediate layer is characterized by high stability during 120 h of the nanofiltration experiment.

The effect of chemical treatment on the performance of the developed membrane after oxidation with sodium hypochlorite solution and acid treatment was investigated (Table 11). It was found that after chemical treatment, the water permeation at ΔP = 0.5MPa slightly increased for the reference NF-PSF–2 and NF-PSF–4 membranes but remained unchanged for TFC NF membranes with a PANI interlayer (Table 11). Moreover, the rejection coefficients for $MgSO_4$ were also the same (>99.99%) for modified TFC NF-PSF/PANI 0.05–0.5–2 and NF-PSF/PANI 0.1–0.5–4 membranes but slightly declined for reference

NF-PSF–2 and NF-PSF–4 membranes (Table 11). Thus, it was found that the developed TFC NF membranes possessed better chemical resistance compared to the reference membranes.

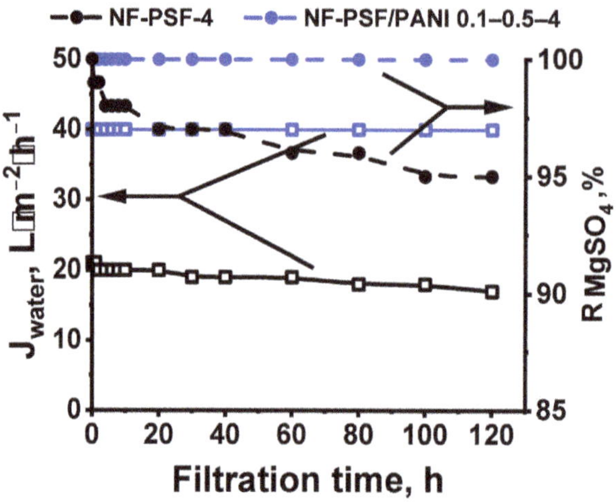

Figure 13. Long-term performance of NF-PSF–4 and NF-PSF/PANI 0.1–0.5–4 membranes (ΔP = 0.5 MPa).

Table 11. Membrane performance of the developed TFC NF membranes after chemical treatment.

Membrane Code	Water Permeation (ΔP = 0.5 MPa), L·m^{-2}h^{-1}			Rejection of $MgSO_4$, %		
	Before	After Chlorine Treatment	After Acid Treatment	Before	After Chlorine Treatment	After Acid Treatment
NF-PSF–2	64	65	67	>99.99	99	98
NF-PSF/PANI 0.05–0.5–2	64	64	64	>99.99	>99.99	>99.99
NF-PSF–4	21	23	25	>99.99	99	99
NF-PSF/PANI 0.1–0.5–4	40	40	40	>99.99	>99.99	>99.99

The nanofiltration performance of the developed TFC NF membranes was compared to the transport characteristic of membranes reported so far (Table 12). It was demonstrated that the pure water permeance, $MgSO_4$ and Na_2SO_4 rejection, and Mg^{2+}/Li^+ separation factor of PSF/PANI/PA nanofiltration membranes surpass most of TFC NF membranes with an interlayer reported previously.

Table 12. Comparison of nanofiltration performance of the developed TFC NF membranes with membranes with an interlayer reported in the literature.

Composition			Membrane Permeance, L·m^{-2}·h^{-1}·bar^{-1}	R ($MgSO_4$), %	R (Na_2SO_4), %	S (Mg^{2+}/Li^+)	Ref.
Membrane Support	Interlayer	PA Layer					
PSF	PDA/PEI	PIP/TMC	7.5	95	97	-	[53]
PSF	TA	PEI/TMC	1.7	95	72	22	[65]
	TA-Cu		2.9	97	76	26.5	
PSF	PEI	TMC/DETA	3.2	94	-	11	[66]
PES	COF-A	MPD + PIP/ Tp + TMC	16.3	98	98	40.4	[38]

Table 12. *Cont.*

Membrane Support	Composition: Interlayer	Composition: PA Layer	Membrane Permeance, $L \cdot m^{-2} \cdot h^{-1} \cdot bar^{-1}$	R ($MgSO_4$), %	R (Na_2SO_4), %	S (Mg^{2+}/Li^+)	Ref.
PES	Hyaluronic acid	PIP/TMC	29.5	91	95	-	[67]
PES	PDADMAC+ carboxylated cellulose nanocrystal	PEI/TMC	3.4	-	-	5.8	[68]
PES/PVP/ PEG 400	GO-hyperbranched polyamide-amine	PIP/TMC	14.3	88	97	-	[69]
PSF/PVP K30	PANI	PIP/TMC	12.8	>99.99	84	37	This work
			8	>99.99	90	58	

4. Conclusions

The principles of PA selective layer formation via interfacial polymerization on PSF ultrafiltration membrane substrates with a PANI interlayer were revealed. It was discovered that the conditions of PANI interlayer application—such as the duration of PANI layer formation and the concentration of aniline—significantly affect the structure and performance of the resulting TFC nanofiltration membranes. It was found that deposition of a PANI layer via oxidative polymerization on the surface of a PSF ultrafiltration membrane yields partial pore blocking, an increase in surface roughness, and hydrophilization of the membrane selective layer. When a higher aniline concentration and longer time of PANI layer formation are applied, a rougher and more hydrophilic selective layer of the membrane substrate with smaller pores is formed. These substantial changes of structure, porosity, and hydrophilic–hydrophobic properties of the membrane selective layer affected the PA layer formation via IP during TFC membrane preparation. It was supposed that the hydrophilic PANI layer covering the pore walls of the membrane substrate enhances the capillary effect of the substrate, and more amine monomers can be absorbed within the pores. Moreover, the release of PIP from the highly hydrophilic membrane substrate coated with a PANI layer is retarded, leading to a lower rate of IP reaction, which results in the formation of more uniform, thinner, and smoother PA layer compared to the reference TFC membrane.

The developed method improves TFC membrane water permeation up to 45–64 $L \cdot m^{-2} \cdot h^{-1}$ at 0.5 MPa and increases selectivity, with rejection coefficients for $MgSO_4$ exceeding 99.99%, for LiCl ranging from 5 to 25%, and for sodium sulfadimetoxine ranging from 80 to 95%. Moreover, the Mg^{2+}/Li^+ separation factor of the developed TFC NF membranes with a PANI interlayer was found to be 37 and 58 compared to 8 and 9 for reference membranes. Additionally, this approach ensures improved long-term operational stability of TFC nanofiltration membranes with a PANI interlayer.

Supplementary Materials: The following supporting information can be downloaded at: https://www.mdpi.com/article/10.3390/polym17091199/s1. Figure S1: The scheme of oxidative polymerization of aniline. Figure S2: FTIR-spectra of reference UF-0 and UF PSF/PANI membranes. Figure S3: Water permeation (J_{water}) of TFC NF membranes with a PANI intermediate layer depending on the concentration of monomers in the IP reaction (ΔP = 0.5 MPa).

Author Contributions: K.S.B.: investigation, writing—original draft preparation, formal analysis, visualization; data curation, and methodology; T.V.P.: conceptualization, methodology, visualization, formal analysis, investigation, writing—original draft preparation, supervision, project administration, funding acquisition, and writing—review and editing; A.V.P.: writing—review and editing; B.Y.:

formal analysis and investigation; S.S.E.: writing—review and editing; A.V.B.: conceptualization, methodology, formal analysis, and writing—review and editing. All authors have read and agreed to the published version of the manuscript.

Funding: This research was funded by Russian Science Foundation, grant number No. 22-73-10034.

Institutional Review Board Statement: Not applicable.

Data Availability Statement: The data are contained within the manuscript.

Acknowledgments: The experimental work was facilitated by equipment from the Resource Centers for Nanotechnology, the Center of Nanofabrication of Photoactive Materials (Nanophotonics), and the Centre for Physical Methods of Surface Investigation and Computer Center at the St. Petersburg State University.

Conflicts of Interest: The authors declare no conflicts of interest.

References

1. Chen, Y.; Liu, X.; Zhang, R.; Wang, W.; Wang, H.; Ren, Y.; Duan, X. Engineering Mineralized Interlayers for Enhanced Nanofiltration: Synergistic Modulation of Polyamide Layer Structure and Catalytic Self-Cleaning Performance. *J. Membr. Sci.* **2025**, *713*, 123319. [CrossRef]
2. Gui, H.; Chen, Z.; Sun, Z.; Sun, L.; Peng, S.; Wang, Y.; Wang, Y.; Tang, J.; Huang, L. GO-PANI Nanofiltration Membrane with pH Response for High Efficiency Water Purification. *J. Water Process Eng.* **2025**, *69*, 106889. [CrossRef]
3. Athinarayanan, B.; Jeong, D.-Y.; Kang, J.-H.; Koo, B.-H. Fabrication of nanoporous aluminum-oxide composite membranes. *J. Korean Phys. Soc.* **2015**, *67*, 1970–1976. [CrossRef]
4. Du, J.; Xue, Y.; Xu, Y.; Zhan, X.; Zhang, D. High Water Permeability Interfacial Polymerized PVDF/EAA Composite Nanofiltration Membrane for Dye Separation. *ACS Appl. Polym. Mater.* **2024**, *6*, 13102–13109. [CrossRef]
5. Abdullahi, A.A.; Saleh, T.A. Amine Grafted Graphene Oxide Embedded in Polyamide Membranes for Simultaneous Removal of Oil, Metals, and Salts from Oily Water. *J. Mol. Liq.* **2024**, *408*, 125352. [CrossRef]
6. Ang, M.B.M.Y.; Wu, Y.-L.; Chu, M.-Y.; Wu, P.-H.; Chiao, Y.-H.; Millare, J.C.; Huang, S.-H.; Tsai, H.-A.; Lee, K.-R. Nanofiltration Membranes Formed through Interfacial Polymerization Involving Cycloalkane Amine Monomer and Trimesoyl Chloride Showing Some Tolerance to Chlorine during Dye Desalination. *Membranes* **2022**, *12*, 333. [CrossRef]
7. Huang, S.; Wu, M.-B.; Zhu, C.-Y.; Ma, M.-Q.; Yang, J.; Wu, J.; Xu, Z.-K. Polyamide Nanofiltration Membranes Incorporated with Cellulose Nanocrystals for Enhanced Water Flux and Chlorine Resistance. *ACS Sustain. Chem. Eng.* **2019**, *7*, 12315–12322. [CrossRef]
8. Liu, S.; Wang, M.; Dong, L.; Cui, Z.; He, B.; Li, J.; Yan, F. Incorporation of Crown Ether into PEI-Polyamide Nanofiltration Membrane for Efficient Mg2+/Li+ Separation. *Sep. Purif. Technol.* **2025**, *360*, 131054. [CrossRef]
9. Liu, J.; Li, Q.; Ji, Y.; Cui, Z.; Mohammad, Y.; Li, J.; He, B. Double Polyamide Layers with $CaCO_3$ Nanoparticles as Scaffolds for High Performance Nanofiltration Membranes. *ACS Appl. Nano Mater.* **2022**, *5*, 8279–8287. [CrossRef]
10. Yu, S.; Ma, M.; Liu, J.; Tao, J.; Liu, M.; Gao, C. Study on Polyamide Thin-Film Composite Nanofiltration Membrane by Interfacial Polymerization of Polyvinylamine (PVAm) and Isophthaloyl Chloride (IPC). *J. Membr. Sci.* **2011**, *379*, 164–173. [CrossRef]
11. Kong, P.; Sun, Z.; Gui, H.; Chen, Z.; Song, Y.; Wang, Y.; Wang, Y.; Kipper, M.J.; Tang, J.; Huang, L. Advances in the Application of Graphene Oxide Composite Loose Nanofiltration Membranes for Dye and Salt Separation. *J. Environ. Chem. Eng.* **2024**, *12*, 114278. [CrossRef]
12. Li, H.; Xiao, S.; Zhao, X.; Yuan, J.; Yu, S. Preparation of High-Flux Loose Nanofiltration Membranes for Efficient Dye/Salt Separation by Controlling Interface Polymerization through Physical and Chemical Dual Constraints. *Sep. Purif. Technol.* **2025**, *362*, 131720. [CrossRef]
13. Xu, G.-R.; An, Z.-H.; Zhao, H.-L.; Liu, Q. Polyamide Layer Modulation for PA-TFC Membranes Optimization: Developments, Mechanisms, and Implications. *Sep. Purif. Technol.* **2023**, *311*, 123200. [CrossRef]
14. Zhang, J.; Yang, X.; Zhang, N.; Tao, Z.; Han, L.; Wei, B.; Qi, R.; Fu, Y.; Wang, Z. Polyamide Membrane Based on Hydrophilic Intermediate Layer for Simultaneous Wetting and Fouling Resistance in Membrane Distillation. *J. Membr. Sci.* **2025**, *717*, 123596. [CrossRef]
15. Xu, G.; Chiao, Y.-H.; Fu, W.; Deng, L.; Hu, M.; Guan, K.; Gonzales, R.R.; Matsuyama, H. Temperature-Modulated Formation of Polyamide Layer for Enhanced Organic Solvent Reverse Osmosis (OSRO) Performance. *J. Membr. Sci.* **2023**, *682*, 121793. [CrossRef]
16. Xiao, F.; Wang, B.; Hu, X.; Nair, S.; Chen, Y. Thin Film Nanocomposite Membrane Containing Zeolitic Imidazolate Framework-8 via Interfacial Polymerization for Highly Permeable Nanofiltration. *J. Taiwan Inst. Chem. Eng.* **2018**, *83*, 159–167. [CrossRef]

17. Li, Y.; You, X.; Li, R.; Li, Y.; Yang, C.; Long, M.; Zhang, R.; Su, Y.; Jiang, Z. Loosening Ultrathin Polyamide Nanofilms through Alkali Hydrolysis for High-Permselective Nanofiltration. *J. Membr. Sci.* **2021**, *637*, 119623. [CrossRef]
18. Minhas, F.T.; Bhanger, M.I. Interfacial Polymerization: A Facile Technique for Developing Solvent Resistant Nanofiltration (SRNF) Membrane. *Procedia Manuf.* **2018**, *17*, 680–687. [CrossRef]
19. Burts, K.; Plisko, T.; Makarava, M.; Krasnova, M.; Penkova, A.; Ermakov, S.; Grigoryev, E.; Komolkin, A.; Bildyukevich, A. The Effect of PEG-Content and Molecular Weight of PEG-PPG-PEG Block Copolymers on the Structure and Performance of Polyphenylsulfone Ultra- and Nanofiltration Membranes. *J. Membr. Sci.* **2024**, *704*, 122869. [CrossRef]
20. Al-Nahari, A.; Liu, S.; Zhang, Q.; Su, B. Ultraselective and Highly Permeable Thin-Film Composite Nanofiltration Membrane Enhanced by Hydrophilic Polydopamine Interlayer. *J. Membr. Sci.* **2025**, *718*, 123686. [CrossRef]
21. Zhao, K.; Cheng, J.; Wang, P.; Hou, H.; An, L.; Yao, J.; Han, Y.; Liu, D.; Ma, J.; Liu, J.; et al. Constructing High-Performance Nanofiltration Membranes Using Nanofiber Supports: Effects of Structural Stability and Polyvinyl Alcohol Interlayers. *Sep. Purif. Technol.* **2024**, *345*, 127445. [CrossRef]
22. Zhang, L.-L.; Zaoui, A.; Sekkal, W. Exfoliated Montmorillonite Nanofillers Role on the High Performance and the Permeability of Polyamide Nanofiltration Membrane. *J. Membr. Sci.* **2025**, *713*, 123303. [CrossRef]
23. Jia, M.-M.; Feng, J.-H.; Shao, W.; Chen, Z.; Yu, J.-R.; Sun, J.-J.; Wu, Q.-Y.; Li, Y.; Xue, M.; Chen, X.-M. In-Situ Interfacial Synthesis of Metal-Organic Framework/Polyamide Thin-Film Nanocomposite Membranes with Elevated Nanofiltration Performances. *J. Membr. Sci.* **2024**, *694*, 122418. [CrossRef]
24. Ahmad, H.; Waheed, A.; Chrystie, R.; Aljundi, I.H. Effect of Surfactants on the Permeation Rate and Selectivity of Polyamide Composite Membranes for Organic Solvent Nanofiltration. *Process Saf. Environ. Prot.* **2024**, *190*, 1052–1066. [CrossRef]
25. Liu, Z.; Wang, Z.; Zhao, L.; Xu, J.; Chen, J.P.; Niu, Q.J. Rod-Coated PVA Interlayer Induced Nanofiltration Membrane: Method Development and Application for Separation of Dye/NaCl with Greater Performance. *J. Membr. Sci.* **2024**, *694*, 122406. [CrossRef]
26. Chi, M.; Zheng, P.; Wei, M.; Zhu, A.; Zhong, L.; Zhang, Q.; Liu, Q. Polyamide Composite Nanofiltration Membrane Modified by Nanoporous TiO2 Interlayer for Enhanced Water Permeability. *J. Ind. Eng. Chem.* **2022**, *115*, 230–240. [CrossRef]
27. Li, W.-L.; Lin, W.-T.; Zhu, C.-Y.; Fu, P.; Zhou, D.; Huang, X.-J.; Xu, Z.-K.; Wan, L.-S. High-Performance Thin Film Composite Nanofiltration Membranes with MoS2 Nanosheet Interlayer. *J. Membr. Sci.* **2023**, *685*, 121956. [CrossRef]
28. Gao, J.; Zhang, M.; Wang, J.; Liu, G.; Liu, H.; Jiang, Y. Bioinspired Modification of Layer-Stacked Molybdenum Disulfide (MoS_2) Membranes for Enhanced Nanofiltration Performance. *ACS Omega* **2019**, *4*, 4012–4022. [CrossRef]
29. Fang, Y.; Zhu, C.-Y.; Han, X.; Ma, Z.-Y.; Yang, H.-C.; Zhang, C.; Liang, H.-Q.; Yang, X.; Xu, Z.-K. Tailoring Polyamide Nanofiltration Membranes by Switching Charge of Nanocellulose Interlayers. *Langmuir* **2024**, *40*, 18233–18241. [CrossRef]
30. Zhu, C.; Zhang, X.; Xu, Z. Polyamide-based Membranes Consisting of Nanocomposite Interlayers for High Performance Nanofiltration. *J. Appl. Polym. Sci.* **2021**, *138*, 49940. [CrossRef]
31. Li, J.; Gong, J.-L.; Fang, S.-Y.; Cao, W.-C.; Tang, S.-Q.; Qin, M.; Wang, Y.-W.; Zhou, H.-Y.; Long, M. Antifouling and Chemical-Resistant Nanofiltration Membrane Modulated by a Functionalized MWCNT Interlayer for Efficient Dye/Salt Separation at High Salinity. *ACS Appl. Mater. Interfaces* **2023**, *15*, 34272–34289. [CrossRef]
32. Long, L.; Wu, C.; Yang, Z.; Tang, C.Y. Carbon Nanotube Interlayer Enhances Water Permeance and Antifouling Performance of Nanofiltration Membranes: Mechanisms and Experimental Evidence. *Environ. Sci. Technol.* **2022**, *56*, 2656–2664. [CrossRef] [PubMed]
33. Gong, G.; Wang, P.; Zhou, Z.; Hu, Y. New Insights into the Role of an Interlayer for the Fabrication of Highly Selective and Permeable Thin-Film Composite Nanofiltration Membrane. *ACS Appl. Mater. Interfaces* **2019**, *11*, 7349–7356. [CrossRef] [PubMed]
34. Cheng, P.; Zhu, T.; Wang, X.; Fan, K.; Liu, Y.; Wang, X.; Xia, S. Enhancing Nanofiltration Selectivity of Metal–Organic Framework Membranes via a Confined Interfacial Polymerization Strategy. *Environ. Sci. Technol.* **2023**, *57*, 12879–12889. [CrossRef] [PubMed]
35. Yao, A.; Hua, D.; Hong, Y.; Pan, J.; Cheng, X.; Tan, K.B.; Zhan, G. Using Cu-TCPP Nanosheets as Interlayers for High-Performance Organic Solvent Nanofiltration Membranes. *ACS Appl. Nano Mater.* **2022**, *5*, 18718–18729. [CrossRef]
36. Cui, X.; Kong, G.; Wei, S.; Zhang, Z.; Kang, Z.; Guo, H. PolyMOF Interlayers Modulated Interfacial Polymerization of Ultra-Thin Nanofiltration Membranes with Efficient and Stable Desalination Performance. *J. Membr. Sci.* **2024**, *702*, 122780. [CrossRef]
37. Fang, Y.-X.; Lin, Y.-F.; Xu, Z.-L.; Pandaya, D.; Xu, S.-J. Interfacial Polymerization Modulated by COFs Interlayer to Fabricate Efficient Mono/Divalent Ions Separation Nanofiltration Membrane. *Desalination* **2024**, *588*, 117977. [CrossRef]
38. Fang, Y.-X.; Xu, Z.-L.; Xu, S.-J. Polyamide Nanofiltration Membrane Modified with Defective Covalent Organic Framework Interlayer for Selective Ion Sieving. *Sep. Purif. Technol.* **2025**, *355*, 129747. [CrossRef]
39. Wang, L.; Fang, M.; Liu, J.; He, J.; Li, J.; Lei, J. Layer-by-Layer Fabrication of High-Performance Polyamide/ZIF-8 Nanocomposite Membrane for Nanofiltration Applications. *ACS Appl. Mater. Interfaces* **2015**, *7*, 24082–24093. [CrossRef]
40. Khandge, R.S.; Nguyen, T.-T.; Qiang, Z.; Yu, H.-K.; Ma, X. Polyamide-Crystalline Covalent Organic Framework Dual-Layer Nanofiltration Membrane with Improved Ion Selectivity. *ACS Appl. Polym. Mater.* **2024**, *6*, 13877–13885. [CrossRef]

41. Jiang, H.; Yang, N.; Hao, Y.; Zhang, L.; Xiao, X.; Sun, Y.; Jiang, B.; Zhang, L. Highly Permeable Polyamide Nanofiltration Membranes Fabricated via the Construction of Anionic Covalent Organic Frameworks/Polydopamine Composite Interlayer. *Sep. Purif. Technol.* **2024**, *347*, 127590. [CrossRef]
42. Zhong, Z.; Li, L.; Yu, S.; Li, X.; Li, H.; Jia, J.; Huang, J.; Cai, M.; Huang, W. Investigation on the Enhancement of SOLVENT-RESISTANT Nanofiltration Membrane Performance Utilizing PDA-UIO-66@ CNT as an Interlayer. *J. Appl. Polym. Sci.* **2024**, *141*, e55274. [CrossRef]
43. Yang, Z.; Wang, F.; Guo, H.; Peng, L.E.; Ma, X.; Song, X.; Wang, Z.; Tang, C.Y. Mechanistic Insights into the Role of Polydopamine Interlayer toward Improved Separation Performance of Polyamide Nanofiltration Membranes. *Environ. Sci. Technol.* **2020**, *54*, 11611–11621. [CrossRef] [PubMed]
44. Zhang, X.; Zhang, X.; Xu, L.; Zhang, G.; Meng, Q. Amphiphilic Interlayer Regulated Interfacial Polymerization for Constructing Polyamide Nanofiltration Membranes with High Perm-Selectivity of Mono-/Divalent Salts. *ACS Appl. Mater. Interfaces* **2024**, *16*, 14321–14332. [CrossRef]
45. He, T.; Chen, F. Enhanced Separation Performance of Composite Nanofiltration Membranes via Electrostatic Air Spray PSS/PEI Interlayer. *Desalination* **2024**, *573*, 117221. [CrossRef]
46. Hui, H.; Ding, H.; Liu, Q.; Zhao, H.; Hou, J.; Zhang, L.; Pei, H.; Li, X. Enhanced Nanofiltration Membranes for Mg2+/Li+ Separation: Development of a Polyethyleneimine Positively Charged Interlayer. *Sep. Purif. Technol.* **2025**, *361*, 131337. [CrossRef]
47. Tian, R.; Zhang, H.; Wang, J.; Dilxat, D.; Xie, T.; Qi, Q.; Wang, Y. Nanofiltration Membrane Functionalization with Enhanced Hardness Cation Removal Using a Mono-Component Interlayer. *Desalination* **2024**, *586*, 117870. [CrossRef]
48. Chen, K.; Li, F.; Wei, T.; Zhou, H.; Zhang, T.; Zhao, S.; Xie, T.; Sun, H.; Li, P.; Niu, Q.J. An Interlayer-Based Positive Charge Compensation Strategy for the Preparation of Highly Selective Mg2+/Li+ Separation Nanofiltration Membranes. *J. Membr. Sci.* **2023**, *684*, 121882. [CrossRef]
49. Yang, X. Controllable Interfacial Polymerization for Nanofiltration Membrane Performance Improvement by the Polyphenol Interlayer. *ACS Omega* **2019**, *4*, 13824–13833. [CrossRef]
50. Zhang, X.; Lv, Y.; Yang, H.-C.; Du, Y.; Xu, Z.-K. Polyphenol Coating as an Interlayer for Thin-Film Composite Membranes with Enhanced Nanofiltration Performance. *ACS Appl. Mater. Interfaces* **2016**, *8*, 32512–32519. [CrossRef]
51. Zha, Z.; He, P.; Zhao, S.; Guo, R.; Wang, Z.; Wang, J. Interlayer-Modulated Polyamide Composite Membrane for Organic Solvent Nanofiltration. *J. Membr. Sci.* **2022**, *647*, 120306. [CrossRef]
52. Peng, H.; Chen, P.; Wu, Y.; Chen, C.; Wang, D. Synergistic Preparation of Organic Solvent Nanofiltration Membranes from Polydopamine/Polyethyleneimine Interlayers with Aminated Attapulgite. *Sep. Purif. Technol.* **2025**, *355*, 129503. [CrossRef]
53. Yang, X.; Du, Y.; Zhang, X.; He, A.; Xu, Z.-K. Nanofiltration Membrane with a Mussel-Inspired Interlayer for Improved Permeation Performance. *Langmuir* **2017**, *33*, 2318–2324. [CrossRef]
54. Wang, T.; Wang, J.; Zhao, Z.; Zheng, X.; Li, J.; Liu, H.; Zhao, Z. Bio-Inspired Fabrication of Anti-Fouling and Stability of Nanofiltration Membranes with a Poly(Dopamine)/Graphene Oxide Interlayer. *Ind. Eng. Chem. Res.* **2021**, *60*, 14868–14883. [CrossRef]
55. Wu, Y.; Yao, C.; Wang, Y.; Ding, M.; Xu, H.; Yan, X.; Gao, L. Water Recovery from the High Salinity Brine: Effect of the Interlayer Structure in the Polyamide Nanofiltration Membrane. *J. Environ. Chem. Eng.* **2024**, *12*, 111963. [CrossRef]
56. Liang, C.; Wang, Q.; Pan, Z.; Yu, S.; Wu, D.; Gao, C. Enhancing Nanofiltration Performance with Tannic Acid and Polyvinyl Alcohol Interlayers for Improved Water Permeability and Selective Solute Rejection. *Desalination* **2025**, *594*, 118277. [CrossRef]
57. Jia, R.; Jin, X.-G.; Xu, Z.-L.; Wu, L.-K.; Tong, Y.-H.; Li, H.-X.; Ping, H.-H.; Ma, X.-H.; Xu, S.-J. Surfactant-Interlayer Assisted Interfacial Polymerization for Constructing Janus Nanofiltration Membranes: Enhanced Li^+/Mg^{2+} Separation Efficiency. *J. Membr. Sci.* **2024**, *712*, 123235. [CrossRef]
58. Wang, X.; Mao, Y.; Gao, N.; Liao, Y.; Zhang, Y. Ceramic Supported Polyamide Composite Nanofiltration Membrane with a Glutaraldehyde Cross-Linked Chitosan Interlayer. *J. Water Process Eng.* **2024**, *68*, 106566. [CrossRef]
59. Guo, Y.; Ji, C.; Ye, Y.; Chen, Y.; Yang, Z.; Xue, S.; Niu, Q.J. High Performance Nanofiltration Membrane Using Self-Doping Sulfonated Polyaniline. *J. Membr. Sci.* **2022**, *652*, 120441. [CrossRef]
60. Guo, Y.; Wei, S.; Chen, Y.; Ye, H.; Xue, S.; Niu, Q.J. Sulfonated Polyaniline Interlayer with Controllable Doping Conditions for High-Performance Nanofiltration. *J. Membr. Sci.* **2023**, *672*, 121478. [CrossRef]
61. Zhang, X.; Zhang, Y.; Zhao, G.; Zhang, Y.; Zhang, Y.; Chang, Y.; Zhang, X.; Yuan, B.; Hu, P.; Zhao, K.; et al. Super Li^+/Mg^{2+} sieving and regenerable potentials of a nanofiltration membrane with intermediate layers of positively-charged hyperbranched polyamide. *Desalination* **2025**, *604*, 118737. [CrossRef]
62. Burts, R.S.; Plisko, T.V.; Davydova, M.V.; Makarava, M.S.; Yuan, B.; Penkova, A.V.; Ermakov, S.S.; Bildyukevich, A.V. The effect of polydiallyldimethylammonium chloride molecular weight in the intermediate layer on the structure and performance of thin film composite membranes for nanofiltration prepared via interfacial polymerization. *Colloids Surf. A Physicochem. Eng. Asp.* **2025**, *716*, 136752. [CrossRef]

63. Wang, Z.; Liang, S.; Kang, Y.; Zhao, W.; Xia, Y.; Yang, J.; Wang, H.; Zhang, X. Manipulating Interfacial Polymerization for Polymeric Nanofilms of Composite Separation Membranes. *Prog. Polym. Sci.* **2021**, *122*, 101450. [CrossRef]
64. Karan, S.; Jiang, Z.; Livingston, A.G. Sub–10 Nm Polyamide Nanofilms with Ultrafast Solvent Transport for Molecular Separation. *Science* **2015**, *348*, 1347–1351. [CrossRef]
65. Fang, S.; Guan, K.; Zhou, S.; Song, Q.; Shi, Y.; Fu, W.; Li, Z.; Xu, P.; Hu, M.; Mai, Z.; et al. Ternary-Coordination-Regulated Polyamide Nanofiltration Membranes for Li^+/Mg^{2+} Separation. *Desalination* **2024**, *581*, 117577. [CrossRef]
66. Li, H.; Li, Y.; Li, M.; Jin, Y.; Kang, G.; Cao, Y. Improving Mg^{2+}/Li^+ Separation Performance of Polyamide Nanofiltration Membrane by Swelling-Embedding-Shrinking Strategy. *J. Membr. Sci.* **2023**, *669*, 121321. [CrossRef]
67. Liu, M.; Chen, W.; Fu, J.; Wang, A.; Ding, M.; Zhang, L.; Han, L.; Gao, L. Hyaluronic Acid-Modified Nanofiltration Membrane for Ultrahigh Water Permeance and Efficient Rejection of PFASs. *Process Saf. Environ. Prot.* **2022**, *166*, 214–221. [CrossRef]
68. Guo, C.; Li, N.; Qian, X.; Shi, J.; Jing, M.; Teng, K.; Xu, Z. Ultra-Thin Double Janus Nanofiltration Membrane for Separation of Li^+ and Mg^{2+}: "Drag" Effect from Carboxyl-Containing Negative Interlayer. *Sep. Purif. Technol.* **2020**, *230*, 115567. [CrossRef]
69. Yang, L.; Zhang, K.; Zhao, F.; Li, Z.; Chen, H.; Wang, K. Customized Polyamide Nanofiltration Membranes with Hyperbranched Polyamide-Amine-Modified Graphene Oxide Interlayer for Desalination. *Desalination* **2025**, *600*, 118496. [CrossRef]

polymers

Article

Effect of Polymer Molecular Weight on the Structure and Properties of Ultra-High-Molecular-Weight Polyethylene Membranes Prepared via Controlled Swelling

Andrey V. Basko [1], Konstantin V. Pochivalov [1,*], Tatyana N. Lebedeva [1], Mikhail Y. Yurov [1], Alexander S. Zabolotnov [2], Sergey S. Gostev [2], Alexey A. Yushkin [3], Alexey V. Volkov [3] and Sergei V. Bronnikov [4]

1 G A Krestov Institute of Solution Chemistry, Russian Academy of Sciences, ul. Akademicheskaya, 1, Ivanovo 153045, Russia; avb@isc-ras.ru (A.V.B.)
2 N N Semenov Institute of Chemical Physics, Russian Academy of Sciences, ul. Kosygina, 4, Moscow 119991, Russia
3 A V Topchiev Institute of Petrochemical Synthesis, Russian Academy of Sciences, Leninsky pr., 29-2, Moscow 119071, Russia
4 Branch of Petersburg Nuclear Physics Institute named by B.P. Konstantinov of National Research Centre "Kurchatov Institute" Institute of Macromolecular Compounds, Bolshoy pr. of V.O., 31, Saint Petersburg 199004, Russia
* Correspondence: pkv@isc-ras.ru

Abstract: A recently proposed method called "controlled swelling of monolithic films" was implemented to prepare ultra-high-molecular-weight polyethylene (UHMWPE) ultrafiltration membranes. For the first time, the effect of UHMWPE molecular weight (MW) on the structure and properties of the membranes prepared via this special case of thermally induced phase separation was studied in detail. The morphology and properties of the membranes were studied using SEM, DSC, liquid–liquid displacement porometry, and standard methods for the evaluation of mechanical properties, permeance, rejection, and abrasion resistance. High-quality membranes with a tensile strength of 5.0–17.8 MPa, a mean pore size of 25–50 nm, permeance of 17–107 L m^{-2} h^{-1} bar^{-1}, rejection of model contaminant (blue dextran) of 72–98%, and great abrasion resistance can be prepared only if the MW of the polymer in the initial monolithic film is sufficiently high. The properties of the membranes can effectively be controlled by changing the MW of the polymer and the mass fraction of the latter in the swollen film. Shrinkage is responsible for the variation in the membrane properties. The membranes prepared from a higher-MW polymer are more prone to shrinking after the removal of the solvent. Shrinkage decreases before rising again and minimizes with an increase in the polymer content in the swollen film.

Keywords: ultra-high-molecular-weight polyethylene; swelling; ultrafiltration membrane; thermally induced phase separation

1. Introduction

Membrane technologies have been applied in different areas, varying from gas separation [1], liquid filtration [2], and water treatment [3,4] to battery separators [5], membrane crystallization devices [6], and the preparation of microparticles [7]. Among the polymers used for membrane preparation, polyvinylidene fluoride [8], polysulphone [9], cellulose acetate [10], polyetherketone [11], and polyolefins, particularly polyethylene [12], polypropylene [13], and poly-4-methyl-1-pentene [14], are the most widespread.

In recent years, an interest in preparing membranes using ultra-high-molecular-weight polyethylene (UHMWPE) has been growing [15]. This is due to its high tensile and puncture strength [16], chemical resistance, and wear resistance [17].

One of the first papers in which attention was paid to the abrasion resistance of membranes used for the filtration of suspensions was published in 1999 [18]. It was shown that magnesium and calcium phosphate crystals can cause abrasion of ceramic membranes in bioreactors, thus leading to a decrease in their permeance. The problem of an increase in permeance, along with a decrease in selectivity and mechanical properties in the polymeric membranes due to their abrasion, has been intensively studied [19,20]. Later, several dozen papers, in which polymeric membranes were prepared and characterized in terms of wear resistance, were published [21–33].

In [22,23], it was shown that the abrasion resistance of the membranes, assessed via the mass loss after treatment with sandpaper or storage in a continuously mixed suspension of silicon carbide, may be improved by adding a small amount of nanoclay to the dope solution. A similar method was also applied in [24], but the amount of nanoclay added was much higher (up to 10%). In ref. [25], it was shown that prefiltration of molasses distillery wastewater through filter paper before the main filtration process using a polymeric membrane negates the abrasion effect of the feed mixture components on membrane stability. In order to prepare abrasion-resistant polyvinylidene fluoride membranes for membrane distillation, adding microsilica to the dope solution for electrospinning was proposed [26]. The addition of ~3 wt.% microsilica enabled the production of superhydrophobic membranes that maintained their properties after 160 h of continuous filtration. A positive effect of nanoparticle (graphene oxide and nanodiamonds) addition on the abrasion resistance of polyvinyl chloride membranes was also observed in [27].

It should be noted that nanoparticle addition does not always improve the abrasion resistance of a membrane. For example, in ref. [28], the authors showed that the addition of titanium dioxide to the dope solution increases the permeance of a membrane but decreases its abrasion resistance. In ref. [29], composite membranes based on polyvinyl chloride and polycarbonate were prepared, and it was shown that an increase in the content of the latter in the membranes not only improved their strength and abrasion resistance but also increased their permeance. An increase in polyurethane additive in composite membranes based on polyvinyl chloride was shown to increase their abrasion resistance [30]. It was also proposed to protect the active, easily abraded material (polyacrylamide membrane) by sandwiching it between the cotton fabric materials [31]. Zirconium-based metal–organic frameworks were added to the collagen fiber membrane to form a highly effective and abrasion-resistant membrane from genuine leather [32]. Superhydrophobic membranes were prepared by forming a layer of spherulitic structures on the surface of the membranes, and it was shown that their hydrophobicity was maintained even after abrasion resistance tests and ultrasound treatment [33]. More details on the abrasion resistance and overall durability of the membranes for liquid filtration were discussed in a review paper [34]. At the same time, despite UHMWPE having an advantage in abrasion resistance, as shown in a literature analysis, to date, there are no papers in which the abrasion resistance of UHMWPE membranes is reported.

A few methods can be used for UHMWPE membrane preparation, including stretching, powder sintering, and thermally induced phase separation (TIPS). The TIPS method is the most widely developed method for UHMWPE membrane formation. But this method is complicated since it is difficult to form a homogeneous dope solution with the required concentration [15]. That is why, in recent years, several original methods for preparing UHMWPE membranes have been developed [35–37].

In our recent paper, we also proposed a new method of UHMWPE membrane preparation based on the controlled swelling of monolithic films and discussed the peculiarities of the structure formation mechanism [38]. This method consists of monolithic film annealing at an elevated temperature in a large amount of solvent (xylene), followed by cooling the swollen film to room temperature, extracting the solvent, and, finally, drying. Membranes' morphologies and properties can effectively be controlled by changing the parameters of the membrane formation process, such as the cooling rate, thermodynamic affinity of dope solution components, and polymer MW [15]. In works published in the 2000s [39–41], it was shown that increasing polymer MW decreases the pore size in polypropylene membranes prepared via TIPS. Later, this tendency was also observed for polyvinylidene fluoride membranes [42]. In ref. [43], it was shown that increasing MW not only decreases the pore size but also decreases the depth of non-solvent diffusion into the dope solution in the process of non-solvent/thermally induced phase separation and corresponding changes in the formed membrane structure.

The effect of polymer MW on the structure and properties of nonporous UHMWPE materials [44,45] and peculiarities of polymer behavior [46,47] were reported. Meanwhile, the effect of polymer MW on the structure and properties of UHMWPE membranes is scarce in the literature. Particularly, it was shown [48] that an increase in UHMWPE MW from 2.5 to $4{\cdot}10^6$ g mol^{-1} leads to a decrease in permeance of membranes prepared via TIPS and an increase in rejection due to the pore size decrease [48]. In ref. [49], we found that an increase in polymer MW results in a change in the phase separation type in the TIPS process from solid–liquid to liquid–liquid.

The goal of the present work is to further develop the understanding of the controlled swelling method for the preparation of UHMWPE membranes from monolithic films, to assess the prospects of using decalin as a swelling agent, and to investigate the effect of polymer MW on the morphology, transport, physico-mechanical, and tribological properties of UHMWPE membranes prepared via controlled swelling.

2. Materials and Methods

2.1. Materials

In this study, 99% decalin (Macklin, Shanghai, China), with a boiling temperature of 186 °C and a density of 0.896 g cm^{-3} at 25 °C, was used as a swelling agent, and 99.6% iso-propanol (Ekos-1, Moscow, Russia), with a boiling temperature of 82.5 °C and a density of 0.785 g cm^{-3} at 25 °C, was used as an extractant.

UHMWPE samples with different MWs were synthesized according to the technique described previously [49]. The samples were prepared via polymerization in situ using a catalyst deposited on graphite nanoplates acting as a filler in the synthesized composite UHMWPE powder. The graphite nanoplate concentration in the synthesized polymer was 0.11 wt.%. Controlling the synthesis conditions (temperature and hydrogen gas addition), three samples with different MWs were prepared. The principal characteristics of the UHMWPE samples are listed in Table 1.

Table 1. Principal characteristics of the synthesized UHMWPE samples and hot-pressed monolithic films.

Sample	Weight Average MW, g mol^{-1}	Melting Temperature *	Crystallinity Degree (α) *, %
LMW	700,000	139.0	85.6
MMW	3,700,000	135.3	53.7
HMW	4,700,000	135.6	55.3

* Melting temperature and crystallinity degree was measured for the prepared monolithic films.

The MW of the samples was detected via gel-permeation chromatography (HLC-8321 GPC/HT Tosoh, Tokyo, Japan) according to the technique described previously [49]. The melting temperature and crystallinity degree (α) of the samples were assessed using a differential scanning calorimetry (DSC) technique.

2.2. Membrane Preparation

The first stage of the membrane preparation was the hot pressing of the synthesized powders into a thin monolithic film. The required quantity of the powder was evenly deposited within a 150 μm thick frame on the surface of a polyethylene terephthalate film lying on a polished stainless steel plate, covered with a second polyethylene terephthalate film and a second stainless steel plate. The package was then placed between the plates of a hydraulic press heated to 170 °C. Then, a pressure of ~10 bar was applied for 10 min to ensure uniform heating of the sample. After that, the pressure was increased to ~160 bar and left for 10 min again. Finally, the sample was cooled down in the press by changing the steam circulating in the instrument plates with cold water. The mean cooling rate in the range 90–120 °C was ca. 30 °C min^{-1}. As a result, monolithic films of 25 × 25 cm were obtained.

The prepared films were cut into squares and placed into a hot weighing bottle with an excess of decalin (bath modulus 1:200) for swelling. The swelling temperatures were chosen based on preliminary experiments so that the required swelling degree could be attained in a reasonable time. At lower temperatures, the swelling was too slow, whereas at higher temperatures, it tended to be nonuniform. The swelling time was adjusted in order to obtain samples with the required polymer mass fraction in swollen film (PMFSF). The latter was determined by taking into account the variation in the sample size due to its swelling. It was calculated as a ratio of the polymer mass in the swollen film to the mass of the swollen film according to Equation (1):

$$PMFSF = m_p/m_{sf} = \left(m_f - m_e\right)/\left(\left(m_f - m_e\right) + \left(\left(a_{sf}b_{sf}d_{sf} - a_0b_0d_0\right)\rho_{dec}\right)\right), \quad (1)$$

where m_p is the polymer mass in the swollen film; m_{sf} is the mass of the swollen film; m_f is the mass of the initial monolithic film; m_e is the mass of the polymer extracted in the swelling process; a_{sf}, b_{sf}, and d_{sf} are the dimensions of the swollen film; a_0, b_0, and d_0 are the dimensions of the initial film; and ρ_{dec} is the decalin density. The value of d_{sf} was not measured but taken as increased by the same factor as a_{sf} and b_{sf} compared to a_0 and b_0.

After the required PMFSF was reached, the bottle with the sample was cooled, and thus, the swelling process was stopped. The membrane samples were transferred to iso-propanol to extract decalin for 12 h. After that, the samples were then transferred into a fresh iso-propanol bath to remove traces of decalin for 24 h and then air-dried to a constant mass.

As many as 12 samples were chosen for the following studies, with the main preparation conditions listed in Table 2.

Table 2. The membrane sample preparation conditions.

Sample Code	Polymer MW, 10^6 g mol^{-1}	Polymer Mass raction in Swollen Film (PMFSF), g g^{-1}	Swelling Temperature, °C	Swelling Time, min
LMW-11	0.7	0.11	105	20
MMW-11	3.7	0.11	124	100
MMW-17	3.7	0.17	122	40
MMW-25	3.7	0.25	115	90

Table 2. *Cont.*

Sample Code	Polymer MW, 10^6 g mol^{-1}	Polymer Mass raction in Swollen Film (PMFSF), g g^{-1}	Swelling Temperature, °C	Swelling Time, min
MMW-29	3.7	0.29	110	60
MMW-34	3.7	0.34	110	55
MMW-37	3.7	0.37	110	50
HMW-11	4.7	0.11	116	80
HMW-17	4.7	0.17	112	30
HMW-25	4.7	0.25	112	25
HMW-29	4.7	0.29	109	30
HMW-34	4.7	0.34	108	20

2.3. Membrane and Film Characterization

In order to determine the melting temperatures and α of UHMWPE in monolithic films and membranes, DSC thermograms were recorded at a scanning rate of 10 °C min^{-1} using a 204F1 Phoenix instrument (NETZSCH, Selb, Germany). The values of α were calculated according to Equation (2):

$$\alpha = \Delta H_m / \Delta H_m^{100\%}, \quad (2)$$

where ΔH_m is the melting enthalpy of the sample determined in the DSC experiment and $\Delta H_m{}^{100\%}$ = 293 J g^{-1} is the melting enthalpy of a hypothetical polyethylene sample, with α = 100% [50]. The melting temperature was taken as the maximum temperature of the endothermic peak in the DSC thermogram.

A DSC study of the UHMWPE mixtures with decalin was performed according to the following procedure. The required amounts of UHMWPE and decalin were placed into an aluminum pan, sealed and heated in a calorimeter from room temperature to 180 °C, annealed at this temperature for 30 min for homogenization, cooled to room temperature, thermostated for 3 min, and heated again to 180 °C. The tensile strength (σ) and relative elongation at break (ε) of the films and membranes were determined using an I-11M instrument (Tochpribor-KB, Ivanovo, Russia). The 25 × 3 mm samples with a working length of 15 mm were studied at an extension rate of 50 mm min^{-1}. The mean value of σ was calculated from five individual measurements according to Equation (3):

$$\sigma = F_{max} / S_0, \text{ MPa} \quad (3)$$

where F_{max} is the breaking force and S_0 is the cross-section area of the sample including (for the membranes) the area of pores. The mean value of ε was calculated from the same five measurements according to Equation (4):

$$\varepsilon = (l_{br} - l_0) / l_0 \quad (4)$$

where l_{br} is the final length of the sample at break and l_0 is the initial length of the sample.

The morphology of the surfaces and cross-section surfaces of the samples was investigated using a Quattro S scanning electron microscope (SEM) (Thermo Fischer Scientific, Brno, Czech Republic). The images were recorded from secondary electrons using a lower detector at the accelerating voltage of 1 kV. In order to investigate the cross-section morphology of the samples, they were cleaved in liquid nitrogen. The samples were coated with gold using a Quorum Q150es plus (Quorum Technologies, Lewes, UK) sputter coater prior to this study.

The porosity of the membranes was calculated using Equation (5):

$$P = \left(V_m - V_f\right)/V_m \quad (5)$$

where V_m and V_f are the volumes of the membrane and initial film, respectively.

Filtration experiments were carried out in a dead-end stirred filtration cell. For the filtration experiments, the membrane coupon was placed onto porous stainless steel disks and sealed with a rubber O-ring. The active membrane area in the cell was 7.9 cm^2. The system was pressurized with helium. The transmembrane pressure was 5 bar. Water was used as a solvent. Blue dextran (M_w = 500 kg mol^{-1}) was used as a solute at a concentration of 10 mg L^{-1}. Due to the hydrophobic nature of UHMWPE, ethanol was filtered through every membrane coupon for 1 h at 5 bar to fill all the membrane pores. Pure water was filtered for 2 h at 5 bar to determine the pure water permeance. Then, the blue dextran solution in water was filtered until 100 mL of permeate was collected. In the case of solution filtration, the feed was stirred at 550 rpm to minimize the concentration polarization effect. The membrane permeance L (L m^{-2} h^{-1} bar^{-1}) was determined as:

$$L = {}^{m}/_{(\rho \cdot S \cdot t \cdot \Delta p)} \quad (6)$$

where Δp is the transmembrane pressure (bar), m is the mass of the permeate (g), ρ is the density of the permeate (g L^{-1}), S is the active membrane area (m^2), and t is the filtration time (h).

The concentration of blue dextran in the feed and permeate was measured with a spectrophotometer at a wavelength of 620 nm. The rejection R was calculated using the following relation:

$$R = (1 - {}^{C_p}/_{C_f}) \cdot 100\% \quad (7)$$

where C_f and Cp denote the solute concentrations in the feed and permeate, respectively.

The mean flow pore size (MFP) was measured by liquid–liquid displacement porosimetry using a POROLIQ 1000 ML porometer (Porometer, Nazareth, Belgium) according to the procedure described in ref. [51]. MFP is such a pore size that half the flux goes through pores of a higher diameter and half the flux goes through pores of a lower diameter. The measurements were carried out at room temperature using a pair of liquids prepared by demixing a mixture of isobutanol and water (1/4, v/v). The coupons (2.5 cm in diameter) were cut from the membrane and then placed into the beaker with a wetting liquid (alcohol-rich phase) for 12 h at room temperature before testing.

The abrasion resistance of the films and membranes was assessed according to the modified technique described in refs. [26,52]. Briefly, the membrane sample of a given mass was attached to a glass by double-sided tape, placed on sandpaper, and moved back and forth with a weight on top of the glass. We used the sandpaper P400 (grain size of 28–40 μm) (Zolder, Yanino-1, Russia). The loading weight was 1 kg, and the number of cycles was 200 since, at a standard weight of 200 g and 50 cycles of movement, there was almost no mass loss due to abrasion.

3. Results and Discussion

3.1. Thermal Behavior of the Decalin Mixtures of UHMWPE of Different MWs

Decalin is known to be a good solvent for UHMWPE [53–55]. However, in ref. [56], a phase diagram was obtained for a mixture of UHMWPE with decalin, containing a binodal with an upper critical solution temperature (Figure 1). The points obtained experimentally were not quite in agreement with the solid lines shown on the same diagram. Furthermore, the porous sample prepared by cooling the mixture containing 40% wt. of UHMWPE and 60% wt. of decalin was shown to have a spherulitic morphology [56]. Meanwhile, the spherulitic morphology is a result of solid–liquid phase separation (i.e., polymer crystallization directly from the homogeneous mixture) rather than the liquid–liquid phase separation associated with binodal on the phase diagram.

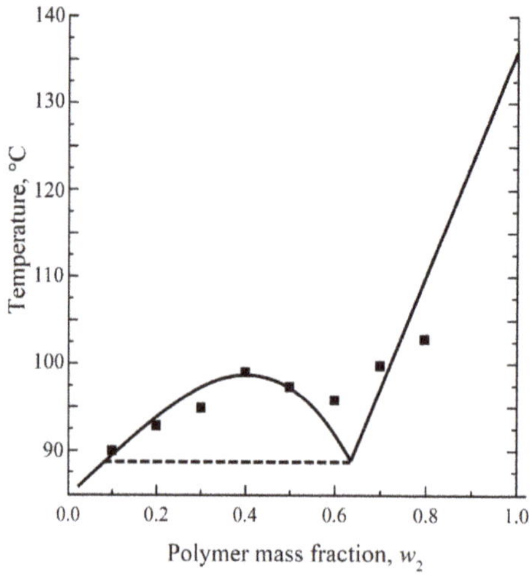

Figure 1. Phase diagram for the UHMWPE–decalin mixture. Adapted with permission from [56].

Figure 2 shows the results of the DSC study of the UHMWPE films and binary mixtures of UHMWPE of different MWs and decalin proportions. For the hot-pressed films, the thermograms recorded in the first heating are shown, while for the mixtures, the second heating endotherms are presented. One can see that all the DSC curves contain a single endothermic peak corresponding to the thermal effect of the UHMWPE amorphization (melting) process. The peak maximum temperature increases with an increase in the polymer content in the mixture. The area of the peak (melting enthalpy related to the overall mass of the mixture) in the mixtures containing LMW increases monotonically and passes through a maximum for the mixtures containing MMW and HMW with increasing polymer concentration in the mixture.

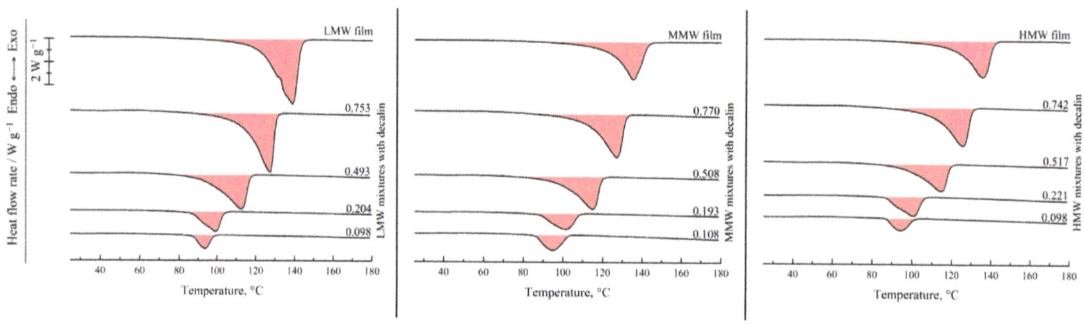

Figure 2. Second heating DSC thermograms of the mixtures of UHMWPE of different MW (LMW—**left**; MMW—**middle**, HMW—**right**) with decalin, together with the first heating DSC thermograms of the corresponding hot-pressed films. The polymer mass fraction in the mixtures is shown above the curves.

Figure 3 shows the DSC thermograms of the same films and mixtures recorded in the first cooling from 180 °C. They contain a single exothermic peak corresponding to the polymer crystallization. Its onset temperature also increases with increasing polymer concentration, and the trend of its area variation is in agreement with that observed on the second heating thermograms.

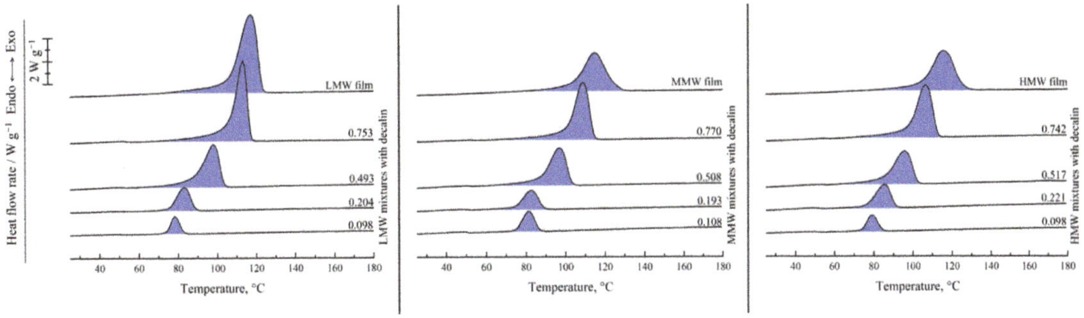

Figure 3. First cooling DSC thermograms of the UHMWPE of different MW (LMW—**left**; MMW—**middle**; HMW—**right**) and their mixtures with decalin. The polymer mass fraction in the mixtures is shown above the curves.

In Figure 4, the dependencies of the endothermic peak maximum and onset of exothermic peaks (black points, left vertical axis), together with polymer crystallinity degrees attained during the mixture's cooling (blue points, right vertical axis), are plotted against the polymer mass fraction in the mixture. The crystallinity degrees were calculated using a modified Equation (2):

$$\alpha = \Delta H_{cr} / \left(w_2 \times \Delta H_m^{100\%} \right) \quad (8)$$

where ΔH_{cr} is an area of the exothermic peak on the DSC thermogram and w_2 is the polymer mass fraction in the mixture.

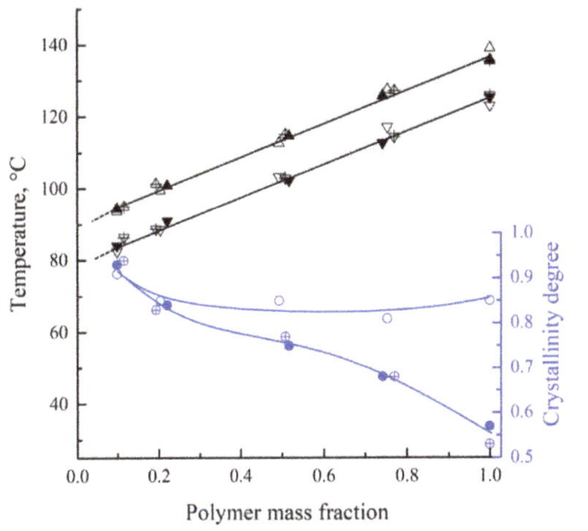

Figure 4. Dependencies of melting (black triangles) and crystallization (inverted black triangles) temperature and crystallinity degree (blue circles) of the LMW (hollow points), MMW (crossed points), and HMW (solid points) UHMWPE.

The dependence of both the melting and crystallization temperatures on the mixture composition represents almost parallel straight lines. The crystallization temperature is 11 °C lower than the melting temperature, which is due to the kinetic sluggishness of the crystallization process. The absence of a horizontal section on these lines clearly indicates that decalin is a good solvent for UHMWPE. Consequently, the phase diagram of the mixture does not contain a liquid–liquid equilibrium binodal. The melting and crystallization temperatures are independent of the polymer MW, which is to be expected for the high-MW polymer samples synthesized in similar conditions.

Interestingly, an increase in the decalin mass fraction in the mixtures containing MMW and HMW leads to a gradual increase in α (from ~55 to ~92%). This is because an increase in the decalin concentration in the mixture promotes higher mobility of the macromolecules and thus enables the formation of a more ideal crystalline structure. At the same time, the α values of the LMW sample show almost no dependence on the mixture composition, being equal to ~85% for the pure polymer and slightly increasing (to ~92%) for the mixture containing ~10% wt. of the polymer. It can be assumed that the shorter macromolecules in this sample are less entangled, which enables them to reach higher values of α during cooling at a rate of 10 °C min^{-1}, even in the absence of plasticizer/diluent.

The dependence of the polymer melting temperature on the mixture (Figure 4) is one of the lines presented on the phase diagram of the binary mixture. In addition, as it was shown previously for various mixtures [57], the phase diagram of the semicrystalline (SC) polymer mixture with a good solvent should contain at least one more boundary curve, reflecting a dependence of the SC polymer swelling degree on temperature (temperature dependence of low molecular liquid solubility in amorphous regions of the SC polymer).

Figure 5 shows the phase diagram containing this curve. In this diagram, the ABC curve reflects a dependence of the polymer melting temperature on the mixture composition, whereas the BD curve corresponds to the temperature dependence of the UHMWPE swelling degree in decalin. The BC line on this diagram was plotted using the DSC data obtained in the present work, while the BD curve was plotted using a dependence of the UHMWPE swelling degree on temperature reported by Zhao et al. [58]. The homogeneous molecular mixtures of the components exist in domain I. In domain II, the systems represent the uniformly swollen SC polymer (gel), while in domain III, such a gel coexists with the almost pure solvent.

The points corresponding to the temperatures of the swelling process of the monolithic films and the solvent concentration in these films at the moment of swelling termination by cooling (w_1 = 1 − PMFSF) are also plotted within the temperature–composition field of the diagram. All these points are located in domain I. This means that termination of swelling the films was actually solutions of decalin in fully amorphous UHMWPE. Nevertheless, the films maintained their shape and integrity due to the very high viscosity of such solutions. Obviously, such "limited" swelling would be unlimited in time, i.e., a homogeneous solution of UHMWPE in the full volume of a decalin bath would be formed with a composition corresponding to the bath modulus (in our case 0.5% wt.) if the swelling process was not ceased by cooling.

An analysis of the phase diagrams shown in Figure 5 reveals that the cooling of the prepared swollen films leads, at first, to their entering domain II, in which the polymer starts to crystallize across the volume of the swollen film, and then to domain III, where, from the swollen SC UHMWPE, almost pure decalin segregates. The details of the structure formation mechanism were discussed in our previous work [38].

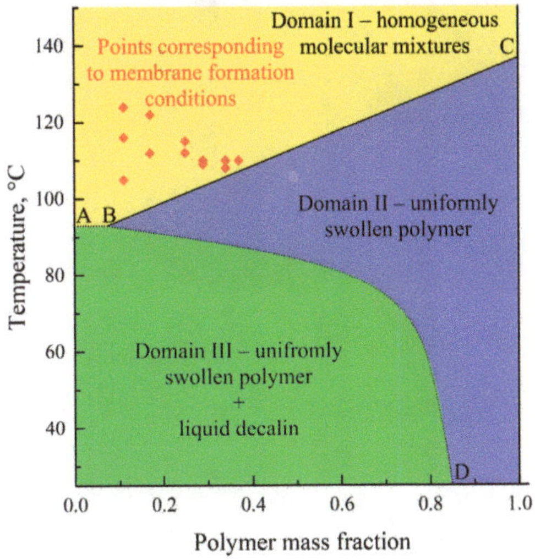

Figure 5. Phase diagram for the UHMWPE mixture with decalin. Red diamonds correspond to the temperature and PMFSF of the samples before termination of the swelling process by cooling.

3.2. *Morphology and Properties of the UHMWPE Monolithic Films*

Prior to discussing the effect of the variation in the process parameters of the membrane formation on their structure and properties, it is useful to consider first how polymer MW affects the characteristics of the hot-pressed monolithic films used for membrane preparation.

Figure 6 shows the SEM images of the surfaces and cross-sections of the monolithic films prepared from LMW, MMW, and HMW. The surfaces of all the films are smooth and homogeneous, which is indicative of their successful preparation. The cross-section surface of the LMW sample is significantly different from those of the MMW and HMW samples. This is due to the inability to obtain a high-quality cross-section of the sample by cleavage due to highly elastic deformation before its break, even at −196 °C, leading to the formation of fibrils and pores. Because of this, the image actually shows a cut surface of the sample instead of a cleavage surface. The MMW and HMW samples have a layered structure (Figure 6). In our opinion, this is because, for these polymers of very high MW, even at a pressure of 160 bar and a temperature of 170 °C, it was not possible to form a homogeneous polymer melt across the thickness of the samples. Meanwhile, the density measurements with a pycnometer showed that all the prepared samples were monolithic and had a density of 0.93–0.94 g cm^{-3}.

Table 3 shows the results of the physico-mechanical tests of the monolithic films. It can be seen that both the σ and ε of the samples decrease with increasing MW, which is in agreement with the results of the SEM study of the films. Indeed, the samples for which we were not able to achieve complete mixing of macromolecules (formation of a homogeneous melt) are characterized by a lower σ and by a more fragile character of the break. The trends of the elasticity modulus change correlate well with the data on the crystallinity degree of the samples (Figure 4). One can see that the LMW sample, with an α value of ~85%, has a higher elasticity modulus than the MMW and HMW samples, with an α value of ca. 55%.

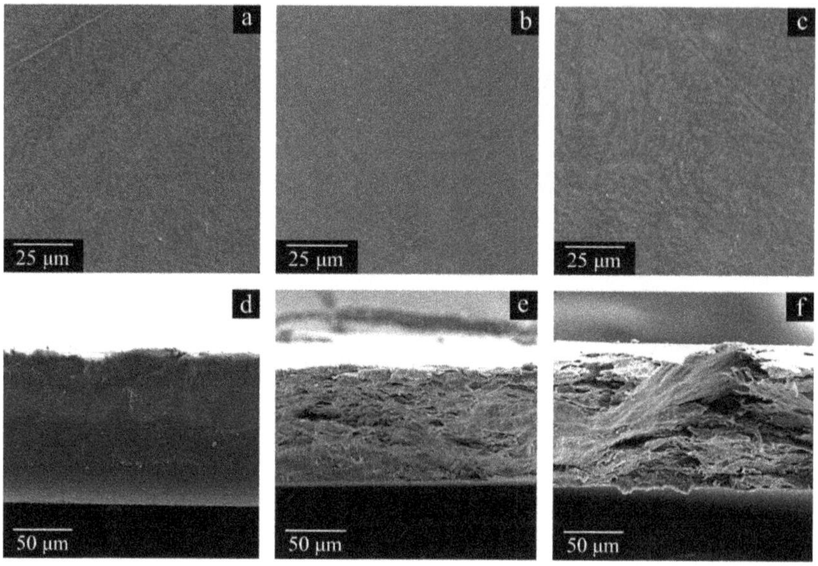

Figure 6. SEM images of the surface (**a**–**c**) and cross-section surface (**d**–**f**) of the monolithic films prepared from LMW (**a**,**d**), MMW (**b**,**e**), and HMW (**c**,**f**).

Table 3. Physico-mechanical properties of the monolithic films.

Sample Code	σ, MPA	ε, %	Elasticity Modulus, MPa	Mass Loss After Abrasion Test, g m^{-2}
LMW	37 ± 0.8	700 ± 40	780 ± 50	12 ± 2
MMW	34.1 ± 0.8	22 ± 2	520 ± 30	20 ± 2
HMW	23.3 ± 0.4	7.5 ± 0.8	570 ± 20	18 ± 2

It is interesting to note that the films prepared from LMW also have the best abrasion resistance. Taking into account the data shown in Figures 4 and 6, this can be explained by a combined effect of two factors. Firstly, this sample has a much higher crystallinity degree, determining a high abrasion resistance. Secondly, its more uniform (monolithic) structure is also favorable for high wear resistance. At the same time, the MMW sample is expectedly characterized by slightly higher mass loss after the abrasion test under the same conditions compared to the HMW sample with a similar crystallinity degree and morphology.

3.3. Morphology and Properties of the Prepared Membranes

It should be noted that we attempted to prepare the membrane samples from the mixtures of decalin with all three types of the UHMWPE samples chosen for study at different PMFSF values. However, it was found that during swelling, the LMW films lost 10 to 30% (depending on the swelling temperature and time) of the initial mass, while for the MMW and HMW films, the mass loss did not exceed ~2%. The samples prepared from LMW often attained a shape significantly different from the initial square, and, in some cases, lost their integrity in either the swelling, extraction, or drying stage. The reason for such a difference in the behavior of the films during swelling is obviously their different MWs.

Indeed, annealing the film in the decalin medium (at temperatures corresponding to the domain I in the phase diagram in Figure 5) makes two processes thermodynamically possible: diffusion of decalin molecules into the film and mass transfer of polymer into the decalin volume. Obviously, the latter process is more pronounced for the films composed of the LMW polymer.

Thus, we were able to prepare only one sample of a membrane suitable for the following studies of morphology, transport, and physico-mechanical properties from the LMW film (at PMFSF = 0.11) and several samples from MMW and HMW with different PMFSF values.

3.3.1. Effect of Polymer MW on Structure and Properties of Membranes at Fixed PMFSF

In order to assess the effect of the polymer MW on the structure and properties of the membranes, the samples prepared with the same PMFSF of 0.11 were compared. Analysis of the data presented in Figures 7–10 and Table 4 regarding the membrane samples prepared at PMFSF = 0.11 allows us to draw the following conclusions.

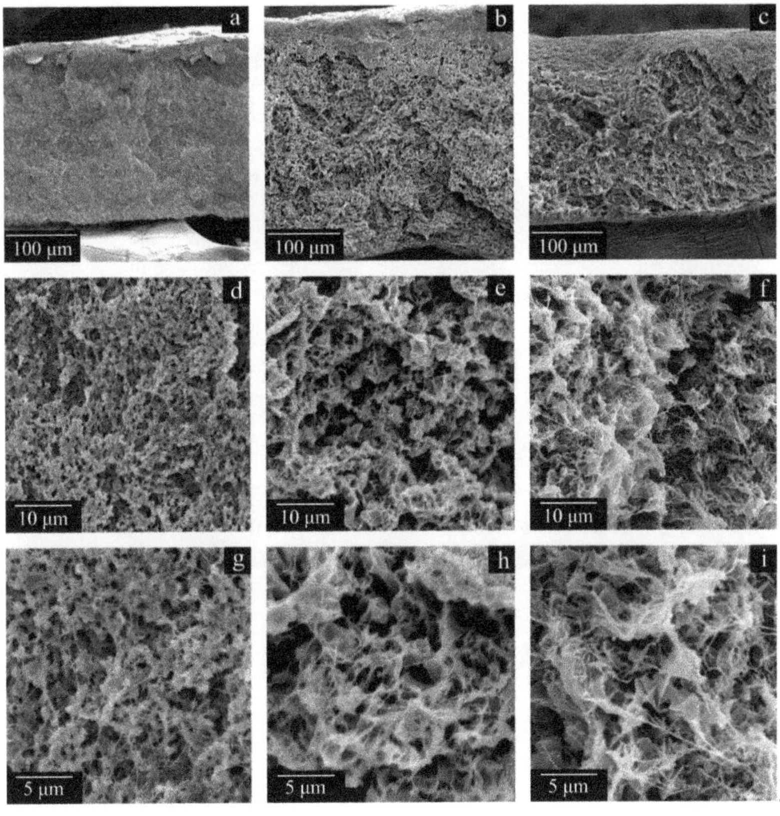

Figure 7. SEM images of the cross-section surface of the LMW-11 (**a**,**d**,**g**), MMW-11 (**b**,**e**,**h**), and HMW-11 (**c**,**f**,**i**) membranes.

Traces of the initial powder particles are visible in the morphology of the cross-section surface of the MMW and HMW monolithic films, but they are absent in the structure of the membranes. Furthermore, the membrane structure itself is uniform across the thickness of the sample. Thus, a treatment of the films with decalin at elevated temperature ensures macromolecule mixing even in the samples where it was not complete at the stage of hot-pressing the UHMWPE powder. Further evidence of this phenomenon is the results of the studies of the membranes' physico-mechanical properties. While the σ values of monolithic films decrease by a factor of ca. 1.5 in an LMW > MMW > HMW series and their ε values decrease in the same series by almost two orders of magnitude, the trends are opposite for the membranes prepared from these polymers. Moreover, the ε of the MMW-11 and HMW-11 samples is even higher than that of the corresponding monolithic films.

The membrane structure in all cases (Figures 7 and 8), expectedly, is a result of polymer crystallization directly from its uniform mixture with decalin. The morphology of the LMW-11 sample consists of globule-like structures interconnected by thin fibrils. Taking into account the data contained in Table 4, it can be concluded that such a structure is characterized by poor mechanical properties. In the morphology of the MMW-11 and HMW-11 samples, the leaf-like formations dominate, similar to those found in other UHMWPE membranes [59]. The morphology of the cross-section of these samples also contains the fibrils that were absent in the initial sample but were formed during its cleavage in liquid nitrogen [60]. When the MMW-11 and HMW-11 samples are compared, the porosity and pore size of the latter sample are significantly lower. It should be noted that since the PMFSF values during these sample preparations were identical, their different porosity must be a result of different shrinkage at the stages of decalin extraction and drying from iso-propanol. Both the transformation of the structure from globule-like to leaf-like and the increase in the proneness of the formed porous structure to shrinkage are consequences of a polymer MW increase. On the one hand, UHMWPEs with MWs higher than $3{\cdot}10^6$ g mol^{-1} are more prone to the formation of leaf-like structures during crystallization. On the other hand, a higher MW leads to increased shrinkage [61].

The structure of the sample surface is also significantly different. When the LMW-11 (Figure 8a,d) and MMW-11 (Figure 8b,e) samples are compared, the quantity and size of open pores of the latter sample are lower. Furthermore, open pores were not even observed on the surface of the HMW-11 sample due to the insufficient resolving power of the SEM device. It can also be noted that the surface of this sample is wrinkled, probably due to nonuniform shrinkage.

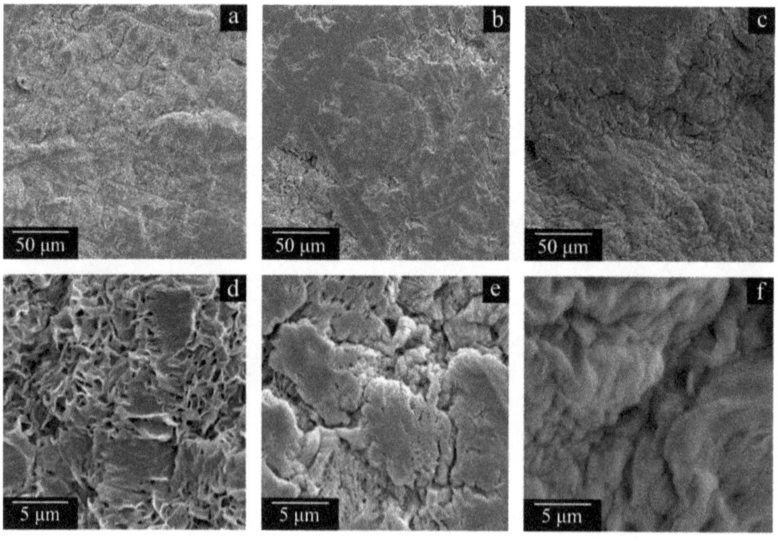

Figure 8. SEM images of surfaces of the LMW-11 (**a**,**d**), MMW-11 (**b**,**e**), and HMW-11 (**c**,**f**) membranes.

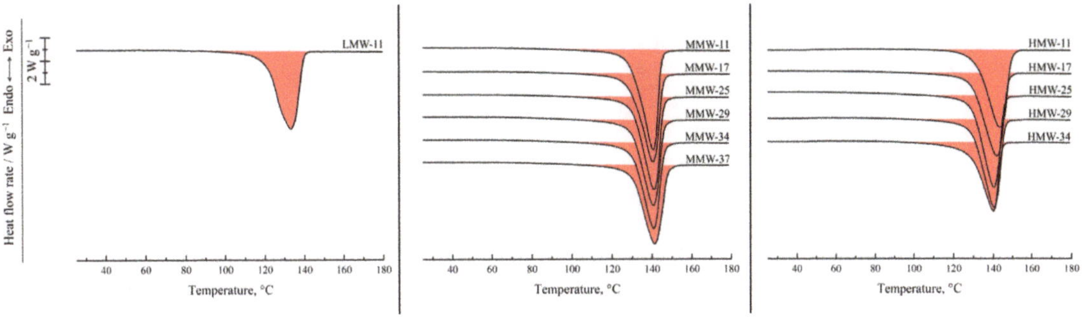

Figure 9. DSC-thermograms of the prepared membranes. Sample codes are shown above the curves.

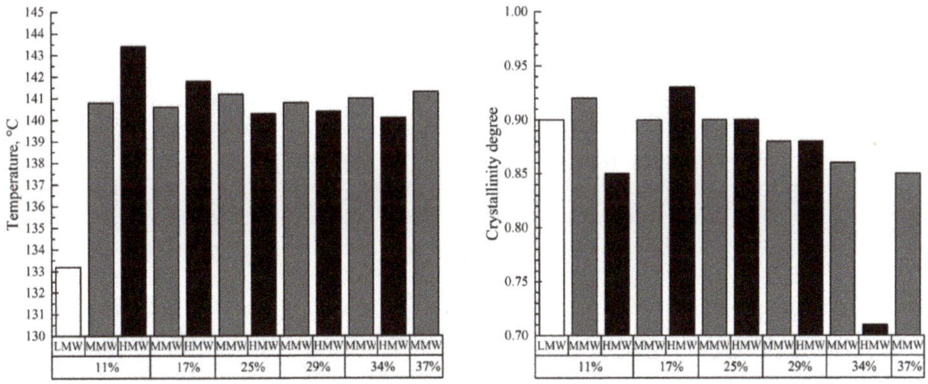

Figure 10. Values of the melting temperatures (**left**) and crystallinity degrees (**right**) of the prepared membranes. The polymer used for the preparation of the sample and PMFSF is indicated under the histograms.

Table 4. Physico-mechanical, transport, and other characteristics of the membranes.

Sample Code	σ, MPA	ε, %	Porosity, %	MFP, nm	Permeance, $L\ m^{-2}\ h^{-1}\ bar^{-1}$	*R*, %	Mass Loss After Abrasion Test, $g\ m^{-2}$
LMW-11	1.0 ± 0.2	9 ± 0.6	60 ± 4	47 ± 5	75 ± 3	73 ± 3	42 ± 5
MMW-11	6.3 ± 0.8	190 ± 25	65 ± 4	36 ± 4	58 ± 3	92.7 ± 1.5	10 ± 1
MMW-17	7.2 ± 0.7	170 ± 20	67 ± 4	39 ± 3	50 ± 2	86 ± 2	11 ± 1
MMW-24	10.1 ± 0.7	180 ± 30	72 ± 5	43 ± 6	66 ± 4	77 ± 3	10 ± 1
MMW-29	11.0 ± 0.8	160 ± 35	74 ± 6	50 ± 4	107 ± 6	72 ± 4	11 ± 1
MMW-34	13.8 ± 1.2	170 ± 30	69 ± 5	41 ± 4	59 ± 5	82 ± 3	11 ± 1
MMW-37	17.8 ± 1.1	130 ± 20	67 ± 7	38 ± 5	47 ± 5	90.4 ± 1.8	11 ± 1
HMW-11	11.9 ± 0.9	125 ± 20	29 ± 3	15 ± 2	1.4 ± 0.2	98.2 ± 0.8	10 ± 1
HMW-17	5.0 ± 0.6	100 ± 15	72 ± 4	39 ± 4	54 ± 3	83 ± 3	11 ± 1
HMW-24	5.6 ± 0.8	90 ± 15	71 ± 4	37 ± 5	50 ± 4	90.7 ± 1.6	11 ± 1
HMW-29	10.6 ± 1.0	105 ± 25	67 ± 4	34 ± 3	44 ± 3	94.7 ± 1.4	10 ± 1
HMW-34	13.6 ± 1.2	45 ± 6	49 ± 5	25 ± 3	17 ± 2	96.9 ± 1.1	14 ± 1

The tendencies of the changes in the transport properties (mean pore size and permeance) are in good agreement with the above-described trends of changes in overall porosity and surface porosity. The LMW-11 sample demonstrates a higher pore size and permeance than the MMW-11 sample, while the latter sample exhibits the smallest pore size and permeance (Table 4). For the samples prepared from the same polymer, the pore size follows porosity, and hence, more porous samples possessed higher MFP and permeance. The rejection of blue dextran also well correlates with MFP. For the dense HMW-11 sample, the rejection was the highest (up to 98%) when the more porous sample LMW-11 possessed a lower rejection of 73%.

Figure 9 shows the DSC thermograms of the membrane samples used to assess the melting temperature of the polymer in these membranes and their crystallinity degree. All the thermograms are characterized by a single endothermic peak reflecting amorphization of UHMWPE. The melting temperature of the membrane samples increases with the increase in polymer MW (Figure 10). Furthermore, if these values are compared with those of UHMWPE in the monolithic films (Figures 2 and 4), it can be concluded that the melting temperature decreases for the LMW sample by ~6 °C, while it increases by ~6 and ~8 °C for the MMW and HMW samples, respectively. The crystallinity degree of UHMWPE in the membranes as a function of polymer MW passes through a quite pronounced maximum (Figure 10). At the same time, the crystallinity degree of UHMWPE in its mixtures with decalin, containing 10% wt. of the polymer (Figures 2–4), is almost the same for all three studied samples. However, it should be noted that the mixtures in the DSC experiment were cooled from 180 °C, while in the membrane preparation, they were cooled from 105 to 124 °C depending on the sample (Table 2). Bearing in mind the data on the morphology of the samples, it can be concluded that in the LMW-11 samples, an entirely new crystalline structure was formed with no traces of the initial crystalline structure of the monolithic film. On the one hand, the presence of decalin creates more favorable conditions for reaching a higher crystallinity degree than that in the initial film. On the other hand, in these conditions, more stressed and defective crystallites were formed that melt at lower temperatures. An increase in the melting temperature with the increase in polymer MW indicates that UHMWPE with a higher MW is able to form a more ideal crystalline structure. It can be assumed that the lower crystallinity degree of the HMW-11 sample is a result of high shrinkage that, probably, led to lamellae fragmentation, as it was shown in ref. [62].

The polymer MW increase leads to an increase in the σ of the membrane samples prepared at PMFSF = 0.11. At the same time, their ε passes through a maximum. The LMW-11 sample has very low mechanical properties, which is attributed to the peculiarities of its structure that significantly differ from the structure of the other samples (see Figure 7a,d,g). The MMW-11 sample has lower tensile strength but higher elongation at break than the HMW-11 sample. Furthermore, the ε of the porous samples is about an order of magnitude higher than that of the corresponding monolithic films. This is due to the fact that the treatment of the films in decalin favors effective mixing of macromolecules originating from different particles of the polymer powder across the film volume and thus the formation of a continuous, uniform matrix. A higher σ value of the MMW-11 sample compared to the HMW-11 sample is a result of a porosity decrease.

The trends observed in the wear resistance studies of the prepared samples are in good agreement with the conclusions discussed above. Particularly, the LMW-11 sample with low mechanical properties quickly collapsed in the abrasion experiments, while the MMW-11 sample and the HMW-11 sample demonstrate excellent wear resistance that is even better than that of the corresponding monolithic films (Table 3). This is obviously due to an increase in the crystallinity degree upon transformation of monolithic films into porous membrane samples. Despite the differences in porosity of the MMW-11 sample and the HMW-11 sample, their wear resistance is almost the same within the experimental error. Apparently, the combined effect of increased MW and lower porosity, which should lead to better abrasion resistance, is compensated for by its deterioration due to lower crystallinity degree.

3.3.2. Effect of PMFSF and Polymer MW on the Structure and Properties of the Membranes

Figures 11 and 12 show the SEM images of the cross-section surfaces of the membranes prepared from decalin mixtures with MMW and HMW at PMFSF higher than 0.11. Analysis of these images and the DSC data shown in Figures 9 and 10 and Table 4 allows drawing a conclusion.

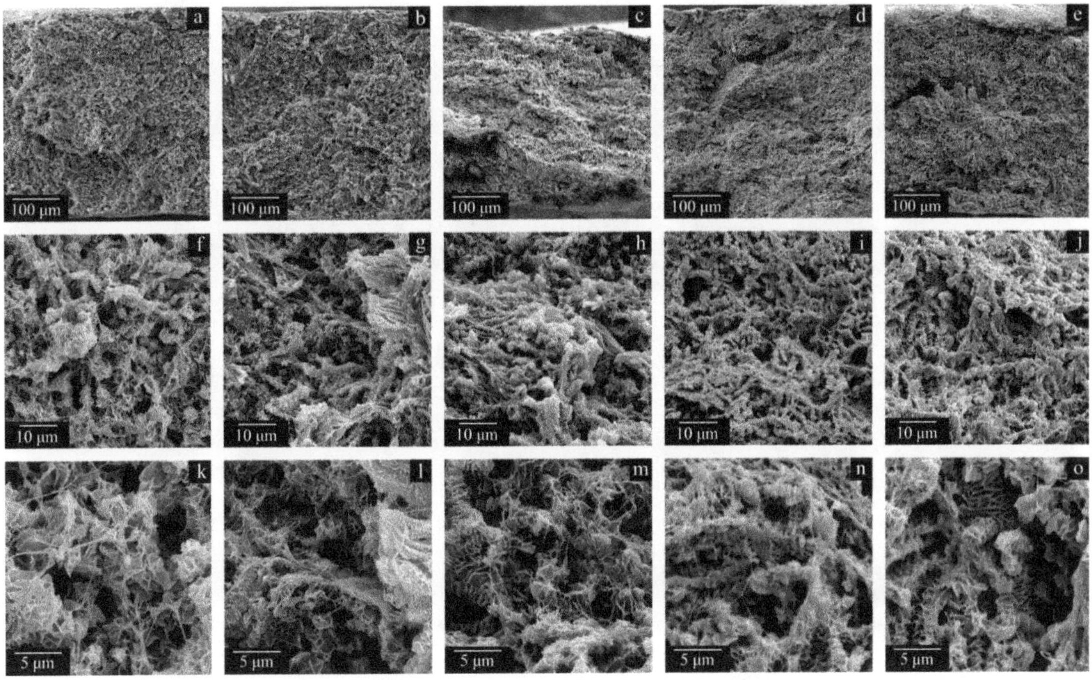

Figure 11. SEM images of the cross-section surfaces of the membranes MMW-17 (**a**,**f**,**k**), MMW-25 (**b**,**g**,**l**), MMW-29 (**c**,**h**,**m**), MMW-34 (**d**,**i**,**n**), and MMW-37 (**e**,**j**,**o**).

The morphology of the cross-section surfaces of all the samples except HMW-34 is similar: they are composed of leaf-like and shish-kebab structures. The experimentally found differences in porosity (Table 4) are almost indiscernible in the SEM images of the cross-section surfaces since even at −196 °C, ideally brittle cleavage of the samples is not possible. A rather unpronounced dependence of porosity on PMFSF is explained by an increase in the sample shrinkage with an increase in PMFSF in the studied range. Only the HMW-34 sample has significantly lower porosity. This means that the effect of lower shrinkage could not compensate for a decrease in porosity due to an increase in PMFSF. In the structure of this sample, both leaf-like and shish-kebab structures are almost absent due to its lower porosity compared to the other samples.

A PMFSF increase in the discussed range does not lead to significant changes in the melting temperature of both samples prepared from MMW and HMW. At the same time, the crystallinity degree decreases as PMFSF increases. In our opinion, this is because a decalin content increase in the swollen films makes macromolecules more mobile and thus makes it possible to achieve a higher crystallinity degree. The significantly lower crystallinity degree of the HMW-34 sample is probably explained by the fact that the swelling conditions were too mild to fully erase the memory of the initial crystalline structure in the sample. Thus, upon its cooling, a network of crystallites, similar to that in the monolithic film, was formed. This assumption also agrees with the abovementioned differences in the morphology of the samples (cf. Figure 12d,h,l and, for example, Figure 12c,g,k).

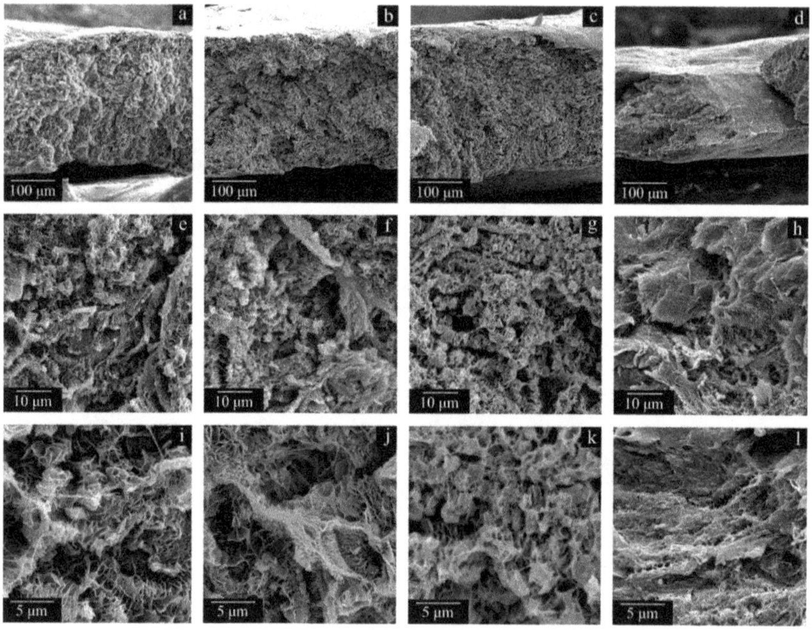

Figure 12. SEM images of the cross-section surface of the membranes HMW-17 (**a**,**e**,**i**), HMW-25 (**b**,**f**,**j**), HMW-29 (**c**,**g**,**k**), and HMW-34 (**d**,**h**,**l**).

The transport properties of the obtained membranes depend on two factors. The first factor is polymer MW. The membranes obtained with the same PMFSF demonstrate a tendency to increase MFP and permeance with decreasing MW (Table 4). The second factor is membrane porosity. More porous samples possess a higher pore size and permeance. The highest permeance of 107 L m^{-2} h^{-1} bar^{-1} was observed for the most porous MMW-29 sample. This sample also possesses a higher pore size MFP of 50 nm. The HMW-34 and HMW-11 samples with a low porosity demonstrate a low pore size of 15 and 25 nm, respectively. The permeance of these samples is expectedly low at 1.4 and 17 L m^{-2} h^{-1} bar^{-1}. Consequently, rejection of the blue dextran dye completely followed expectations based on their pore size. The highest rejection of 98% was observed for the densest HMW-11 sample, while the most porous MMW-29 sample with the highest pore size possessed the lowest rejection of 72%.

In the discussed range of PMFSF, the σ value of the samples prepared from MMW monotonically increases with an increase in PMFSF, while that of the samples prepared from HMW passes through a minimum. Such a character of the tensile strength variation is due to their different porosities and crystallinity degrees. An increase in the crystallinity degree usually leads to higher σ values. In this case, when the crystallinity degree is ca. 90% and higher, its increase probably decreases the number of tie chains that transfer the external stress between the crystallites and thus makes the sample less robust. This is why, in our opinion, in the series of the MMW-17, MMW-25, MMW-29, MMW-34, and MMW-37 samples with almost the same porosity but decreasing crystallinity degree, the tensile strength increases. The same can be said about the HMW-17, HMW-25, and HMW-29 samples. The higher σ values of the HMW-11 and HMW-34 samples compared to those of the HMW-17, HMW-25, and HMW-29 samples are explained by the lower porosity of the former. In turn, the low porosity of the HMW-11 sample is a result of its very high shrinkage, whereas, for the HMW-34 sample, it is a consequence of its initially high PMFSF.

The wear resistance of the samples also correlates well with the data on the crystallinity degree of the samples. Particularly, almost all the discussed samples have the same (within experimental error) values of mass loss. The only exception is the HMW-34 sample with a significantly lower crystallinity degree. It should once again be noted that the wear resistance of the samples was assessed using a modified method (five times higher than the standard pressing load and four times the distance (200 cycles of sample moving)) since the standard method [26,52] yielded almost negligible values of mass loss. This means that the prepared samples have a great wear resistance, much higher than that reported in the literature for the membrane samples prepared from various polymers, since UHMWPE has a substantial advantage over other polymers in this regard.

Before concluding, it is interesting to discuss why the porosity and some transport properties of the samples with an increase in PMFSF pass through a maximum and their tensile strength and rejection pass through a minimum (Table 4). Obviously, the variation in porosity, mean pore size, and permeance is a result of competition between two factors: the quantity of decalin in the swollen film ($w_1 = 1 - \text{PMFSF}$), which increases porosity, and shrinkage, which results in a decrease in porosity. Our experiments show that shrinkage increases with an increase in w_1. The observed trend in the σ variation is also a result of competition between these two factors, and it is also affected by the crystallinity degree.

Indeed, at low PMFSF, despite the high amount of decalin in the swollen film, high shrinkage leads to a significant decrease in overall porosity and thus to high values of tensile strength. An increase in PMFSF makes the shrinkage effect on porosity less pronounced, so that the samples, even at a lower decalin content in the swollen film, are more porous and, consequently, less strong. Although a subsequent increase in the PMFSF value continues to diminish the shrinkage effect on membrane porosity, the latter decreases because of the increase in the decalin content in the swollen film.

This observation can also explain why the membranes possess a uniform pore structure across their thicknesses (see Figures 7a–c, 11a–e and 12a–d). The swelling, as expected, is probably nonuniform, and the polymer concentration actually is the lowest near the surface of the film and the highest in the middle of the film. However, after the swelling process is stopped, in the stage of decalin extraction and drying the membranes of the extractant, the outer (more swollen) layers of the membrane shrink more while the inner (less swollen) layers shrink less. This ultimately leads to uniform porosity across the film thickness.

A decrease in the crystallinity degree of the membranes with the increase in PMFSF probably increases the amount of tie chains transferring the external load and thus increases tensile strength. However, according to the obtained dependencies, σ = f(PMFSF) and P = f(PMFSF), this factor is not a decisive one. It should also be noted that the proneness of the samples to shrinkage was found to increase with an increase in polymer MW. This is due to the higher stresses of the longer macromolecules that participate in a higher number of entanglements and their relaxation after removal of non-compressible liquid (solvent or extractant) from the pores.

4. Conclusions

In this work, the polymer MW effect on the structure and properties of membranes prepared via the controlled swelling of the monolithic UHMWPE film method was assessed for the first time. It was shown that the method allows for preparing membranes with a uniform porous structure, even if the monolithic films prepared via hot pressing have defects due to the very high viscosity of the melt. It was established that the formation of high-quality membranes by this method is impossible at a polymer MW of $0.7{\cdot}10^6$ g mol^{-1} or lower. The abrasion resistance of the UHMWPE membranes was also assessed for the first time. It was shown that all the samples had excellent tribological properties, which supports their recommendation for the filtration of highly abrasive suspensions. It was also found that the abrasion resistance was almost independent of the MW of the polymer and membrane porosity for the studied samples, but it increased with an increase in the crystallinity degree. The effect of the polymer concentration in the swollen films on the structure and properties of the membranes was studied. It was established that with an increase in PMFSF, the tensile strength of the membranes passes through a minimum, whereas porosity, mean pore size, and permeance pass through a maximum, while the shrinkage and crystallinity degree decrease. The explanation of such dependencies was proposed. Analysis of the SEM images of the prepared samples showed that all the samples have a morphology consisting of leaf-like formations and shish-kebab structures.

Author Contributions: Conceptualization, A.V.B. and K.V.P.; formal analysis, M.Y.Y., A.S.Z. and S.S.G.; investigation, T.N.L., M.Y.Y. and A.A.Y.; resources, A.S.Z.; writing—original draft preparation, A.V.B. and K.V.P.; writing—review and editing, K.V.P., A.A.Y., A.V.V. and S.V.B.; visualization, A.V.B. All authors have read and agreed to the published version of the manuscript.

Funding: The study was supported by RSF, grant No. 24-29-00433.

Data Availability Statement: The raw data supporting the conclusions of this article will be made available by the authors on request.

Acknowledgments: The authors are grateful for the opportunity to use the equipment of the Upper Volga Region Center of Physico-Chemical Research.

Conflicts of Interest: The authors declare no conflicts of interest.

References

1. Kang, G.D.; Cao, Y.M. Application and Modification of Poly(Vinylidene Fluoride) (PVDF) Membranes—A Review. *J. Membr. Sci.* **2014**, *463*, 145–165. [CrossRef]
2. Volkov, A.; Yushkin, A.; Kachula, Y.; Khotimsky, V.; Volkov, V. Application of Negative Retention in Organic Solvent Nanofiltration for Solutes Fractionation. *Sep. Purif. Technol.* **2014**, *124*, 43–48. [CrossRef]
3. Ezugbe, E.O.; Rathilal, S. Membrane Technologies in Wastewater Treatment: A Review. *Membranes* **2020**, *10*, 89. [CrossRef]
4. Wen, Y.; Yuan, J.; Ma, X.; Wang, S.; Liu, Y. Polymeric Nanocomposite Membranes for Water Treatment: A Review. *Environ. Chem. Lett.* **2019**, *17*, 1539–1551. [CrossRef]
5. Li, L.; Duan, Y. Engineering Polymer-Based Porous Membrane for Sustainable Lithium-Ion Battery Separators. *Polymers* **2023**, *15*, 3690. [CrossRef]
6. Jiang, X.; Shao, Y.; Sheng, L.; Li, P.; He, G. Membrane Crystallization for Process Intensification and Control: A Review. *Engineering* **2021**, *7*, 50–62. [CrossRef]
7. Vladisavljević, G.T. Preparation of Microparticles and Nanoparticles Using Membrane-Assisted Dispersion, Micromixing, and Evaporation Processes. *Particuology* **2024**, *84*, 30–44. [CrossRef]
8. Pochivalov, K.V.; Basko, A.V.; Lebedeva, T.N.; Yurov, M.Y.; Yushkin, A.A.; Bronnikov, S.V.; Volkov, A.V. PVDF Membrane Formation via NIPS in Isothermal and Non-Isothermal Conditions: Thermodynamics, Structure, and Properties. *Membr. Membr. Technol.* **2024**, *6*, 473–490. [CrossRef]
9. Serbanescu, O.S.; Voicu, S.I.; Thakur, V.K. Polysulfone Functionalized Membranes: Properties and Challenges. *Mater. Today Chem.* **2020**, *17*, 100302. [CrossRef]

10. Wang, H.H.; Jung, J.T.; Kim, J.F.; Kim, S.; Drioli, E.; Lee, Y.M. A Novel Green Solvent Alternative for Polymeric Membrane Preparation via Nonsolvent-Induced Phase Separation (NIPS). *J. Membr. Sci.* **2019**, *574*, 44–54. [CrossRef]
11. Kammakakam, I.; Lai, Z. Next-Generation Ultrafiltration Membranes: A Review of Material Design, Properties, Recent Progress, and Challenges. *Chemosphere* **2023**, *316*, 137669. [CrossRef]
12. Yang, S.; Xiao, C.; Huang, Y.; Ji, D.; Chen, K. Effect of Additive and Coagulation Bath Temperature on Structure and Properties of HDPE Membranes via Thermally Induced Phase Separation. *J. Mater. Sci.* **2022**, *57*, 4834–4849. [CrossRef]
13. Pochivalov, K.; Basko, A.; Yurov, M.; Lebedeva, T.; Shalygin, M.; Lavrentyev, V.; Yushkin, A.; Anokhina, T.; Volkov, A. Polypropylene Membranes Prepared via Non-Solvent/Thermally Induced Phase Separation: Effect of Non-Solvent Nature. *J. Membr. Sci.* **2024**, *703*, 122839. [CrossRef]
14. Tang, Y.; Li, M.; Lin, Y.; Wang, L.; Wu, F.; Wang, X. A Novel Green Diluent for the Preparation of Poly(4-Methyl-1-Pentene) Membranes via a Thermally-Induced Phase Separation Method. *Membranes* **2021**, *11*, 622. [CrossRef]
15. Basko, A.; Pochivalov, K. Current State-of-the-Art in Membrane Formation from Ultra-High Molecular Weight Polyethylene. *Membranes* **2022**, *12*, 1137. [CrossRef]
16. Ding, L.; Li, D.; Du, F.; Zhang, D.; Zhang, S.; Wu, T. Novel Preparation of Lithium-ion Battery Wet-processed Separator Based on the Synergistic Effect of Porous Skeleton Nano-Al_2O_3 in Situ Blending and Synchro-draw. *Polym. Int.* **2023**, *72*, 61–70. [CrossRef]
17. Dayyoub, T.; Maksimkin, A.; Olifirov, L.K.; Chukov, D.; Kolesnikov, E.; Kaloshkin, S.D.; Telyshev, D.V. Structural, Mechanical, and Tribological Properties of Oriented Ultra-High Molecular Weight Polyethylene/Graphene Nanoplates/Polyaniline Films. *Polymers* **2023**, *15*, 758. [CrossRef]
18. Cicek, N.; Dionysiou, D.; Suidan, M.T.; Ginestet, P.; Audic, J.M. Performance Deterioration and Structural Changes of a Ceramic Membrane Bioreactor Due to Inorganic Abrasion. *J. Membr. Sci.* **1999**, *163*, 19–28. [CrossRef]
19. Mozia, S.; Szymański, K.; Michalkiewicz, B.; Tryba, B.; Toyoda, M.; Morawski, A.W. Effect of Process Parameters on Fouling and Stability of MF/UF TiO2 Membranes in a Photocatalytic Membrane Reactor. *Sep. Purif. Technol.* **2015**, *142*, 137–148. [CrossRef]
20. Chesters, S.P.; Pena, N.; Gallego, S.; Fazel, M.; Armstrong, M.W.; del Vigo, F. Results from 99 Seawater RO Membrane Autopsies. *IDA J. Desalin. Water Reuse* **2013**, *5*, 40–47. [CrossRef]
21. Chen, S.; Zhou, J.; Li, K.; Chen, Z.; Li, X.; Su, X.; Ao, X.; Xie, H.; Chen, L.; Wu, X.; et al. Superhydrophobic and Robust Photothermal/Electrothermal PVDF-a/CNT-S@PDMS Membrane for Crude Oil Removal and Dye Adsorption. *Compos. Sci. Technol.* **2023**, *244*, 110267. [CrossRef]
22. Lai, C.Y.; Groth, A.; Gray, S.; Duke, M. Preparation and Characterization of Poly(Vinylidene Fluoride)/Nanoclay Nanocomposite Flat Sheet Membranes for Abrasion Resistance. *Water Res.* **2014**, *57*, 56–66. [CrossRef] [PubMed]
23. Lai, C.Y.; Groth, A.; Gray, S.; Duke, M. Enhanced Abrasion Resistant PVDF/Nanoclay Hollow Fibre Composite Membranes for Water Treatment. *J. Membr. Sci.* **2013**, *449*, 146–157. [CrossRef]
24. Ji, J.; Zhou, S.; Lai, C.Y.; Wang, B.; Li, K. PVDF/Palygorskite Composite Ultrafiltration Membranes with Enhanced Abrasion Resistance and Flux. *J. Membr. Sci.* **2015**, *495*, 91–100. [CrossRef]
25. Arimi, M.M.; Namango, S.S.; Götz, G.; Zhang, Y.; Kiriamiti, K.; Geißen, S.U. The Abrasion Effects of Natural Organic Particles on Membrane Permeability and the Size Distribution of Recalcitrants in a Colored Effluent. *J. Membr. Sci.* **2016**, *509*, 1–9. [CrossRef]
26. Liao, Y.; Zheng, G.; Huang, J.J.; Tian, M.; Wang, R. Development of Robust and Superhydrophobic Membranes to Mitigate Membrane Scaling and Fouling in Membrane Distillation. *J. Membr. Sci.* **2020**, *601*, 117962. [CrossRef]
27. Khakpour, S.; Jafarzadeh, Y.; Yegani, R. Incorporation of Graphene Oxide/Nanodiamond Nanocomposite into PVC Ultrafiltration Membranes. *Chem. Eng. Res. Des.* **2019**, *152*, 60–70. [CrossRef]
28. Behboudi, A.; Jafarzadeh, Y.; Yegani, R. Preparation and Characterization of TiO_2 Embedded PVC Ultrafiltration Membranes. *Chem. Eng. Res. Des.* **2016**, *114*, 96–107. [CrossRef]
29. Behboudi, A.; Jafarzadeh, Y.; Yegani, R. Polyvinyl Chloride/Polycarbonate Blend Ultrafiltration Membranes for Water Treatment. *J. Membr. Sci.* **2017**, *534*, 18–24. [CrossRef]
30. Wang, N.; Raza, A.; Si, Y.; Yu, J.; Sun, G.; Ding, B. Tortuously Structured Polyvinyl Chloride/Polyurethane Fibrous Membranes for High-Efficiency Fine Particulate Filtration. *J. Colloid Interface Sci.* **2013**, *398*, 240–246. [CrossRef] [PubMed]
31. Xiao, Q.R.; Sun, S. An Anti-Abrasion Sandwich Structured Membrane for Highly Efficient Oil/Water Separation. *Polym. Adv. Technol.* **2023**, *34*, 3659–3666. [CrossRef]
32. Ye, X.; Huang, R.; Zheng, Z.; Liu, J.; Chen, J.; Lv, Y. Anti-Abrasion Collagen Fiber-Based Membrane Functionalized by UiO-66-NH2 with Ultra-High Efficiency and Stability for Oil-in-Water Emulsions Separation. *Chin. J. Chem. Eng.* **2024**, *66*, 285–297. [CrossRef]
33. Han, M.; Liu, G.; Fang, C.; Zhu, L. Robust Superhydrophobic Poly(Vinylidene Fluoride) Membranes with Spherulitic Surface Morphology against Wetting and Scaling in Membrane Distillation. *J. Membr. Sci.* **2024**, *708*, 123027. [CrossRef]
34. Lin, X.; Hong, J. Recent Advances in Robust Superwettable Membranes for Oil–Water Separation. *Adv. Mater. Interfaces* **2019**, *6*. [CrossRef]
35. Cao, X.; Li, Y.; He, G. Fabrication of Self-Lubricating Porous UHMWPE with Excellent Mechanical Properties and Friction Performance via Rotary Sintering. *Polymers* **2020**, *12*, 1335. [CrossRef] [PubMed]

36. Zhang, Q.; Lan, L.; Zheng, Z.; Liu, P.; Wu, H.; Guo, S.; Lin, C.; He, G. Constructing Highly Oriented and Condensed Shish-Kebab Crystalline Structure of HDPE/UHMWPE Blends via Intense Stretching Process: Achieving High Mechanical Properties and in-Plane Thermal Conductivity. *Polymer* **2022**, *241*, 124532. [CrossRef]
37. Quan, J.; Yu, J.; Wang, Y.; Hu, Z. Construction of Intrinsic Superhydrophobic Ultra-High Molecular Weight Polyethylene Composite Membrane for DCMD. *J. Membr. Sci.* **2022**, *648*, 120353. [CrossRef]
38. Pochivalov, K.; Basko, A.; Lebedeva, T.; Yurov, M.; Yushkin, A.; Volkov, A.; Bronnikov, S. Controlled Swelling of Monolithic Films as a Facile Approach to the Synthesis of UHMWPE Membranes. *Membranes* **2023**, *13*, 422. [CrossRef]
39. Matsuyama, H.; Maki, T.; Teramoto, M.; Asano, K. Effect of Polypropylene Molecular Weight on Porous Membrane Formation by Thermally Induced Phase Separation. *J. Membr. Sci.* **2002**, *204*, 323–328. [CrossRef]
40. Atkinson, P.M.; Lloyd, D.R. Anisotropic Flat Sheet Membrane Formation via TIPS: Atmospheric Convection and Polymer Molecular Weight Effects. *J. Membr. Sci.* **2000**, *175*, 225–238. [CrossRef]
41. Yave, W.; Quijada, R.; Serafini, D.; Lloyd, D. Effect of the Polypropylene Type on Polymer–Diluent Phase Diagrams and Membrane Structure in Membranes Formed via the TIPS ProcessPart I. Metallocene and Ziegler–Natta Polypropylenes. *J. Membr. Sci.* **2005**, *263*, 146–153. [CrossRef]
42. Hassankiadeh, N.T.; Cui, Z.; Kim, J.H.; Shin, D.W.; Sanguineti, A.; Arcella, V.; Lee, Y.M.; Drioli, E. PVDF Hollow Fiber Membranes Prepared from Green Diluent via Thermally Induced Phase Separation: Effect of PVDF Molecular Weight. *J. Membr. Sci.* **2014**, *471*, 237–246. [CrossRef]
43. Zhang, P.; Fang, C.; Rajabzadeh, S.; Liu, W.; Jia, Y.; Shen, Q.; Zhang, L.; Wang, S.; Kato, N.; Matsuyama, H. Effect of Polymer Molecular Weight on Structure and Performance of PVDF Hollow Fiber Membranes Prepared via TIPS Process with Co-Extrusion of Solvent Using Triple Orifice Spinneret. *J. Membr. Sci.* **2021**, *620*, 118854. [CrossRef]
44. Deplancke, T.; Lame, O.; Rousset, F.; Seguela, R.; Vigier, G. Mechanisms of Chain Reentanglement during the Sintering of UHMWPE Nascent Powder: Effect of Molecular Weight. *Macromolecules* **2015**, *48*, 5328–5338. [CrossRef]
45. Vadivel, H.S.; Bek, M.; Šebenik, U.; Perše, L.S.; Kádár, R.; Emami, N.; Kalin, M. Do the Particle Size, Molecular Weight, and Processing of UHMWPE Affect Its Thermomechanical and Tribological Performance? *J. Mater. Res. Technol.* **2021**, *12*, 1728–1737. [CrossRef]
46. Cao, C.; Jiang, W.; Lin, Y.; Chen, X.; Qian, Q.; Chen, Q.; Yu, D.; Chen, X. Sensitive Phase Separation Behavior of Ultra-High Molecular Weight Polyethylene in Polybutene. *Polym. Test.* **2020**, *81*, 106243. [CrossRef]
47. Yang, Y.; Sheng, L.; Zhang, H.; Li, M.L.; Xu, R.; Bai, Y.; Song, S.; Liu, G.; Wang, T.; Huang, X.; et al. Investigation on the Solid–Liquid (S–L) Phase Separation of the PE/LP Blend with Different Molecular Weight Polyethylene. *Polym. Bull.* **2022**, *79*, 1521–1534. [CrossRef]
48. Liu, R.; Wang, X.; Yu, J.; Wang, Y.; Zhu, J.; Hu, Z. Development and Evaluation of UHMWPE/Woven Fabric Composite Microfiltration Membranes via Thermally Induced Phase Separation. *RSC Adv.* **2016**, *6*, 90701–90710. [CrossRef]
49. Basko, A.V.; Lebedeva, T.N.; Yurov, M.Y.; Zabolotnov, A.S.; Gostev, S.S.; Gusarov, S.S.; Pochivalov, K.V. Thermally Induced Phase Separation of UHMWPE Mixture with Dioctyl Adipate: Competition of Liquid–Liquid Phase Separation and Polymer Crystallization. *Thermochim. Acta* **2024**, *738*, 179787. [CrossRef]
50. Mirabella, F.M.; Bafna, A. Determination of the Crystallinity of Polyethylene/α-Olefin Copolymers by Thermal Analysis: Relationship of the Heat of Fusion of 100% Polyethylene Crystal and the Density. *J. Polym. Sci. Part B Polym. Phys.* **2002**, *40*, 1637–1643. [CrossRef]
51. Yushkin, A.A.; Efimov, M.N.; Malakhov, A.O.; Karpacheva, G.P.; Bondarenko, G.; Marbelia, L.; Vankelecom, I.F.J.; Volkov, A.V. Creation of Highly Stable Porous Polyacrylonitrile Membranes Using Infrared Heating. *React. Funct. Polym.* **2021**, *158*, 104793. [CrossRef]
52. Yin, Z.; Yuan, F.; Li, M.; Xue, M.; Zhou, D.; Chen, Y.; Liu, X.; Luo, Y.; Hong, Z.; Xie, C.; et al. Self-Cleaning, Underwater Writable, Heat-Insulated and Photocatalytic Cellulose Membrane for High-Efficient Oil/Water Separation and Removal of Hazardous Organic Pollutants. *Prog. Org. Coatings* **2021**, *157*, 106311. [CrossRef]
53. Wang, Y.; Fu, J.; Song, Q.; Yu, J.; Wang, Y.; Hu, Z. Regulating the Dissolving System of Ultra-high Molecular Weight Polyethylene to Enhance the High-strength and High-modulus Properties of Resultant Fibers. *J. Appl. Polym. Sci.* **2022**, *139*. [CrossRef]
54. Brem, A.; Lhost, O.; Tervoort, T.A. Influence of Solvent Quality and Crystallization Conditions on the Drawability of Ultra-High Molecular Weight Polyethylene Cast from Solution. *Macromolecules* **2020**, *53*, 5957–5970. [CrossRef]
55. Lemstra, P.J. Chapter 1: High-Performance Polyethylene Fibers. *Adv. Ind. Eng. Polym. Res.* **2022**, *5*, 49–59. [CrossRef]
56. Ding, H.; Tian, Y.; Wang, L.; Liu, B. Preparation of Ultrahigh-Molecular-Weight Polyethylene Membranes via a Thermally Induced Phase-Separation Method. *J. Appl. Polym. Sci.* **2007**, *105*, 3355–3362. [CrossRef]
57. Pochivalov, K.V.; Basko, A.V.; Kudryavtsev, Y.V. Binary Mixtures of Semicrystalline Polymers with Low-Molecular-Mass Compounds: Thermal Behaviour and Phase Structure. *Russ. Chem. Rev.* **2020**, *89*, 311–338. [CrossRef]
58. Yu, Q.; Gao, J.; Wang, Z.; Feng, X.; Zhao, Y.; Chen, L. Excellent Interfacial Bonding in UHMWPE Fiber/ Epoxy Resin Composites Fabricated via a Swelling Approach. *J. Polym. Res.* **2023**, *30*, 275. [CrossRef]

59. Basko, A.; Pochivalov, K.; Yurov, M.; Lebedeva, T.; Novikov, I.; Pakalnis, V.; Kuryndin, I.; Bronnikov, S.; Zabolotnov, A.; Yushkin, A.; et al. Preparation of Porous Ultrahigh Molecular Weight Polyethylene Separators for Li-Ion Batteries via Thermally Induced Phase Separation Method. *J. Ind. Eng. Chem.* 2025; in press. [CrossRef]
60. Sui, Y.; Li, J.; Qiu, Z.; Cui, Y.; Cong, C.; Meng, X.; Ye, H.; Zhou, Q. Effects of the Sintering Temperature on the Superior Cryogenic Toughness of Ultra-High Molecular Weight Polyethylene (UHMWPE). *Chem. Eng. J.* **2022**, *444*, 136366. [CrossRef]
61. Capaccio, G.; Ward, I.M. Shrinkage, Shrinkage Force and the Structure of Ultra High Modulus Polyethylenes. *Colloid Polym. Sci.* **1982**, *260*, 46–55. [CrossRef]
62. Pakhomov, P.; Khizhnyak, S.; Reuter, H.; Lechner, M.; Tshmel, A. Gel-to-Solid Transition in Polyethylene from the Viewpoint of the Crystallization Process. *Macromolecules* **2003**, *36*, 4868–4873. [CrossRef]

polymers

Article

Study on Deposition of Coaxial Electrospinning Fibers by Coaxial Auxiliary Flow Field

Rongguang Zhang [1,2], Xun Chen [1,2,*], Han Wang [1,2], Jianfeng Sun [1,2], Shize Huang [2], Xuanzhi Zhang [1,2] and Jiecai Long [1,2]

1 State Key Laboratory of Precision Electronic Manufacturing Technology and Equipment, Guangdong University of Technology, Guangzhou 510006, China
2 School of Electromechnical Engineering, Guangdong University of Technology, Guangzhou 510006, China
* Correspondence: xunchen@gdut.edu.cn

Abstract: Gas-assisted coaxial electrospinning (GACES) is a simple and general method for the mass preparation of coaxial nanofiber membranes, which has great industrial potential. However, in the manufacturing process, due to the bending instability of the jet in the electric field and the pulling effect of the gas flow field, the deposition uniformity of the fiber is still a big problem. Through finite element simulation analysis of the flow field in the manufacturing process and the construction of the jet mechanics model after adding the flow field, the influence mechanism of coaxial auxiliary flow on the fiber deposition area and its uniformity was successfully revealed in this research. Finally, the deposition area and thickness uniformity of coaxial fibers are increased by 3 times (the deposition area: 19.63 $cm^2 \rightarrow 78.50$ cm^2) and 2.34 times (the standard variance: 3 $\mu m^2 \rightarrow 10$ μm^2) by gas-assisted coaxial electrospinning. At the same time, the coaxial auxiliary gas flow also reduces the coaxial fiber diameter by 36.9% (the average fiber diameter: 241 nm $\pm$ 5 nm $\rightarrow$ 152 nm $\pm$ 23 nm) and the distribution range by 66% (the standard variance: 1.5×10^2 $nm^2 \rightarrow 51$ nm^2). This research provides a reliable idea and experimental basis for homogeneous preparation of coaxial nanofiber membranes.

Keywords: gas-assisted coaxial electrospinning (GACES); coaxial auxiliary flow field; uniformity

1. Introduction

Coaxial nanofibers have attracted much attention from all fields due to their unique micro-/nanostructure [1–5]. Common coaxial nanofiber fabrication processes include template fibers [6], self-assembly [7], template synthesis [8], coaxial electrospinning [9–13], etc. Among them, coaxial electrospinning is a simple and versatile method for preparing coaxial nanofiber membranes, which can be divided into needleless coaxial electrospinning and needle coaxial electrospinning depending on the spinneret [14–17]. However, so far, the electrospinning technology that can prepare coaxial nanofibers on a large scale is still subject to many limitations. Gas-assisted electrospinning is a new technology for the preparation of nanofibers that has emerged in recent years and has the advantage of efficient large-scale preparation of nanofibers [18,19]. Therefore, the gas-assisted coaxial electrospinning (GACES) technology has great industrial application potential in the large-scale preparation of coaxial nanofibers, but it has received less attention and has been developed slowly.

Meanwhile, current methods for controlling coaxial nanofiber deposition mainly rely on simple and inaccurate techniques that vary electrospinning time, which often produce

https://doi.org/10.3390/polym17030396

coaxial nanofiber membranes of non-uniform thickness. Moreover, the bending instability of the jet may also lead to the non-uniform deposition of nanofibers, resulting in local thin regions of the coaxial nanofiber membrane [20,21]. This non-uniformity can reduce the functionality of coaxial nanofiber membrane as a filter or scaffold, as it can affect the filtration efficiency or mechanical properties, affecting the application of electrospinning coaxial membranes in various fields [22,23]. So, in order to achieve uniform deposition of coaxial fibers in electrospinning, it is necessary to research the fiber deposition in the gas-assisted coaxial electrospinning process.

The objective of this research is to investigate the effect of coaxial gas flow on the deposition uniformity of coaxial nanofibers. Through the finite element simulation analysis of the gas-assisted coaxial electrospinning process and the construction of the mechanical model of the gas-assisted coaxial electrospinning jet, the influence mechanism of the coaxial flow field on the deposition of coaxial fibers was successfully revealed. At the same time, the quantitative analysis of the experimental results was carried out to realize the quantitative characterization of the fiber deposition area, deposition thickness uniformity, fiber average diameter, and distribution interval under different gas field parameters. Finally, the deposition area and thickness uniformity of coaxial fibers are increased by 3 times and 2.34 times by gas-assisted coaxial electrospinning. At the same time, the coaxial auxiliary gas flow also reduces the coaxial fiber diameter by 36.9% and the distribution range by 66%.

2. Materials and Methods

2.1. Materials

Polyvinyl alcohol (PVA-220, Kuraray, Tokyo, Japan) and polyethylene oxide (PEO-N3000) were obtained from Aladdin Chemical Co. (Shanghai, China). All reagents were of analytical grade and used as received.

2.2. Fabrication of Nanofiber Membrane

A 2 wt% solution of PVA and a 6.5 wt% solution of PEO in ion water were prepared. The experimental device is self-built, as shown in Figure 1. In this research, we selected a coaxial gas duct nozzle with an inner diameter of 2.78 mm and a coaxial electrospinning needle with an inner/outer diameter of 1.6/2.1 mm and 0.7/1.07 mm, respectively. The needle was charged to 15 kV using a power supply, and the PEO shell solution and PVA core solution were pumped from the gas-assisted coaxial electrospinning nozzle at a rate of 1.6 mL/h and 0.1 mL/h, respectively, via a double-channel pump. The pressure regulating valve control of auxiliary gas flow pressure is set for 0/0.01/0.02/0.03/0.04/0.05 MPa, and the needle was positioned 25 cm above a grounded collector. A jet of the polymer solution was launched from the needle tip and was collected on the grounded surface, and an ultra-fine coaxial fiber membrane was collected on the collector.

2.3. Characterization and Deposition Evaluation of Nanofibers

The surface morphology of the nanofibers was observed by scanning electron microscopy (SEM, TESCAN MIRA3, Brno-Kohoutovice, Czech Republic). The distribution of nanofiber diameters was calculated from the SEM images using ImageJ software 1.51W (NIH, Bethesda, MD, USA). We randomly selected 100 nanofibers from the SEM images of each nanofiber membrane sample for diameter measurement and analyzed and displayed the diameter distribution of the nanofibers.

2.4. Numerical Simulation of Flow Field

COMSOL 6.2 Multiphysics software was used to construct the gas flow field model of an auxiliary flow coaxial electrospinning for finite element simulation analysis, and the two-dimensional lines with different axis distances were parametrically scanned to obtain the relevant gas flow field velocity distribution data set.

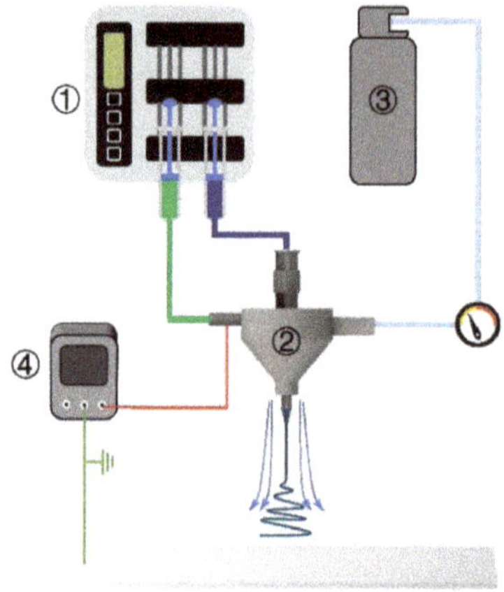

Figure 1. Schematic diagram of coaxial electrospinning by coaxial auxiliary flow field: ① double-channel pump, ② auxiliary flow coaxial electrospinning nozzle, ③ compressed gas, and ④ high-voltage power supply.

3. Results and Discussion

3.1. *Theoretical Research of Electrospinning Instability by Introducing Coaxial Gas Flow*

3.1.1. Simulation Analysis of Flow Field in Auxiliary Flow Coaxial Electrospinning

By using COMSOL 6.2 Multiphysics software, the gas flow field model of auxiliary flow coaxial electrospinning was constructed and simulated.

As shown in Figure 2a, the boundary conditions of the gas flow field are set. The gas flow import is set to the normal flow rate of 2 m/s. The gas flow export condition is set to 0 Pa. The resulting flow speed distribution cloud map and speed vector diagram are presented in Figure 2b–e. From Figure 2b–d, we can see that the overall flow rate gradually decreases along the negative direction of the Z axis from the nozzle, while the action range of the gas flow gradually increases. At the same time, it can be seen from Figure 2e that at the front of the needle, there was an obvious reverse flow phenomenon. This phenomenon had also been reported in the paper of the predecessors [24,25].

As shown in Figure 3, the gas flow is sprayed from the +z direction to the zero point. To create the coordinate system, the center point of the needle outlet is the center point, and +z and +r are defined as the positive directions of the velocity of the axial and radial flow fields, respectively.

In Figure 3, we define the distance with the axis of Dr, and two-dimensional lines were drawn in Dr = 0/0.125/0.25/0.375/0.5/0.625/0.75/0.875/1/1.25/1.5/1.75/2 mm. By collecting the highest point of axial/radial components of the velocity field on the two-dimensional line, the axial/radial components of the velocity field are analyzed statistically, as shown in Figure 4a,b.

From Figure 4a,b, it can be seen that the overall flow field rate at the diffusion boundary of the flow field gradually decreases with the diffusion. By fitting the data, the following relationship can be obtained:

$$\begin{cases} V_z \propto e^{-0.2z} V_o + 100 \\ V_r \propto 0.01 e^{-0.2z} V_o + 1 \end{cases} \tag{1}$$

where V_z (m/s) and V_r (m/s) represent the axial/radial components of the gas field velocity, respectively, V_0 (m/s) is the initial velocity of gas flow import, and z (mm) is the axial coordinate.

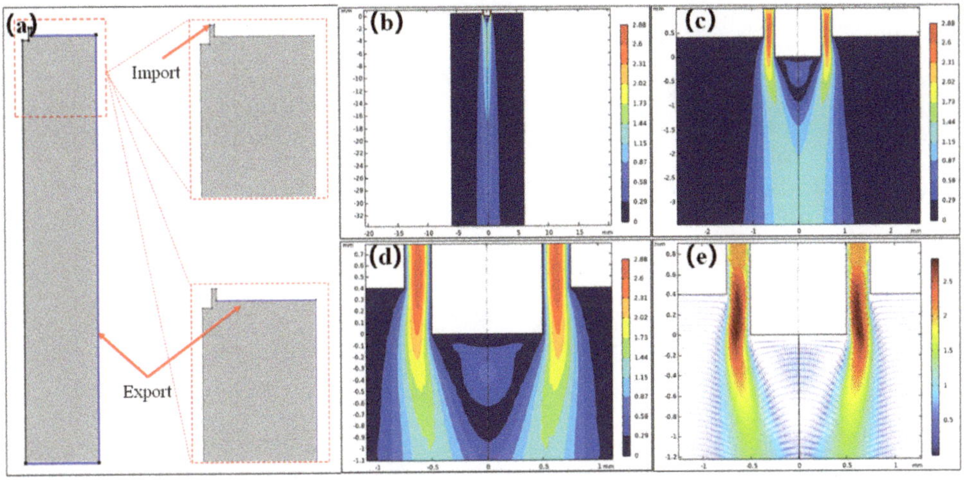

Figure 2. Diagram of flow field model establishment and simulation results: (**a**) boundary condition settings of the gas flow import and export; (**b**) overall velocity distribution cloud map; (**c**,**d**) local velocity distribution cloud map at different scales; (**e**) speed vector diagram of the needle tip.

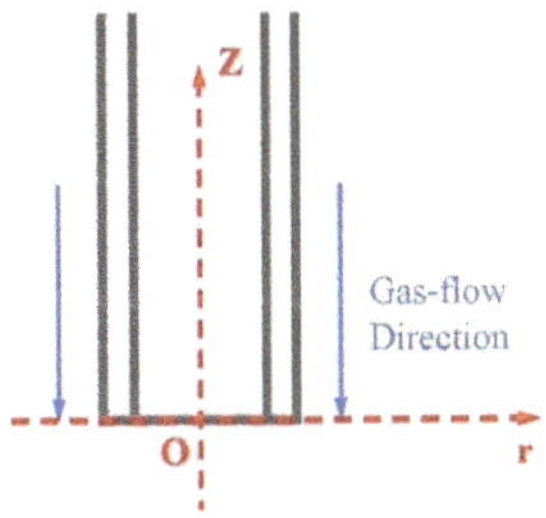

Figure 3. A coordinate system diagram of the gas flow at the needle.

3.1.2. Mechanical Modeling and Analysis of Auxiliary Flow Coaxial Electrospinning Jet

With reference to the one-dimensional steady model [26,27], the motion process of the electrospinning jet is analyzed. Under the action of an electric field, the charged jet accelerates along a straight line, and the viscous resistance increases. When the viscous resistance is greater than the electric field force, the jet enters the unstable stage, the acceleration of the jet becomes zero, and the small disturbance in the gas will cause the jet to deviate from the equilibrium position. Therefore, the motion radius of the jet is an important parameter to characterize the degree of instability.

In Figure 5a, the motion behavior of any particle M in the unsteady phase of the jet can be divided into linear motion in the vertical direction and approximate circular motion in the horizontal direction. Force analysis is carried out on particle M, as shown in Figure 5b.

Based on the centripetal force equation, Newton's second law of motion, and the relevant force decomposition rules, we can obtain Equation (2) as follows:

$$\begin{cases} F_n = \frac{mV^2}{R} \\ \tau = AV + BV^2 \\ \tau_1 = \tau \sin\theta \\ \tau_1 = F_n \end{cases} \quad (2)$$

where F_n represents the centripetal force of jet motion, m represents the mass of the jet particle, V represents the velocity of particle M, R represents the motion radius of the jet, τ represents viscous resistance, A and B are constants, τ_1 is the component force of τ in the radial direction, and θ is the angle between τ and τ_1.

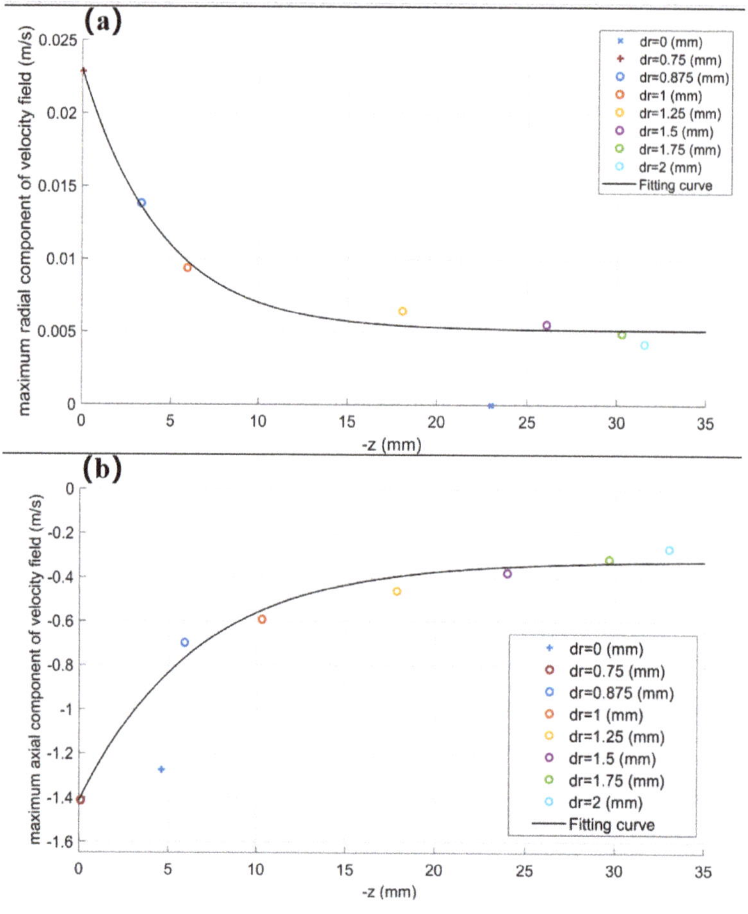

Figure 4. The distribution and fitting curve of the maximum axial/radial component of the velocity field in the axial direction: (**a**) variation diagram of maximum radial component of velocity field; (**b**) variation diagram of maximum axial component of velocity field.

From Yu-Qin Wan et al.'s conclusion [28], we can obtain Equation (3):

$$\begin{cases} V \propto r^{-2} \\ r \propto z^{-\frac{1}{4}} \end{cases} \quad (3)$$

where r is the motion radius of the jet, which is also R (mm) in this research, and z is the axial coordinate.

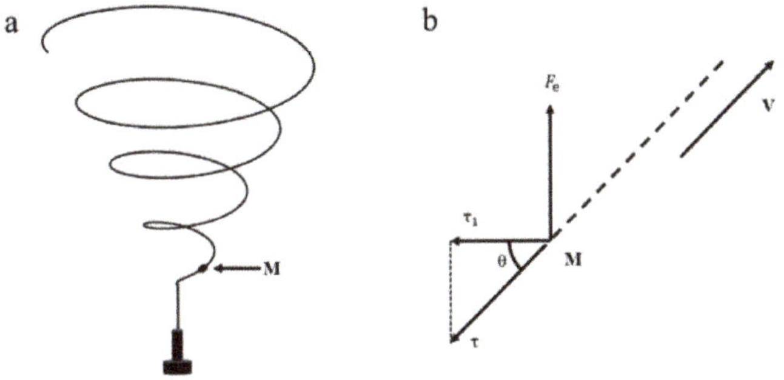

Figure 5. Force analysis of spinning jet without gas field: (**a**) electrospinning jet without gas field diagram; (**b**) force analysis diagram.

Equation (4) can be obtained by combining Equations (2) and (3).

$$R \propto \frac{1}{z^{-\frac{1}{2}}+1} \tag{4}$$

Therefore, Equation (5) gives the relationship between the radius of the motion of the jet (R) and the axial coordinate (z) without a gas field, and the trend diagram is shown in Figure 6. This relationship demonstrates that the fiber after the draft of the jet has a fixed deposition region and the size of the region is related to the axial coordinate. This conclusion is consistent with the experimental process captured by Reneker et al. through high-speed cameras [29].

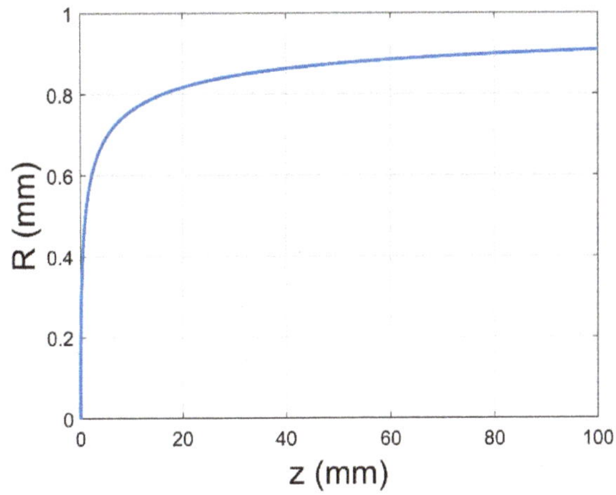

Figure 6. The trend diagram of the relationship between the jet radius (R) and the axial coordinate (z) without a gas field.

When the gas field force is involved, particle M is affected not only by viscous resistance and electric field force, but also by gas field force. Here, the magnitude and direction of viscous resistance and electric field force remain unchanged, as shown in Figure 7.

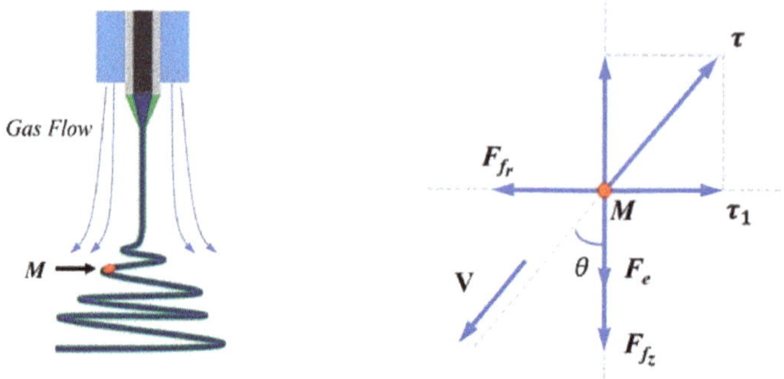

Figure 7. Force analysis of spinning jet with gas field.

Similarly to Equation (2), we can obtain Equation (5) as follows:

$$\begin{cases} F_n = \frac{mV^2}{R_0} \\ \tau = AV + BV^2 \\ \tau_1 = \tau \sin\theta \\ F_{f_r} = \frac{1}{2}\rho S {V_r}^2 \\ \tau_1 - F_{f_r} = F_n \end{cases} \tag{5}$$

where R_0 (mm) represents the ideal motion radius of the jet, F_{f_r} represents the radial component of the gas field force, ρ and S are constants, and V_r represents the radial component of the gas field velocity.

Equation (6) can be obtained by combining Equations (4) and (5).

$$R_0 \propto \frac{1}{z^{-\frac{1}{2}} + 1 - \frac{1}{2}{V_r}^2 z^{-1}} \tag{6}$$

Therefore, from Equation (6), we can know the relationship between the ideal radius of the motion of the jet and the axial coordinate and the radial component of the gas field velocity after the introduction of the gas field. The ideal radius of the motion of the jet continues to increase either when increasing the axial coordinate or radial component of the gas field velocity.

Meanwhile, combining these results with Equation (1), we can obtain the trend in R_0 with V_0 as shown in Figure 8. However, R_0 is the deposition radius in the ideal situation without the consideration of the spinning height limitation. And this relation equation cannot correctly describe the relationship between the real deposition radius R and V_0. Therefore, by reintroducing the influence brought by the axial physical field, we can obtain Equations (7) and (9) as follows.

When the axial wind field component is not considered,

$$\begin{cases} z = \frac{1}{2}\bar{a}\bar{t}^2 \\ \tau = AV + BV^2 \\ \tau_2 = \tau \cos\theta \\ F_e - \tau_2 = m\bar{a} \end{cases} \tag{7}$$

where z is the axial coordinate, which is also the spinning distance, $\bar{a}$ represents the axial acceleration without consideration of the axial gas field component force, $\bar{t}$ (s) represents the motion time without consideration of the axial gas field component force, τ_2 is the component force of τ in the axial direction, and F_e (N) represents the electric field force.

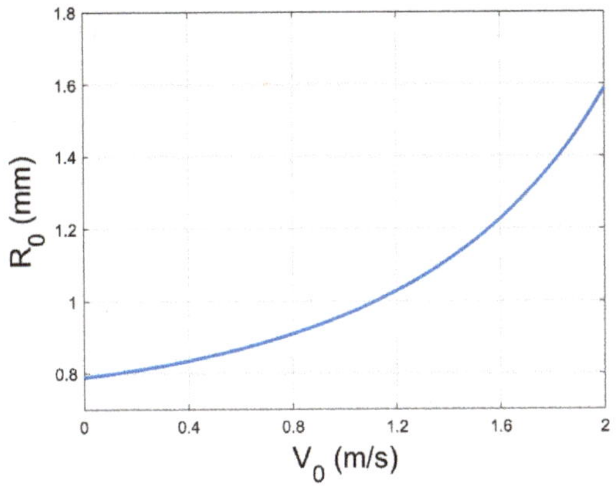

Figure 8. The trend diagram of the relationship between the jet radius R0 and the initial velocity of gas flow V_0 considering the radial component of the gas field.

Equation (8) can be obtained by combining Equations (4) and (7).

$$\bar{t} \propto \sqrt{\frac{2}{F_e z^{-1} - z^{-\frac{1}{2}} - 1}} \tag{8}$$

When the influence of the axial gas field component force is considered,

$$\begin{cases} z = \frac{1}{2}at^2 \\ \tau = AV + BV^2 \\ \tau_2 = \tau\cos\theta \\ F_{f_z} = \frac{1}{2}\rho S {V_z}^2 \\ F_e + F_{f_z} - \tau_2 = ma \end{cases} \tag{9}$$

where a represents the axial acceleration when considering the axial gas field component force, $t(\mathrm{s})$ represents the jet motion time after considering the axial gas field component force, F_{f_z} represents the axial component of the gas field force, and V_z represents the axial component of the gas field velocity.

Equation (10) can be obtained by combining Equations (4) and (9).

$$t \propto \sqrt{\frac{2}{F_e z^{-1} + \frac{1}{2}{V_z}^2 z^{-1} - z^{-\frac{1}{2}} - 1}} \tag{10}$$

And then, we set a change rate for the jet radius as $\overline{R}$ (mm/s), obtained from Equation (11).

$$\overline{R} = \frac{R_0}{\bar{t}} \tag{11}$$

Therefore, the radius of the jet after considering the effects of radial/axial physical field can be obtained from Equation (12).

$$R = \overline{R}t \tag{12}$$

Finally, by coupling the above equations, the following can be obtained:

$$R \propto \sqrt{\frac{F_e Z^{-1} - (z^{-\frac{1}{2}} + 1)}{\left[F_e z^{-1} + \frac{1}{2}{V_z}^2 z^{-1} - (z^{-\frac{1}{2}} + 1)\right]\left[(z^{-\frac{1}{2}} + 1) - \frac{1}{2}{V_r}^2 z^{-1}\right]^2}} \tag{13}$$

By combining Equation (1), we can obtain a trend in R with V_0 as shown in Figure 9.

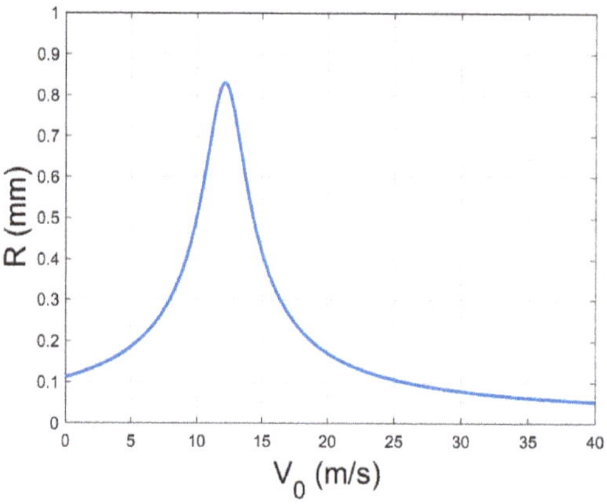

Figure 9. The trend diagram of the relationship between the jet radius R and the initial velocity of gas flow V_0 considering the radial component of the gas field.

Consequently, the jet, which was originally only affected by the vertical electric field force, was affected by the axial and radial wind forces due to the intervention of the gas field, making the fiber deposition break through the original natural area and the deposition circle larger, which we believe is one of the keys to make the fiber deposition more uniform. However, the axial gas field force causes the jet to accelerate towards the collector, thereby imposing a constraint on the continued increase in the diameter of the deposition circle. Therefore, with an increase in the gas field flow rate, the area of the fiber deposition will demonstrate a tendency to increase and then decrease. So far, the mechanical model of the single nozzle case has been fully described. When the mass production of multiple nozzles is really carried out, the related effects of the flow field and electric field between the nozzles must be considered in the subsequent work.

3.2. *Experimental Verification*

3.2.1. Effect of Coaxial Gas Flow on Coaxial Fiber Deposition Area and Uniformity

By setting different auxiliary gas pressures, the results of fiber deposition can be achieved, as shown in Figure 10a–f. By measuring the diameter and thickness of the fiber deposition circle, we obtained the scatter diagram of the diameter distribution and thickness distribution of the fiber deposition circle under different pressures, as shown in Figure 10g,h.

In Figure 10a–g, it can be clearly observed that as the gas pressure increases (0 MPa → 0.05 MPa), the diameter of the fiber deposition circle presents a tendency to increase and then decrease (5 cm/0 MPa → 10 cm/0.03 MPa → 7 cm/0.05 MPa). Compared with the results without auxiliary gas flow, the diameter of the fiber deposition circle we are most concerned about increases from 5 cm (0 MPa) to 10 cm (0.03 MPa) at most, which means that the fiber deposition area has increased by three times (19.63 cm^2 → 78.50 cm^2). Then, according to Bernoulli's principle, we can see that, with other parameters constant, the gas pressure is positively correlated with the initial flow rate of the nozzle, i.e., $V_0 \propto P$. So, this experimental result verifies the relationship between the real deposition radius R and the initial velocity of gas flow V_0.

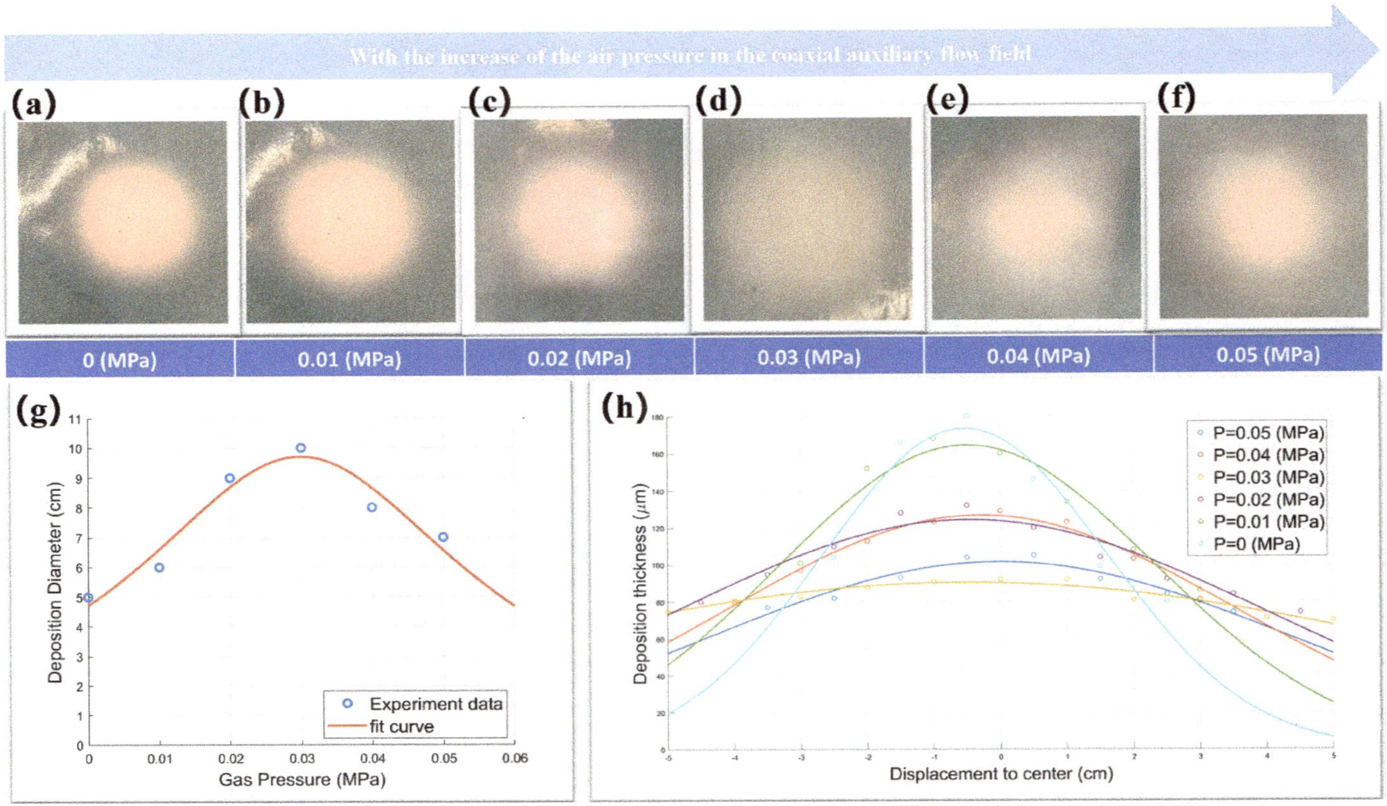

Figure 10. Fiber deposition effect under different gas pressures: (**a**–**f**) fiber deposition circle diagram; (**g**) diameter change and fitting curve of the fiber deposition circle; (**h**) comparison of scatter maps and fitting curves of fiber deposition thickness.

At the same time, according to Figure 10h, it can be seen that when the gas pressure is 0.03 MPa, the diameter of the fiber deposition circle is the largest and the thickness uniformity is the best. In order to further quantify the thickness uniformity, the thickness distribution data of the fiber deposition circle under different gas pressures were simulated by Gaussian fitting, and six fitting curves were compared and analyzed.

The one-dimensional form of the standard Gaussian is shown in Equation (14):

$$y = ae^{-\frac{(x-b)^2}{2c^2}} \tag{14}$$

where a represents the height of the peak of the curve, b represents the coordinates of the center of the peak, and c represents the standard variance.

As can be seen from Figure 10h, compared to without auxiliary gas flow, the standard variance, which we are most concerned about, increases from 3 μm^2 (0 MPa) to 10 μm^2 (0.03 MPa) at the highest, which means that the thickness uniformity increases by 2.34 times.

3.2.2. Effect of Coaxial Gas Flow on Coaxial Fiber Morphology and Diameter Distribution

Due to the addition of coaxial gas flow, the jet fibrosis environment may change. Therefore, we conducted a statistical analysis on fiber morphology and diameter distribution before and after the intervention of coaxial gas flow.

According to Figure 11a,b, when the gas pressure is lower than 0.05 MPa, the fiber morphology basically maintains a good fiber shape, without the appearance of beads and lumps. And even the fiber diameter becomes thinner, because the addition of gas flow increases the tractive force on the jet, which is conducive to the thinning of the fiber. However, when the gas pressure reached 0.05 MPa, the fiber deposition showed obvious lump and fiber adhesion, which was because the excessive gas pressure made the fiber reach the collector too fast, and the fiber curing was not sufficient. In order to further quantify the change in fiber diameter distribution, Gaussian fitting was performed on fiber diameter distribution data under different gas pressures, and six fitting curves and their average diameters were compared and analyzed, as shown in Figure 12.

$$y = y_0 + \frac{A}{w\sqrt{\frac{pi}{2}}}e^{-2*(\frac{x-x_c}{w})^2} \tag{15}$$

By comparing Equations (14) and (15), it can be seen that a is positively correlated with $\frac{A}{w}$, and b and c are positively correlated with x_c and w, respectively. And y_0 is an auxiliary fitting parameter, so this term will not be compared later.

As can be seen from Figure 12a,b, with the increase in gas pressure, the average fiber diameter decreases first and then increases (241 nm/0 MPa $\rightarrow$ 152 nm/0.03 MPa $\rightarrow$ 252 nm/0.05 MPa). Compared to the results without auxiliary gas flow, the average fiber diameter was reduced from 241 nm $\pm$ 50 nm (0 MPa) to 152 nm $\pm$ 23 nm (0.03 MPa), representing a 36.9% reduction in fiber diameter. The standard variance of the fiber diameter distribution can be reduced from 1.5×10^2 nm^2 (0 MPa) to 51 nm^2 (0.03 MPa), which means that the fiber diameter distribution range is reduced by 66%. Unexpectedly, the fiber diameter can be more concentrated in a specific range.

Therefore, by adding suitable coaxial auxiliary gas flow, the fiber deposition area can be increased by 3 times, and the uniformity of fiber deposition thickness can be increased by 2.34 times. At the same time, because the gas flow increases the tractive force on the jet, suitable coaxial auxiliary gas flow can reduce the average diameter of the fiber by 36.9% and the standard variance of the fiber distribution by 65.1%.

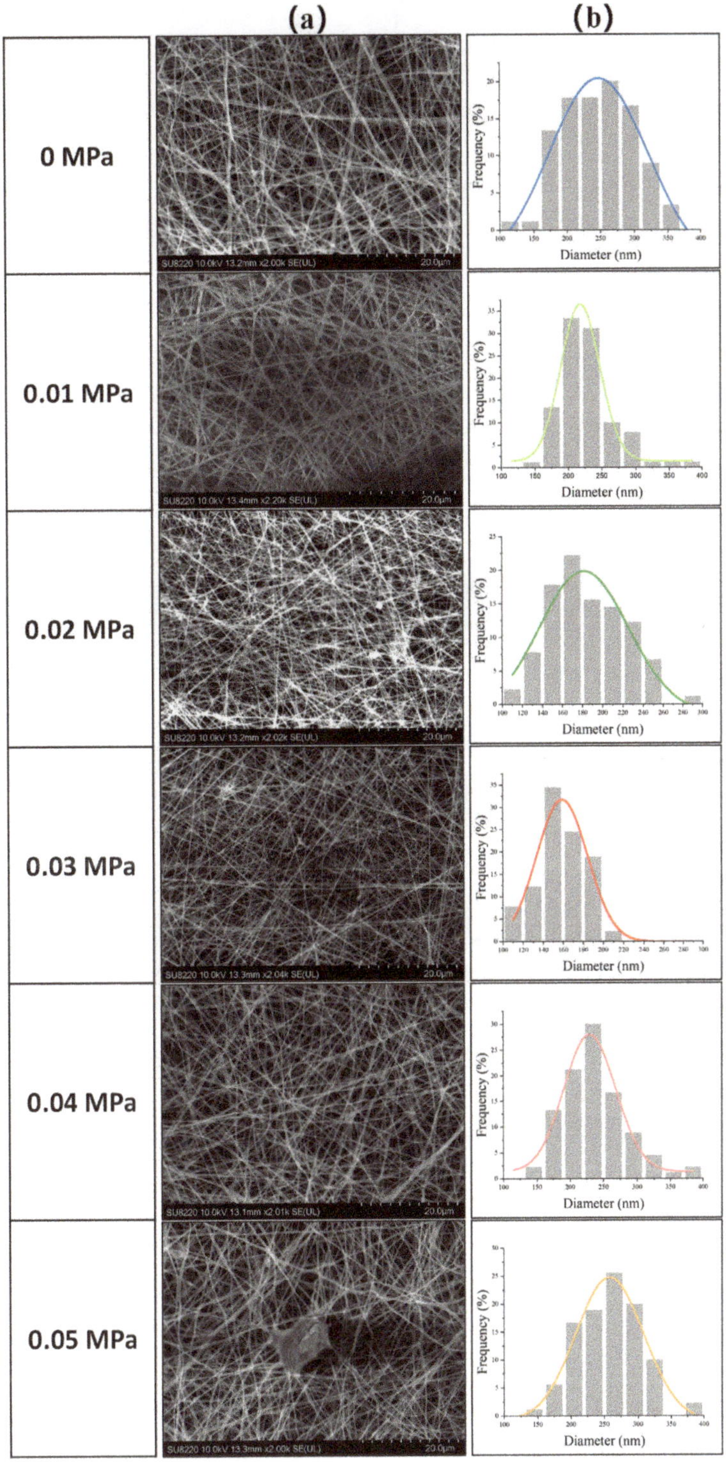

Figure 11. Electrospinning effect under different gas pressures: (**a**) SEM image of fiber morphology; (**b**) nanofiber diameter distribution graph and fitting curve.

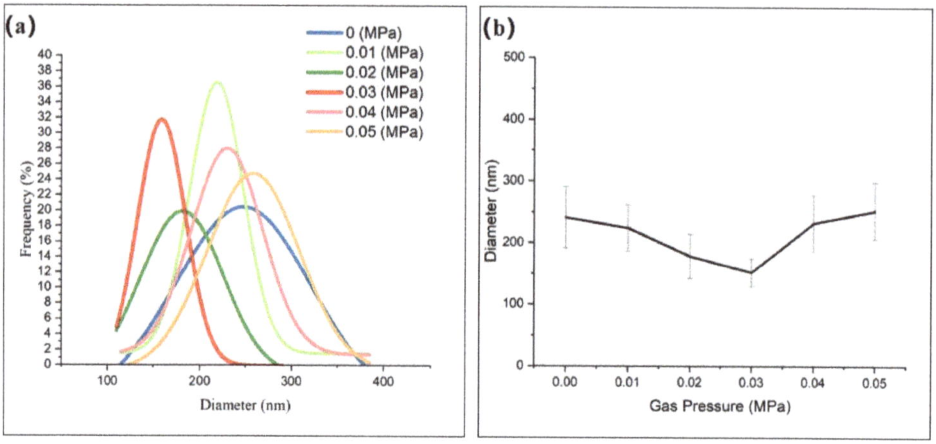

Figure 12. Quantitative analysis of electrospinning effect under different gas pressures: (**a**) the fitting curve of the nanofiber diameter distribution; (**b**) comparison of average diameters.

4. Conclusions

In this research, we simulate the gas flow field in the process of gas-assisted coaxial electrospinning by finite element analysis and establish the jet mechanics model of this process. Through a series of formula derivation and experiments, the influence mechanism of coaxial auxiliary gas flow on the fiber deposition area and its uniformity has been successfully revealed and verified. Finally, in the experiment of gas-assisted coaxial electrospinning, the deposition area and thickness uniformity of coaxial fibers are increased by 3 times (19.63 $cm^2 \rightarrow$ 78.50 cm^2) and 2.34 times (3 $\mu m^2 \rightarrow$ 10 μm^2), respectively. At the same time, the average diameter and distribution range of coaxial fibers decreased by 36.9% (241 nm $\pm$ 50 nm $\rightarrow$ 152 nm $\pm$ 23 nm) and 66% (1.5×10^2 $nm^2 \rightarrow$ 51 nm^2), respectively. Therefore, the addition of a coaxial gas field can effectively increase the deposition area of the coaxial fiber, improve the uniformity of coaxial fiber deposition, refine fiber diameter, and make fiber distribution more concentrated, which provides theoretical guidance and experimental basis for the uniform preparation of coaxial electrospinning fibers. In a follow-up study, we focused on the large-scale preparation of coaxial fibers by gas-assisted coaxial electrospinning. There are still, at least, the following problems that have to be overcome: ① the development of a multi-nozzle module with uniform liquid and gas supply; ② the improvement of gas flow environmental impacts and structural optimization of the multi-nozzle module; and ③ the development of a coaxial nanofiber quality evaluation system and related equipment for large-scale preparation.

Author Contributions: R.Z.: methodology, validation, investigation, data curation, visualization, writing—original draft. X.C.: supervision, conceptualization, writing—reviewing and editing. H.W.: funding acquisition, supervision, conceptualization, writing—reviewing and editing. J.S.: methodology, resources, writing—reviewing and editing. S.H.: investigation, data curation, visualization. X.Z.: investigation, data curation, visualization. J.L.: investigation, visualization. All authors have read and agreed to the published version of the manuscript.

Funding: This research was funded by the National Natural Science Foundation of China, grant numbers 52475435, 52305456, and U20A6004; the Science and Technology Projects in Guangzhou, grant number 2023A04J1610; and the Postdoctoral Fellowship Program (Grade C) of the China Postdoctoral Science Foundation under grant number GZC20230569.

Institutional Review Board Statement: Not applicable.

Data Availability Statement: The data presented in this study are available on request from the corresponding author. The data are not publicly available due to [Some of the data relate to the core information of production].

Acknowledgments: We would like to offer special thanks to Han Wang for giving us access to the scanning electron microscopy.

Conflicts of Interest: The authors declare no conflicts of interest.

References

1. Zhi, D.; Li, T.; Qi, Z.; Li, J.; Tian, Y.; Deng, W.; Meng, F. Core-shell heterogeneous graphene-based aerogel microspheres for high-performance broadband microwave absorption via resonance loss and sequential attenuation. *Chem. Eng. J.* **2022**, *433*, 134496. [CrossRef]
2. Wang, Y.; Liu, L.; Zhu, Y.; Wang, L.; Yu, D.G.; Liu, L.Y. Tri-Layer Core–Shell Fibers from Coaxial Electrospinning for a Modified Release of Metronidazole. *Pharmaceutics* **2023**, *15*, 2561. [CrossRef] [PubMed]
3. Lu, Y.; Xiao, X.; Fu, J.; Huan, C.; Qi, S.; Zhan, Y.; Zhu, Y.; Xu, G. Novel smart textile with phase change materials encapsulated core-sheath structure fabricated by coaxial electrospinning. *Chem. Eng. J.* **2019**, *355*, 532–539. [CrossRef]
4. Pant, B.; Park, M.; Park, S.-J. Drug Delivery Applications of Core-Sheath Nanofibers Prepared by Coaxial Electrospinning, A Review. *Pharmaceutics* **2019**, *11*, 305. [CrossRef] [PubMed]
5. Han, D.A.J. Coaxial Electrospinning Formation of Complex Polymer Fibers and their Applications. *ChemPlusChem* **2019**, *84*, 1453–1497. [CrossRef]
6. Bognitzki, M.; Hou, H.; Ishaque, M.; Frese, T.; Hellwig, M.; Schwarte, C.; Schaper, A.; Wendorff, J.H.; Greiner, A. Polymer, metal, and hybrid nano and mesotubes by coating degradable polymer template fibers (TUFT process). *Adv. Mater.* **2000**, *12*, 637–640. [CrossRef]
7. Liu, G.; Qiao, L.; Guo, A. Diblock copolymer nanofibers. *Macromolecules* **1996**, *29*, 5508–5510. [CrossRef]
8. Jang, J.; Lim, B.; Lee, J.; Hyeon, T. Fabrication of a novel polypyrrole/poly (methyl methacrylate) coaxial nanocable using mesoporous silica as a nanoreactor. *Chem. Commun.* **2001**, *1*, 83–84. [CrossRef]
9. Rashedi, S.; Afshar, S.; Rostamic, A.; Ghazalian, M.; Nazockdaste, H. Co-electrospinning poly (lactic acid)/gelati nanofibrous scaffold prepared by a new solvent system, morphological, mechanical and in vitro degradability properties. *Int. J. Polym. Mater. Polym. Biomater.* **2021**, *70*, 545–553. [CrossRef]
10. Mahmood, K.; Kamilah, H.; Karim, A.A.; Ariffin, F. Enhancing the functional properties of fish gelatin mats by dual encapsulation of essential oils in β-cyclodextrins/fish gelatin matrix via coaxial electrospinning. *Food Hydrocoll.* **2023**, *137*, 108324. [CrossRef]
11. Anaya Mancipe, J.M.; Boldrini Pereira, L.C.; de Miranda Borchio, P.G.; Dias, M.L.; da Silva Moreira Thiré, R.M. Novel polycaprolactone (PCL)-type I collagen core-shell electrospinning nanofibers for wound healing applications. *J. Biomed. Mater. Res. Part B Appl. Biomater.* **2023**, *111*, 366–381. [CrossRef] [PubMed]
12. Ji, W.; Yang, F.; Beucken, J.J.v.D.; Bian, Z.; Fan, M.; Chen, Z.; Jansen, J.A. Fibrous scaffolds loaded with protein prepared by blend or coaxial electrospinning. *Acta Biomater.* **2010**, *6*, 4199–4207. [CrossRef] [PubMed]
13. Movahedi, M.; Asefnejad, A.; Rafienia, M.; Khorasani, M.T. Potential of novel electrospinning core-shell structured polyurethane/starch (hyaluronic acid) nanofibers for skin tissue engineering, In vitro and in vivo evaluation. *Int. J. Biol. Macromol.* **2020**, *146*, 627–637. [CrossRef]
14. Vysloužilová, L.; Valtera, J.; Pejchar, K.; Beran, J.; Lukáš, D. Design of coaxial needleless electrospinning electrode with respect to the distribution of electric field. *Appl. Mech. Mater.* **2014**, *693*, 394–399. [CrossRef]
15. Vysloužilová, L.; Buzgo, M.; Pokorný, P.; Chvojka, J.; Míčková, A.; Rampichová, M.; Kula, J.; Pejchar, K.; Bílek, M.; Lukáš, D.; et al. Needleless coaxial electrospinning, A novel approach to mass production of coaxial nanofibers. *Int. J. Pharm.* **2017**, *516*, 293–300. [CrossRef]
16. Buzgo, M.; Mickova, A.; Rampichova, M.; Doupnik, M. Blend electrospinning, coaxial electrospinning, and emulsion electrospinning techniques. In *Core-shell Nanostructures for Drug Delivery and Theranostics*; Woodhead Publishing Series in Biomaterials: Duxford, UK, 2018; pp. 325–347. [CrossRef]
17. Batka, O.; Skrivanek, J.; Beran, J. Optimizing the Shape of the Spinning Electrode for Needleless Coaxial Electrospinning. *Appl. Sci.* **2024**, *14*, 4638. [CrossRef]
18. Magaz, A.; Roberts, A.D.; Faraji, S.; Nascimento, T.R.L.; Medeiros, E.S.; Zhang, W.; Greenhalgh, R.D.; Mautner, A.; Li, X.; Blaker, J.J. Porous, Aligned and Biomimetic Fibers of Regenerated Silk Fibroin Produced by Solution Blow Spinning. *Biomacromolecules* **2018**, *19*, 4542–4553. [CrossRef]
19. Tandon, B.; Kamble, P.; Olsson, R.T.; Blaker, J.J.; Cartmell, S.H. Fabrication and Characterisation of Stimuli Responsive Piezoelectric PVDF and Hydroxyapatite-Filled PVDF Fibrous Membranes. *Molecules* **2019**, *24*, 1903. [CrossRef]

20. Hwang, S.H.; Song, J.Y.; Ryu, H.I.; Oh, J.H.; Lee, S.; Lee, D.; Park, D.Y.; Park, S.M. Correction to: Adaptive Electrospinning System Based on Reinforcement Learning for Uniform-Thickness Nanofiber Air Filters. *Adv. Fiber Mater.* **2023**, *5*, 695. [CrossRef]
21. Zhang, R.; Chen, X.; Wang, H.; Zeng, J.; Zhang, X.; Chen, X. Study on uniformity of multi-needle electrostatic spinning by auxiliary flow field. *Micro Nano Letters.* **2024**, *19*, e12200. [CrossRef]
22. Ryu, H.I.; Koo, M.S.; Kim, S.; Kim, S.; Park, Y.A.; Park, S.M. Uniform-thickness electrospinning nanofibers mat production system based on real-time thickness measurement. *Sci. Rep.* **2020**, *10*, 20847. [CrossRef] [PubMed]
23. Shao, Z.; Wang, Q.; Gui, Z.; Shen, R.; Chen, R.; Liu, Y.; Zheng, G. Electrospun bimodal nanofibrous membranes for high-performance, multifunctional, and light-weight air filtration, A review. *Sep. Purif. Technol.* **2024**, *358*, 130417. [CrossRef]
24. Atif, R.; Combrinck, M.; Khaliq, J.; Hassanin, A.H.; Shehata, N.; Elnabawy, E.; Shyha, I. Solution Blow Spinning of High-Performance Submicron Polyvinylidene Fluoride Fibres, Computational Fluid Mechanics Modelling and Experimental Results. *Polymers* **2020**, *12*, 1140. [CrossRef] [PubMed]
25. Xu, G.; Chen, M.; Gao, Y.; Chen, Y.; Luo, Z.; Wang, H.; Fan, J.; Luo, J.; Ou, W.; Zeng, J.; et al. Gas-assisted electrospinning of high-performance ceramic fibers, Optimal design modelling and experimental results of the gas channel of the nozzle. *Front. Mater.* **2023**, *10*, 1113168. [CrossRef]
26. Wan, Y. Research on Electrospinning Process Behavior and Vibration Electrospinning Technology. Ph.D. Thesis, Donghua University, Shanghai, China, 2006.
27. Wu, Y. Research on the Jet Instability of Magnetio-Electrospinning. Ph.D. Thesis, Donghua University, Shanghai, China, 2008.
28. Zhao, Y.; Jiang, J.; Li, W.; Wang, X.; Zhang, K.; Zhu, P.; Zheng, G. Electrospinning jet behaviors under the constraints of a sheath gas. *AIP Adv.* **2016**, *6*, 115022. [CrossRef]
29. Reneker, D.H.; Yarin, A.L.; Fong, H.; Koombhongse, S. Bending Instability of Electrically Charged Liquid Jets of Polymer Solutions in Electrospinning. *J. Appl. Phys.* **2000**, *87*, 4531–4547. [CrossRef]

polymers

MDPI

Article

Synthesis of Jicama (*Pachyrhizus erosus*) Starch Particles by Electrospraying: Effect of the Hydrolysis Degree

Fatima Sarahi Serrano-Villa [1], Eduardo Morales-Sánchez [2], José Alfredo Téllez-Morales [1], Verónica Cuellar-Sánchez [1], Reynold R. Farrera-Rebollo [1] and Georgina Calderón-Domínguez [1,*]

1 Departamento de Ingeniería Bioquímica, Escuela Nacional de Ciencias Biológicas, Instituto Politécnico Nacional, Ciudad de México 07738, Mexico; fserranov2001@alumno.ipn.mx (F.S.S.-V.)
2 Centro de Investigación en Ciencia Aplicada y Tecnología Avanzada, Instituto Politécnico Nacional, Querétaro 76090, Mexico; emoraless@ipn.mx
* Correspondence: gcalderon@ipn.mx

Abstract: Electrohydrodynamic atomization (EHDA) has significant advantages for microencapsulating compounds in various structures using biopolymers, where more research using pure starch is required. Concerning this, jicama starch and its hydrolysates have not yet been tested, despite their unique characteristics, which come from an alternative low-value-added crop source. Rapid acid hydrolysis of jicama starch with H_2SO_4 resulted in dextrins with a degree of hydrolysis (DE) from 0.4 to 19% within 1–12 h, and syrup solids at 24 h (DE = 42%). This process modifies the water retention capacity of jicama starch, gel viscosity, surface tension, and electrical conductivity. Hydrolyzed starch particles obtained by electrospraying (10 kV, L = 10 cm, Q = 2 mL/h) showed Feret diameters and roundness significantly influenced ($p \leq 0.05$) by the degree of hydrolysis rather than the concentration of solids. It was found that hydrolyzed jicama starch with a DE < 6.3% can be used as the sole wall material to form particles by electrospraying, as they facilitate the formation of stable and rounded like-microspheres particles; this was not feasible above this threshold. The results suggest that the jicama starch's ability to be used as a wall material in the electrospray synthesis of particles or microspheres appears to be determined by the degree of hydrolysis.

Keywords: electrospraying; jicama starch; acid hydrolysis; particle synthesis; thermal; rheological and structural starch properties

1. Introduction

In the food industry, it is common practice to include functional compounds in formulations, looking to develop foods with different technological or nutraceutical properties. For example, probiotics, as well as other compounds with nutritional, therapeutic, and protective purposes, or those that improve food sensory desirable characteristics such as texture, flavor, aroma, and color, among others, are commonly included [1]. In this regard, microencapsulation (including microspheres and microcapsules) is commonly used to fulfill this purpose [2,3].

There are different techniques for preparing microcapsules, including electrohydrodynamic atomization (EHDA), spray drying, extrusion, freeze drying, ultra-high pressure, coagulation, fluidized bed coating, molecular inclusion, chemical, and enzymatic methods [4]. These techniques have disadvantages, such as the use of high temperatures and/or the need for more complex equipment and materials [5].

Compared to other techniques, EHDA, also known as electrospraying, has significant advantages such as ease of operation, low cost, straightforward structural adjustments, and the ability to incorporate nutrients and complex molecules easily [6] in the absence of heat treatment, gaining special importance in the food and pharmaceutical areas during the last decade [7,8].

The process's behavior and the electrospraying product's final morphology are influenced by the interaction of Coulombic, inertial, electric field, viscoelastic, and surface tension forces [9]. Consequently, electrohydrodynamic atomization can generate different structures based on process conditions (applied voltage, flow rate, distance to the collector, equipment configuration), the properties of the raw materials used in the polymeric solution including the type, structure, and concentration of the polymer, as well as electrical conductivity, surface tension, and viscosity, among others [10,11]. Cuellar-Sánchez et al. [9] cited that EHDA products can be synthesized with raw materials from different origins, among which proteins (gelatin, pectin, zein) and polysaccharides (chitosan, alginate, starch) are widely used.

Starch is a biopolymer formed by α-D-glucose units linked by α (1 → 4) and α (1 → 6) glycosidic bonds, arranged in two types of high molecular weight molecules: amylose and amylopectin [12]. The internal structure of starch granules is described by the growth rings model, which is composed of alternating semi-crystalline (amylopectin) and amorphous (amylose) regions [13]. Starch is used as a standard wall material in the food industry to produce delivery systems for functional compounds, nutrition enhancers, and food preservatives, among others [4]. Starch, as wall material for encapsulation technology, has advantages such as being biocompatible, stable, widely available, low cost, tasteless, completely biodegradable, and can be modified into derivatives with different desired properties by functionalization or alteration of its granular structure [5].

The methods used to modify the physicochemical and functional properties of starch include physical, biochemical, genetic, and chemical mechanisms or a combination of them [14]. Among these methods, starch gelatinization is a physical modification that allows the creation of an encapsulating material, which envelopes the compound of interest when starch gels retrograde [15]. On the other hand, acid hydrolysis is a chemical modification and the simplest and most common method applied to modify starch [16,17]. Acid hydrolysis produces different starch structures and derivatives as a function of the process conditions applied, which include temperature, hydrolysis duration, acid concentration, type of acid, agitation speed, and starch structural characteristics [18].

Starch treatment with hydrochloric or sulfuric acid has been shown to cause depolymerization of the amylose and amylopectin chains [19,20]. Thus, dextrins, maltodextrins, solid syrups [21], and nanocrystals are produced by varying parameters such as hydrolysis time, acid type, and concentration, among others [15,22]. Consequently, hydrolyzed starch exhibits an increase in the enthalpy change (ΔH), as well as in the temperature rank of gelatinization endotherms (or the peak gelatinization temperature). This treatment also increases starch solubility and crystallinity, swelling power, and a loss of pasting viscosity [13,17].

Starch is commonly extracted from vegetables, with maize and potatoes serving as the primary sources for industrial use, and at a lower proportion, tuberous roots [9,12]. Over the past decade, the interest in utilizing non-conventional sources to produce starch has increased, particularly agro-industrial residues or low-added-value crops that do not compete as a crucial food source for human populations and exhibit different properties [23]. In this context, jicama root (*Pachyrhizus erosus*) is an example. It contains significant amounts of starch (40–60% on a dry basis) at commercial maturity [24].

Jicama starch presents polygonal, spherical, and irregular granules with a smooth surface, ranging in size from 3 to 21 μm [25]. Its structural arrangement is predominantly of the CA type polymorphism, with a relative crystallinity of 29 to 44%. Regarding its chemical composition, it has been reported to present a variable content of amylose (11–25%) depending on the genetic variety [23] and less than 1% of proteins, lipids, and ashes, in which calcium, potassium, and phosphate are higher than in potato starch [24,25].

To our knowledge, jicama starch has not been tested in the electrohydrodynamic technology. As a wall material of importance in the electrospraying process, starch is commonly used as a modified compound. Cuellar-Sánchez et al. (2022) [9] cited in this regard that gelatinized and chemically modified starches are a very common material for capsule and fiber synthesis. However, studies using starch as the only polymer are limited and require further investigation in the food field [26]. Research on the formation of starch-based capsules and particles from electrohydrodynamic spraying techniques has focused on the use of modified maize starch, such as resistant [27–29] debranched [30], resistant maltodextrins [31,32], and octenyl succinated starch [33]. The studies often involve the use of copolymers, including gums [31], chitosan, and alginate [29], among others, and surfactants like Tween 80 [32], Span 20, and lecithin [31] to enhance the solution properties to be electrosprayed.

Physical properties of the starch solutions to be electrosprayed, such as viscosity, polymer concentration, and molecular weight of the polymer, play important roles [9]. For example, in some studies, solutions based on potato starch solubilized in formic acid were processed to enhance electrospinning fiber production [34,35]. The authors attribute this to starch formylation, its chemical gelatinization, and a reduction in its molecular weight through hydrolysis, which also modifies the viscosity and influences the role of surface tension.

We hypothesize that the properties of starch solutions can be modified by acid hydrolysis with sulfuric acid to improve the EHDA process. This acid is known to produce more specific cleavages than HCl [14,36], and it could form sulphate esters [22] that allow it to interact with compounds of food interest to be encapsulated, such as proteins or enzymes like glucose oxidase in breadmaking. The present work aims to evaluate the effect of acid hydrolysis on the preparation of microspheres from jicama (*Pachyrhizus erosus*) starch using simple electrohydrodynamic atomization.

2. Materials and Methods

2.1. Materials

Jicama tubers (*P. erosus* L. Urban, variety) from San Juan, Nayarit, Mexico, were used. All reagents were analytical grade. Sodium bisulfite (3556-20) and phenol (2058-01) were bought from JT Baker, acetone (6016) from Fermont, and sulfuric acid (EMSURE®, 1.00731), D-glucose (G8270), Tween 80 (10099956), and glycerol (G9012) from Sigma-Aldrich. The enzymatic kits K-TSTA-100A 02/22 and K-AMYL 06/18 were supplied by Megazyme. Starch cup (998118), SPC rotor (999188), and DIN SST (999292) geometries used in RVA analysis and flow curves were supplied by TA Instruments. Finally, distilled water was used in all experiments and was produced in a local potabilization plant.

2.2. Starch Extraction

The method proposed by González-Lemus et al. [25] was followed, in which the jicama tuber was rinsed and cut into cubes (1 cm^3); it was then soaked for 30 min in a 1500 ppm solution of sodium bisulfite (3556-20, J.T. Baker, Avantor Performance Materials, Center Valley, PA, USA) in a 1:3, root-solution (w/v) ratio. The jicama cubes were placed in a plastic bag, immersed in a dry ice/acetone bath for 10 min, and stored at −24 °C for 24 h.

The cubes were then thawed, defrosted, and stored in a dry ice/acetone bath. They were then thawed (12 h, 25 °C) and ground (jicama-water 1:3 *w/v*) in an Oster PS blender for 60 s. The resulting suspension was processed in an ultrasonic bath (Cole-Parmer 08895-69, 40 kHz, Vernon Hills, IL, USA) for 10 min. It was then filtered through a plastic mesh 250 μm (Monyl, Sefar Nytal PA, Switzerland) and centrifuged (Dynamica Velocity 18R, Metrix, UK) at 4500× *g* for 10 min at 4 °C. The supernatant was removed, and the precipitate dried at 30 °C for 24 h (AFOS Ltd., Afos Mini-Klean, Hull, UK).

2.3. Starch Acid Hydrolysis

The hydrolysis was carried out following the methodology outlined by López-Hernández et al. (2022) [36], with some adaptations: Starch dispersions containing 15% (*w/v*) of native starch were prepared in a 3.16 M sulfuric acid solution (1.00731, EMSURE, SIGMA, St. Louis, MO, USA). These dispersions were then incubated at 40 °C with constant agitation at 150 rpm in a shaking incubator (Barnstead Lab-Line A-Class, Barnstead international, Dubuque, IA, USA) between 1 and 24 h [20]. The time interval was selected based on the results from preliminary tests; times longer than 24 h produced insoluble solids, which do not allow the formation of stable solutions, making them unable to be processed by electrospraying. The samples were washed with distilled water consecutively until a pH as close to neutral as possible was achieved. In each case, the starch was recovered by centrifugation at 4500× *g* at 4 °C for 5 min (Dynamica Velocity 18R, Metrix, UK). The starch was then dried at 30 °C in a convection oven (AFOS, Hull, East Yorkshire, UK) and stored in a desiccator with anhydrous Drierite^{TM} (Drierite, Xenia, OH, USA) at room temperature for later use.

2.4. Hydrolysis Degree

The degree of hydrolysis was developed based on Sánchez de la Concha et al. [14], measured from the first supernatant at the end of the hydrolysis incubation, through the phenol method developed according to González-Vázquez et al. [37], where 1 mL of sample is mixed under dark conditions with 0.6 mL of phenol (2058-01, J.T. Baker, Avantor Performance Materials, Center Valley, PA, USA) at 5% (*w/v*), and 3.6 mL of concentrated H_2SO_4 (1.00731, EMSURE, SIGMA, St. Louis, MO, USA), and incubated for 30 min at 25 ± 1 °C. Subsequently, the reaction is stopped in an ice bath, and the absorbance is read at 490 nm (Multiskan Go spectrophotometer; Thermo Fisher Scientific; Vantaa, Finland). The values are quantified with a standard curve from 0 to 100 μg/mL D-glucose (G8270, SIGMA, St. Louis, MO, USA). Finally, the degree of hydrolysis is expressed as the percentage of total sugars in the first supernatant at a given time to the initial weight of dry starch, also denoted as dextrose equivalent (DE).

2.5. Starch Purity

The purity of the extracted starch was detected using the K-TSTA-100A 02/22 kit (Megazyme International Ireland Ltd., Co Wicklow, Ireland), a process outlined in both AOAC 996.11 and AACC method 76-13.01.

2.6. Amylose and Amylopectin Ratio

The procedure was carried out under the guidelines in the protocol of the K-AMYL 06/18 kit from Megazyme® (Megazyme International Ireland Ltd., Co Wicklow, Ireland). The basis of this kit is the modified method of amylose detection by precipitation of amylopectin with Concanavalin A previously developed by Yun & Matheson (1990) [38].

2.7. Optical Microscopy

Native and modified jicama starch (0.1 g) was suspended in 1 mL of Tween 80 (10099956, SIGMA, St. Louis, MO, USA), then 1 mL glycerol (G9012, SIGMA, St. Louis, MO, USA) was added, vortexed for 5 s, and processed for 60 s in a Branson-1510 ultrasonic bath (Danbury, CT, USA). Images were then captured using a Nikon Eclipse 50i microscope (NIKON Instruments INC., NY, USA), equipped with a DS-U3 digital camera (NIKON Instruments INC., NY, USA), with the aid of NIS-Elements F 2.30 software (NIKON Instruments INC., NY, USA). In contrast, the observation of electrosprayed starch microspheres was conducted directly on the slide. Finally, based on these images, approximately 100 particles were analyzed for some shape descriptors such as perimeter, area, roundness, and Feret (Fe) diameters with ImageJ software (https://imagej.net/ij/download.html, accessed on 14 January 2025, NIH, MD, USA).

2.8. Scanning Electron Microscopy

Native and modified starches were structurally characterized using micrographs obtained with an environmental scanning electron microscope (FEI Quanta, FEG 250, Hillsboro, OR, USA) at 5 kV with the back-scattered electron detector. Sample preparation was performed by directly sprinkling dried native and modified starch onto aluminium trays provided with a double-sided carbon tape.

2.9. Differential Scanning Calorimetry

The thermal properties of starch were evaluated by weighing 2.5 mg of jicama starch in an aluminum capsule, adding 7 μL of distilled water, and then hermetically sealing the capsule [39]. Samples were equilibrated for 24 h at room temperature for subsequent analysis in a differential scanning calorimeter (DSC Q2000, TA Instruments, New Castle, DE, USA), heating at a rate of 5 °C/min [40] over a temperature range of 20 to 120 °C [25]. The glass transition (Tg), characterized by the onset (To), peak (Tp), termination (Te), and enthalpy change (ΔH) temperatures, was obtained from the heat absorption endotherm using TA Universal Analysis 2000 software (TA Instruments, New Castle, DE, USA).

2.10. Fourier Transform Coupled Infrared Spectroscopy

Spectra were obtained by placing the powdered starch (~0.1 g, 25 °C) directly on the surface of a Cary 630 portable FTIR spectrometer (Agilent Technologies Inc., Santa Clara, CA, USA) equipped with a single-bounce attenuated total reflectance (ATR) diamond crystal interface. Spectra were collected in the 4000–650 cm^{-1} spectral region with a resolution of 4 cm^{-1} by pressing the powder sample onto the crystal using a pressure clamp. The absorbance spectrum was obtained by comparing the spectrum of the sample with that of a blank optical path (reference spectrum). The analysis of spectral data was conducted using Spectragryph software (v1.2.16.1, Spectragryph, Germany).

2.11. X-Ray Diffraction

The starch samples were placed in the sample holder, and diffraction patterns were obtained (40 kV, 15 mA, CuK, λ = 0.154 nm, 5–60° (2θ°)). To evaluate the degree of crystallinity, Equation (1) is used according to González-Lemus et al. (2018) [25], where C_R, A_C, and A_a represent the relative crystallinity, the crystalline area, and the amorphous area, respectively.

$$C_R = \frac{A_c}{A_c + A_a} * 100 \quad (1)$$

2.12. Visco-Amylographic Profile

Viscosity profiles were obtained based on Contreras-Jiménez et al. [12] with some modifications. In brief, 2 g of native and modified starch powder was weighed and suspended in 18 mL of distilled water before being placed in the starch cup of the rheometer to perform the test. The curves were obtained in a Discovery HR-3 rheometer (TA Instruments, New Castle, DE, USA) using a Starch cell geometry (Starch cell Aluminium-999188), with a constant rotation at 16 rad.s^{-1}. Initially, the temperature was stabilized at 25 °C for 60 s, after, the temperature was increased at a rate of 10 °C/min to 95 °C, held for 360s, finally cooled at a rate of 10 °C/min to 50 °C, where it was held for 300 s. Data were analyzed in the TA Instruments Trios (4.0.2.30774) software to obtain the pasting parameters shown in Figure 1. As illustrated in the diagram, the pasting temperature (PT) denotes the minimum temperature necessary for the baking of a sample obtained at the onset of the viscosity increase. The peak viscosity (PV) corresponds to the maximum viscosity observed at the equilibrium point between swelling and leaching of the polymer. Furthermore, the peak temperature (PkT) is the temperature at PV. Collectively, these parameters signify the water-binding capacity of the starch. Trough viscosity (TV) is defined as the minimum viscosity obtained after the peak viscosity during the cooking process. The difference between PV and TV is known as breakdown viscosity (BDV), which is a measure of the degree of disintegration of the starch granules when subjected to temperature and shear force. The final viscosity (FV) measures the viscosity of the gel after cooking and cooling, and indicates its stability at the end of this process. Finally, the setback viscosity (SBV) is given by the equation SBV = FV-TV and refers to the retrogradation of the starch molecules during cooling [41,42].

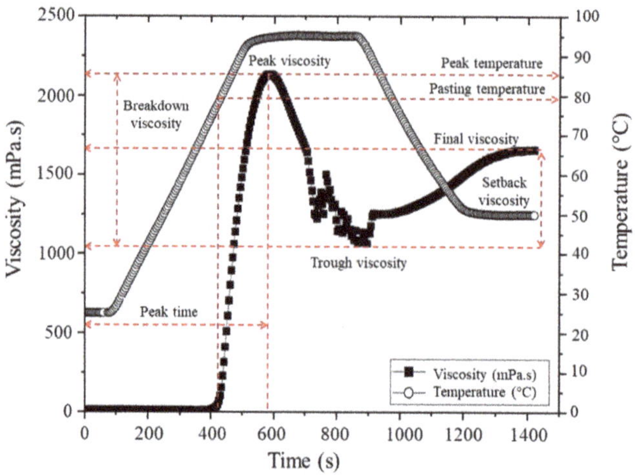

Figure 1. HR-3 rheometer starch viscoamylographic profile. Own data curve. Tags adapted from [41], polymers, 2025.

2.13. Preparation of Electrosprayed Solutions

Based on the design of experiments for microsphere synthesis, starch dispersions were prepared at the required concentration (w/v), subsequently heated to boiling, and maintained at this temperature for 10 min to achieve a complete gelatinization under constant stirring in a heated Super-Nuova Multi-Place stirrer (Thermo Scientific) then was cooled to room temperature and poured into hypodermic syringes for further use.

2.14. *Apparent Viscosity*

Apparent viscosity for electrosprayed solutions was measured according to Li, Kong, and Ziegler (2020) [43], using a Discovery HR-3 hybrid rotational rheometer (TA Instruments, New Castle, DE, USA) using a parallel plate geometry with 40 mm diameter (998807 TA Instruments, Elstree, UK) and determined from flow curves with a shear rate of 0.1 at 100 s^{-1}, at 20 °C. The flow curves were fitted to a model using TRIOS TA Instruments software (4.0.2.30774) and the "best flow fit (Stress vs. Rate)" analysis tool, choosing the model with the best fit and its parameters. Also, the apparent viscosity was introduced in the design of experiments to evaluate the impact of acid hydrolysis and starch concentration on the viscosity of the electrosprayed solutions using Design Expert V9 software (Stat-Ease In., Minneapolis, MN, USA).

2.15. *Surface Tension*

The surface tension (γ) of the electrosprayed solutions was determined using a K6-Krüss tensiometer (Hamburg, Germany), which is based on measuring the tensile strength of a stretched sheet until it breaks. The samples were analyzed in triplicate at room temperature following the method described by the manufacturer. The surface tension value is obtained directly from the equipment in mN/m units. The effect of starch hydrolysis on the surface tension of the electrosprayed solutions was evaluated using Design Expert V9 software (Stat-Ease In., Minneapolis, MN, USA) based on the design of experiments used for microsphere synthesis.

2.16. *Electrical Conductivity*

The quantification of this property was conducted using a ST20 Starter conductivity meter (OHAUS, Parsippany, NJ, USA) with an accuracy of 0 to 20 mS under the manufacturer's instructions. Briefly, the electrode is rinsed in water, dried with a paper towel, and inserted into the measuring solution to obtain the electrical conductivity value.

2.17. *Electrospray Synthesis of Jicama Starch Microspheres*

The synthesis of jicama starch microspheres with varying degrees of hydrolysis was conducted using a basic electrospraying apparatus (CICATA-IPN, Queretaro, Mexico), which comprises a vertical syringe pump, a power source, and a single capillary nozzle, previously employed by Rentería-Ortega et al. (2020) [44]. To evaluate the effect of the degree of hydrolysis, a design of experiments using response surface methodology was developed, following a central composite design with 6 replicates at the central point, resulting in a total of 20 runs (Table 1). The design expert software (v.9) was employed. The parameters evaluated encompassed hydrolysis time (1 to 24 h), within a range designated as rapid hydrolysis [20], starch concentration (6.6 to 23.4% *w*/*v*), and applied voltage (8.30 to 16.70 kV). These parameters were adjusted following a comprehensive literature review on the starch utilization as a wall material in diverse electrospraying methodologies [27–33,45]. Furthermore, the flow rate was set at 2 mL/h for all experiments according to the equipment characteristics.

Table 1. Design of experiments to evaluate the effect of acid hydrolysis on the electrospray synthesis of jicama starch microspheres.

Starch Hydrolysis Time (h)	Starch Concentration (% *w*/*v*)	Voltage (kV)
1	10	10
1	10	15
1	20	10

Table 1. *Cont.*

Starch Hydrolysis Time (h)	Starch Concentration (% *w*/*v*)	Voltage (kV)
1	20	15
6	15	12.5
12.5	6.6	12.5
12.5	15	8.3
12.5	15	12.5
12.5	15	12.5
12.5	15	12.5
12.5	15	12.5
12.5	15	12.5
12.5	15	12.5
12.5	15	16.7
12.5	23.4	12.5
24	10	10
24	10	15
24	20	10
24	20	15

2.18. Statistical Analysis

Results are presented as arithmetic mean plus/minus sample standard deviation from at least three replicates in each experiment. All analyses were made in MYSTAT software (SYSTAT 13, http://systatsoftware.com/downloads/download-mystat/ consulted on 19 October 2021, Inpixon, Palo Alto, CA, USA). On the other hand, the electrospray starch microspheres synthesis data were evaluated using Design Expert 9 software (Stat-Ease Inc., MN, USA).

3. Results and Discussion

3.1. Physicochemical Characterization of Jicama (P. erosus) Native Starch

As illustrated in Table 2, the physicochemical characterization of native jicama starch shows that the extraction process is suitable for this material, as a high-purity material was obtained, along with a low concentration of maltodextrins and glucose. These values of glucose and maltodextrins could have been produced by starch damage during extraction [46]. Conversely, the amylose content is consistent with the range reported for certain jicama varieties [23] and analogous to values obtained in other studies (23.6%, 25.2%, and 26.4%, respectively) [12,24,47] and is considered, due to its amylose content, a normal starch [46]. The percentage of amylose is important because it has been demonstrated to impact the structure of starch, the viscosity of the resultant gel, and the susceptibility or resistance to acid hydrolysis [48,49].

Table 2. Chemical properties of native jicama starch.

Purity (g/100 g Starch)	Maltodextrins (g/100 g Starch)	D-Glucose (g/100 g Starch)	Amylose (g/100 g Starch)	Amylopectin (g/100 g Starch)	Humidity (% RH)
93.2 ± 2.9	7.5 ± 0.6	1.23 ± 0.04	20.2 ± 0.3	79.8 ± 0.3	10.1 ± 0.7

3.2. Morphology and Feret Diameter of Native Jicama Starch

The jicama starch granule (Figure 2) showed spherical, polygonal, and oval compacted shapes, which are similar to the findings reported by Martínez-Bustos et al., (2006) [50]. Regardless of its shape, the native jicama starch granules (Figure 2c) had a closed and dotted hilum; the surface texture was smooth, but exhibited facets resulting from pressure;

some broken granules were also observed [51]. In comparison, the Feret diameter of native jicama starch ranged from 3.5 to 19.5 μm, which is consistent with the findings reported in other studies [24,25]. This indicates that the jicama starch has a narrower size distribution compared to potato starch (5 to 100 μm) and slightly smaller granules than those observed in maize (3 to 30 μm) [46].

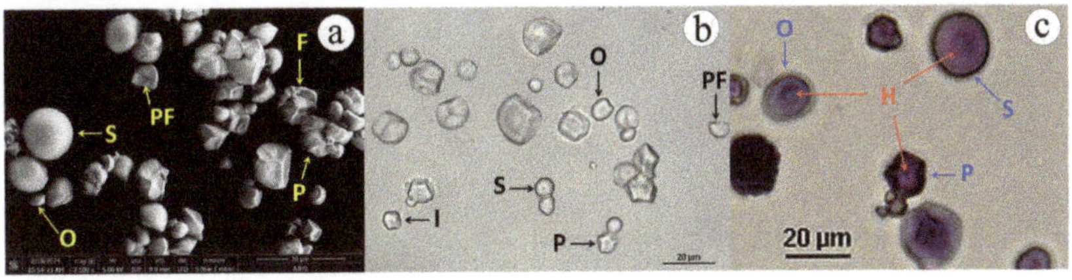

Figure 2. ESEM (**a**) and light microscopy (**b**,**c**) micrographs of jicama (*P. erosus*) starch granules. S = spherical, O = oval, P = polyhedral, I = irregular, PF = pressure facet, F = fragmented or damaged granule, H = hilum.

3.3. Hydrolysis of Jicama Starch

The hydrolysis progression of jicama starch during the first 24 h (Table 3) reveals a maximum of 42.7% hydrolysis degree, following a potential behavior (R^2 = 0.9825), a result analogous to that of mung bean starch under the same hydrolysis conditions, agreeing that both jicama and mung starches possess the same C-type polymorphism [23,49]. It is suggested in the literature that a phase known as fast hydrolysis occurs during this period, involving the breakdown of the outer amorphous region (amylose chains with a degree of polymerization > 300) and the intra-cluster crystalline zone (long amylopectin chains) [14,20,49]. On the other hand, starch hydrolysis products with a dextrose equivalent (DE) less than 20 are classified as maltodextrins; in this study, they were produced within the first 12 h, after which starch hydrolysis products were designated as syrup solids [21].

Table 3. Hydrolysis degree of jicama starch.

Hydrolysis Time (h)	Hydrolysis Degree (DE [1])
1	0.4 ± 0.1^{a}
3	0.8 ± 0.1^{a}
6	6.3 ± 0.3^{b}
9	10.3 ± 0.2^{c}
12	19.3 ± 0.5^{d}
24	42.7 ± 0.7^{e}

[1] DE: dextrose equivalent (%). Different letters in the superscripts indicate significant differences ($p \leq 0.05$).

3.4. Effect of Hydrolysis on Morphology and Jicama Starch Feret Diameter

The micrographs illustrating the changes in the starch granules' morphology after hydrolysis (Figure 3) showed that between 3 and 6 h of hydrolysis, superficial erosion is observed, subsequently followed by the formation of cracks in what could be the amorphous zone of the granules. At 12 h, there is an increase in the presence of granules with cracks, accompanied by the release of granule fragments (2 < Fe < 4 μm) (Figure 4). In addition, spherical and oval granules demonstrate greater integrity than polygonal and irregularly shaped granules.

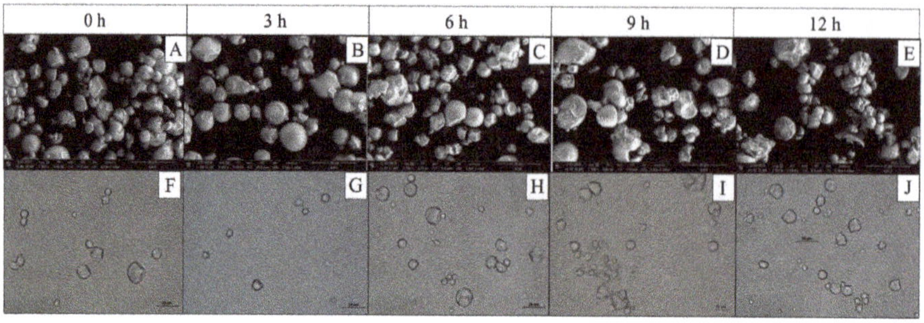

Figure 3. Effect of H_2SO_4 hydrolysis time on morphometric jicama starch characteristics. (**A–E**): SEM micrographs; (**F–J**): brightfield micrographs.

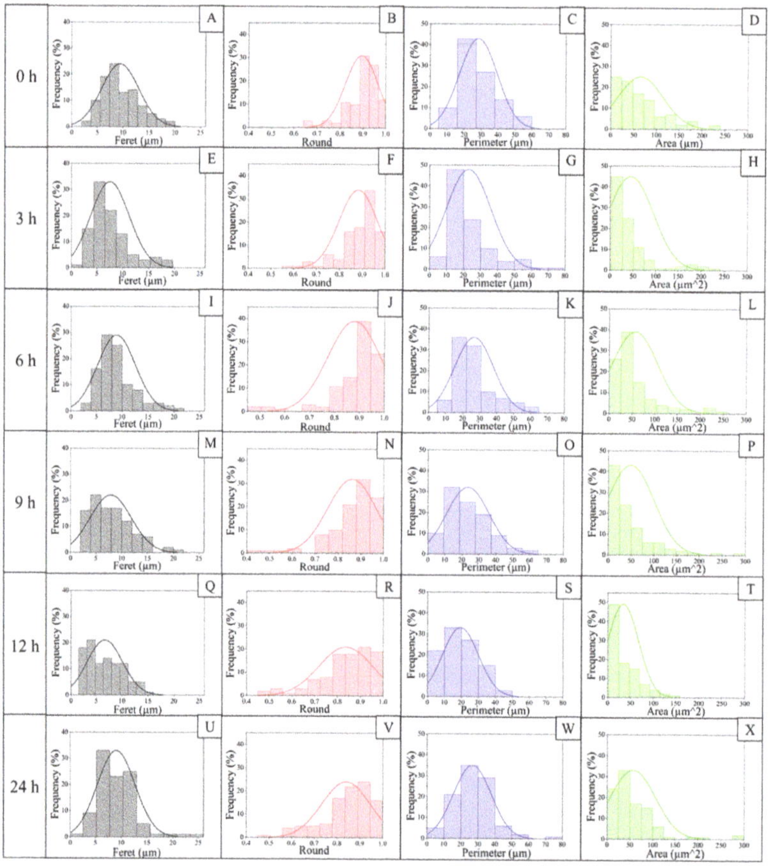

Figure 4. Morphometric parameters according to the hydrolysis degree of jicama starch. Feret distributions are comprised on (**A,E,I,M,Q,U**) from jicama starch native (0 h) and hydrolyzed for 3, 6, 9, 12, and 24 h respectively. Round distributions in (**B,F,J,N,R,V**) from jicama starch native (0 h) and hydrolyzed for 3, 6, 9, 12, and 24 h respectively. Perimeter distributions are presented in (**C,G,K,O,S,W**) from jicama starch native (0 h) and hydrolyzed for 3, 6, 9, 12, and 24 h respectively. Area distributions are comprised in (**D,H,L,P,T,X**) from jicama starch native (0 h) and hydrolyzed for 3, 6, 9, 12, and 24 h respectively. Frequency distributions were expressed as a percentage of about 100 particles analyzed.

Regarding the morphometric parameters (Figure 4), the roundness of the granules decreases during this process, which may be associated with their fragmentation, generating an irregular morphology that also modifies the perimeter. On the other hand, both perimeter and area show a distribution that widens at 3 h and narrows towards 12 h of hydrolysis. This may be associated with swelling and subsequent fragmentation and fracture of the granule. This is followed by a release of fragments between 12 and 24 h of hydrolysis, which may then agglomerate. These changes may be associated with how hydrolysis occurs in the starch structure. The literature indicates that the process starts in the amorphous region of the granule, followed by erosion in the crystalline area [46,52]. There are two mechanisms described for this process; for type A starches, such as corn starch, the channels running from the surface inwards can provide an initiation zone for attack, being different for type B starches such as potato starch, where the process of erosion occurs, resulting in the generation of channels that widen as hydrolysis progresses [20,53]; in this case, the jicama starch displaying characteristics of both types A and B starches must follow both breaking processes, but with some granules remaining unfragmented.

3.5. *Thermal Properties of Jicama Starch and Its Hydrolysates*

The changes in the thermal parameters of jicama starch as the hydrolysis time elapses are shown in Table 4. It is well known that when starch is heated at high temperature with a water content above 60% (w/v), the gelatinization process, detectable from DSC endotherms [54], occurs. In this process, the semi-crystalline structure of starch is transformed into an amorphous one, due to the disruption of the molecular order (rupture of hydrogen bonds) reflecting in the gradual loss of the granular structure in an irreversible way [24,46], as well as the change from a slurry into a semi-solid or gel with significant viscous properties [55]. In the case of native jicama starch, we see that the values of peak temperature (Tp) are very similar to those reported by Ramírez-Miranda et al., (2017) [24] and by González-Lemus et al., (2018) [25], where variations in these values regularly depend on cultivar, extraction method, place of origin, amylose content, degree of amylopectin cross-linking, starch crystallinity, chain length distribution, concentration of components (fats, proteins, etc.) and amount of damaged starch [23]. As the hydrolysis process proceeds, the Tp shows the highest value at 6 h, possibly suggesting the formation of stronger intermolecular bonds or increased molecular order or crystallinity [56]. Subsequently, a decrease is perceived at 12 h and no endotherm is formed at 24 h. On the other hand, the enthalpy change (ΔH) reflects the amount of double helix order [17]. For native starch, its value is around that reported again by González-Lemus et al., (2018) [25] and Shi, et al. (2021) [57]. During the hydrolysis process, ΔH remains unchanged within the first 6 h, decreases significantly at 12 h, and ceases to be perceived at 24 h, given the absence of an endotherm (Table 5). This behavior is characteristic of the acid hydrolysis process of granular starch [56]. This may indicate the gradual loss of order in the crystal structure and the variation in the bond strengths as there is a simultaneous attack on amylose and amylopectin in the initial phase of hydrolysis [23], while the absence of the endotherm could be indicating less amount of double helix and a less-ordered crystal structure [17,48].

Table 4. Effect of acid hydrolysis time of jicama starch on thermal properties and FTIR parameters.

Hydrolysis Time	Tp	ΔH	IR Ratios (FTIR Parameters)		Relative Crystallinity
(h)	(°C)	(J/g)	Double Helix Degree (995/1022)	Order Degree (1047/1022)	(%)
0	66.91 ± 0.12 [a]	2.72 ± 0.06 [a]	0.9985	1.0085	39.97
1	64.35 ± 0.06 [b]	2.85 ± 0.18 [a]	0.9985	1.0271	37.12
6	69.25 ± 0.22 [c]	2.74 ± 0.44 [a]	0.9988	1.0344	37.29
12	65.86 ± 0.17 [d]	1.45 ± 0.21 [b]	0.9986	1.0220	36.8
24	ND	ND	1.0019	1.0230	41.86

Tp: peak temperature; ΔH: gelatinization enthalpy; different letters in the superscripts indicate significant differences ($p \leq 0.05$).

Table 5. Changes in the viscoamylographic profile of jicama starch as function of hydrolysis time.

Hydrolysis Time	PT	PV	PkT	TV	BDV	FV	SBV	Tp
(h)	(°C)	(mPa.s)	(°C)	(mPa.s)	(mPa.s)	(mPa.s)	(mPa.s)	(s)
0	76 ± 1 [a]	2137.4 ± 98.4 [a]	94.6 ± 0.1 [a]	1003.1 ± 43.2 [a]	1134.3 ± 58.9 [a]	1661.5 ± 102.8 [a]	658.4 ± 61.4 [a]	410.2 ± 6.9 [a]
1	75.4 ± 0 [a]	58.5 ± 0 [b]	84.8 ± 1 [bc]	22.7 ± 0 [b]	11.9 ± 0 [b]	33.4 ± 0 [b]	3.6 ± 0 [b]	406.2 ± 0 [a]
3	74.8 ± 0.6 [a]	21 ± 1.9 [b]	83.1 ± 1.2 [bc]	15.2 ± 1.1 [b]	5.8 ± 0.9 [b]	20.9 ± 1.4 [b]	5.7 ± 0.3 [b]	398.1 ± 6.9 [a]
6	76.8 ± 1.6 [a]	16.1 ± 1.9 [b]	83.8 ± 0.3 [b]	11.5 ± 1 [b]	4.5 ± 1 [b]	16.5 ± 1.1 [b]	4.9 ± 0.3 [b]	412.5 ± 10.8 [a]
9	73.7 ± 5.7 [a]	13.3 ± 0.6 [b]	80.8 ± 1.5 [c]	9.9 ± 0.2 [b]	3.4 ± 0.4 [b]	14.4 ± 0.3 [b]	4.5 ± 0.2 [b]	397.3 ± 33.2 [a]
12	74.8 ± 4.5 [a]	13.8 ± 0.3 [b]	81.2 ± 1.9 [bc]	9.9 ± 0.1 [b]	3.8 ± 0.3 [b]	14.7 ± 0.4 [b]	4.7 ± 0.3 [b]	405.1 ± 29.6 [a]
24	62.8 ± 4.2 [b]	14.9 ± 0.2 [b]	69.5 ± 1.2 [d]	9.8 ± 0.1 [b]	5.1 ± 0.3 [b]	14.1 ± 0.1 [b]	4.3 ± 0.2 [b]	326.2 ± 24 [b]

PT: pasting temperature; PV: peak viscosity at PkT; PkT: peak temperature; TV: trough viscosity; BDV: breakdown viscosity is equal to PV-TV; FV: final viscosity; SBV: setback viscosity is equal to FV-TV; Tp: peak time. Different letters in the superscripts indicate significant differences ($p \leq 0.05$).

3.6. Fourier Transform Infrared Spectroscopy

The spectra of both native starch and the hydrolysates analyzed (Figure 5A) show a similar behavior to that reported by Wigati et al., (2022) [58] for native jicama starch. All spectra display a peak at 3258.6 ± 17.1 cm^{-1}, ranging from 3000 to 3500 cm^{-1}, associated with the hydrogen bonds of water (O-H stretching). In both native and hydrolyzed starch, a prominent peak is present around 2889 cm^{-1} in the region corresponding to the C-H bond vibration, where the change in intensity has also been related to the progress of starch retrogradation [59]. Subsequently, a consistent, but small, peak is found in all samples around 1653 cm^{-1}, which may be associated with conjugated carbonyl and carboxyl groups and C-O vibrations, or an interaction between water and starch. On the other hand, a peak is detected around 1340 cm^{-1}, indicating O-H stretching. Its intensity decreases with hydrolysis, reaching its lowest point at 6 h and rising slightly to remain at 12 and 24 h. Regarding the characteristic region of carbohydrates (1200 to 800 cm^{-1}), shown in Figure 5B, both in native and hydrolyzed starch, a peak at 1149 cm^{-1} can be distinguished, which could be associated with a C-O stretching of sucrose confirmed by the presence of a peak around 926 cm^{-1}, this molecule may come from the starch extraction process where there was no complete purification, and would help explain the peak around 860 cm^{-1}, which may be associated with a C-C stretching of fructose produced during the hydrolysis of sucrose or some fructooligosaccharide. However, this contrasts with the fact that these substances are soluble in water and could have been removed by washing during hydrolysis.

Additionally, there is a peak at 1078 cm^{-1} related to C-O and C-C stretching and C-O-H bending, while the most prominent peak is found at 1003 cm^{-1} in native starch; in hydrolyzed jicama starch, it shifts towards 1014 cm^{-1}. This peak is associated with the amorphous zone [60], where its intensity decreases in the hydrolyzed sample during 6 h and increases at 12 and 24 h. On the other hand, the FTIR technique allows for the visualization of the double helix degree through the intensity ratio at 995/1022 cm^{-1} and the degree of order through the intensity ratio at 1047/1022 cm^{-1}. Regarding this, in Table 5 "IR ratios (FTIR parameters)", it is observed that the degree of double helix is maintained up to 12 h increases slightly at 24 h. Finally, the degree of order increases in hydrolyzed

Regarding the morphometric parameters (Figure 4), the roundness of the granules decreases during this process, which may be associated with their fragmentation, generating an irregular morphology that also modifies the perimeter. On the other hand, both perimeter and area show a distribution that widens at 3 h and narrows towards 12 h of hydrolysis. This may be associated with swelling and subsequent fragmentation and fracture of the granule. This is followed by a release of fragments between 12 and 24 h of hydrolysis, which may then agglomerate. These changes may be associated with how hydrolysis occurs in the starch structure. The literature indicates that the process starts in the amorphous region of the granule, followed by erosion in the crystalline area [46,52]. There are two mechanisms described for this process; for type A starches, such as corn starch, the channels running from the surface inwards can provide an initiation zone for attack, being different for type B starches such as potato starch, where the process of erosion occurs, resulting in the generation of channels that widen as hydrolysis progresses [20,53]; in this case, the jicama starch displaying characteristics of both types A and B starches must follow both breaking processes, but with some granules remaining unfragmented.

3.5. Thermal Properties of Jicama Starch and Its Hydrolysates

The changes in the thermal parameters of jicama starch as the hydrolysis time elapses are shown in Table 4. It is well known that when starch is heated at high temperature with a water content above 60% (w/v), the gelatinization process, detectable from DSC endotherms [54], occurs. In this process, the semi-crystalline structure of starch is transformed into an amorphous one, due to the disruption of the molecular order (rupture of hydrogen bonds) reflecting in the gradual loss of the granular structure in an irreversible way [24,46], as well as the change from a slurry into a semi-solid or gel with significant viscous properties [55]. In the case of native jicama starch, we see that the values of peak temperature (Tp) are very similar to those reported by Ramírez-Miranda et al., (2017) [24] and by González-Lemus et al., (2018) [25], where variations in these values regularly depend on cultivar, extraction method, place of origin, amylose content, degree of amylopectin cross-linking, starch crystallinity, chain length distribution, concentration of components (fats, proteins, etc.) and amount of damaged starch [23]. As the hydrolysis process proceeds, the Tp shows the highest value at 6 h, possibly suggesting the formation of stronger intermolecular bonds or increased molecular order or crystallinity [56]. Subsequently, a decrease is perceived at 12 h and no endotherm is formed at 24 h. On the other hand, the enthalpy change (ΔH) reflects the amount of double helix order [17]. For native starch, its value is around that reported again by González-Lemus et al., (2018) [25] and Shi, et al. (2021) [57]. During the hydrolysis process, ΔH remains unchanged within the first 6 h, decreases significantly at 12 h, and ceases to be perceived at 24 h, given the absence of an endotherm (Table 5). This behavior is characteristic of the acid hydrolysis process of granular starch [56]. This may indicate the gradual loss of order in the crystal structure and the variation in the bond strengths as there is a simultaneous attack on amylose and amylopectin in the initial phase of hydrolysis [23], while the absence of the endotherm could be indicating less amount of double helix and a less-ordered crystal structure [17,48].

Table 4. Effect of acid hydrolysis time of jicama starch on thermal properties and FTIR parameters.

Hydrolysis Time (h)	Tp (°C)	ΔH (J/g)	IR Ratios (FTIR Parameters) Double Helix Degree (995/1022)	IR Ratios (FTIR Parameters) Order Degree (1047/1022)	Relative Crystallinity (%)
0	66.91 ± 0.12 [a]	2.72 ± 0.06 [a]	0.9985	1.0085	39.97
1	64.35 ± 0.06 [b]	2.85 ± 0.18 [a]	0.9985	1.0271	37.12
6	69.25 ± 0.22 [c]	2.74 ± 0.44 [a]	0.9988	1.0344	37.29
12	65.86 ± 0.17 [d]	1.45 ± 0.21 [b]	0.9986	1.0220	36.8
24	ND	ND	1.0019	1.0230	41.86

Tp: peak temperature; ΔH: gelatinization enthalpy; different letters in the superscripts indicate significant differences ($p \leq 0.05$).

Table 5. Changes in the viscoamylographic profile of jicama starch as function of hydrolysis time.

Hydrolysis Time (h)	PT (°C)	PV (mPa.s)	PkT (°C)	TV (mPa.s)	BDV (mPa.s)	FV (mPa.s)	SBV (mPa.s)	Tp (s)
0	76 ± 1 [a]	2137.4 ± 98.4 [a]	94.6 ± 0.1 [a]	1003.1 ± 43.2 [a]	1134.3 ± 58.9 [a]	1661.5 ± 102.8 [a]	658.4 ± 61.4 [a]	410.2 ± 6.9 [a]
1	75.4 ± 0 [a]	58.5 ± 0 [b]	84.8 ± 1 [bc]	22.7 ± 0 [b]	11.9 ± 0 [b]	33.4 ± 0 [b]	3.6 ± 0 [b]	406.2 ± 0 [a]
3	74.8 ± 0.6 [a]	21 ± 1.9 [b]	83.1 ± 1.2 [bc]	15.2 ± 1.1 [b]	5.8 ± 0.9 [b]	20.9 ± 1.4 [b]	5.7 ± 0.3 [b]	398.1 ± 6.9 [a]
6	76.8 ± 1.6 [a]	16.1 ± 1.9 [b]	83.8 ± 0.3 [b]	11.5 ± 1 [b]	4.5 ± 1 [b]	16.5 ± 1.1 [b]	4.9 ± 0.3 [b]	412.5 ± 10.8 [a]
9	73.7 ± 5.7 [a]	13.3 ± 0.6 [b]	80.8 ± 1.5 [c]	9.9 ± 0.2 [b]	3.4 ± 0.4 [b]	14.4 ± 0.3 [b]	4.5 ± 0.2 [b]	397.3 ± 33.2 [a]
12	74.8 ± 4.5 [a]	13.8 ± 0.3 [b]	81.2 ± 1.9 [bc]	9.9 ± 0.1 [b]	3.8 ± 0.3 [b]	14.7 ± 0.4 [b]	4.7 ± 0.3 [b]	405.1 ± 29.6 [a]
24	62.8 ± 4.2 [b]	14.9 ± 0.2 [b]	69.5 ± 1.2 [d]	9.8 ± 0.1 [b]	5.1 ± 0.3 [b]	14.1 ± 0.1 [b]	4.3 ± 0.2 [b]	326.2 ± 24 [b]

PT: pasting temperature; PV: peak viscosity at PkT; PkT: peak temperature; TV: trough viscosity; BDV: breakdown viscosity is equal to PV-TV; FV: final viscosity; SBV: setback viscosity is equal to FV-TV; Tp: peak time. Different letters in the superscripts indicate significant differences ($p \leq 0.05$).

3.6. *Fourier Transform Infrared Spectroscopy*

The spectra of both native starch and the hydrolysates analyzed (Figure 5A) show a similar behavior to that reported by Wigati et al., (2022) [58] for native jicama starch. All spectra display a peak at 3258.6 ± 17.1 cm^{-1}, ranging from 3000 to 3500 cm^{-1}, associated with the hydrogen bonds of water (O-H stretching). In both native and hydrolyzed starch, a prominent peak is present around 2889 cm^{-1} in the region corresponding to the C-H bond vibration, where the change in intensity has also been related to the progress of starch retrogradation [59]. Subsequently, a consistent, but small, peak is found in all samples around 1653 cm^{-1}, which may be associated with conjugated carbonyl and carboxyl groups and C-O vibrations, or an interaction between water and starch. On the other hand, a peak is detected around 1340 cm^{-1}, indicating O-H stretching. Its intensity decreases with hydrolysis, reaching its lowest point at 6 h and rising slightly to remain at 12 and 24 h. Regarding the characteristic region of carbohydrates (1200 to 800 cm^{-1}), shown in Figure 5B, both in native and hydrolyzed starch, a peak at 1149 cm^{-1} can be distinguished, which could be associated with a C-O stretching of sucrose confirmed by the presence of a peak around 926 cm^{-1}, this molecule may come from the starch extraction process where there was no complete purification, and would help explain the peak around 860 cm^{-1}, which may be associated with a C-C stretching of fructose produced during the hydrolysis of sucrose or some fructooligosaccharide. However, this contrasts with the fact that these substances are soluble in water and could have been removed by washing during hydrolysis.

Additionally, there is a peak at 1078 cm^{-1} related to C-O and C-C stretching and C-O-H bending, while the most prominent peak is found at 1003 cm^{-1} in native starch; in hydrolyzed jicama starch, it shifts towards 1014 cm^{-1}. This peak is associated with the amorphous zone [60], where its intensity decreases in the hydrolyzed sample during 6 h and increases at 12 and 24 h. On the other hand, the FTIR technique allows for the visualization of the double helix degree through the intensity ratio at 995/1022 cm^{-1} and the degree of order through the intensity ratio at 1047/1022 cm^{-1}. Regarding this, in Table 5 "IR ratios (FTIR parameters)", it is observed that the degree of double helix is maintained up to 12 h increases slightly at 24 h. Finally, the degree of order increases in hydrolyzed jicama starch with a peak at 6 h and decreases to remain stable at 12 and 24 h of hydrolysis. Compared with the ΔH data, we see that the degree of order measured by the double helix dissociation [17] tends to decrease. This could be related to the sensitivity of the method, where DSC may not measure the degree of order of the short chains generated during acid hydrolysis [52].

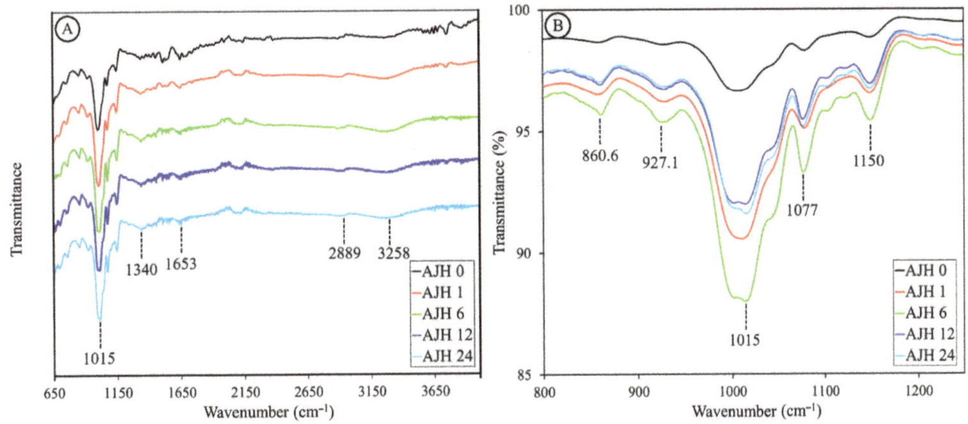

Figure 5. FTIR spectra of native jicama starch (AJH0) and subjected to acid hydrolysis for 1 (AJH1), 6 (AJH6), 12 (AJH12), and 24 h (AJH24). (**A**) shows the spectra from 650 to 4000 cm^{-1}, whereas (**B**) shows the spectra in the carbohydrate region (800 to 1250 cm^{-1}).

3.7. X-Ray Diffraction of Native and Hydrolyzed Jicama Starch

X-ray diffraction results (Figure 6A) of native jicama starch show weak peaks at 5.6°, 10.06°, 11.2°, 20.4°, and 26.16° (2θ), strong peaks at 15.18° and 23.1° (2θ), and a duplex at 17.18° and 18. 06° (2θ). The results mentioned above are characteristic values for a legume starch C-type diffraction pattern [46], coinciding with that reported by several authors [12,24,25]. On the other hand, it is observed that the relative crystallinity in the modified jicama starch shows a decrease until a minimum detected at 12 h of acid hydrolysis, while at 24 h it increases again, while the peaks are conserved in all treatments (Figure 6B). The factors affecting the degree of crystallinity include the amylopectin content and the orientation and size of the double helices in the crystalline area [23]. In turn, how the acid attacks the starch structure may be associated with the diffraction pattern [53]. In this sense, starches with a C-type diffraction pattern tend to have an intermediate susceptibility to hydrolysis compared to those of type A (more resistant) and type B (less resistant) [20,49].

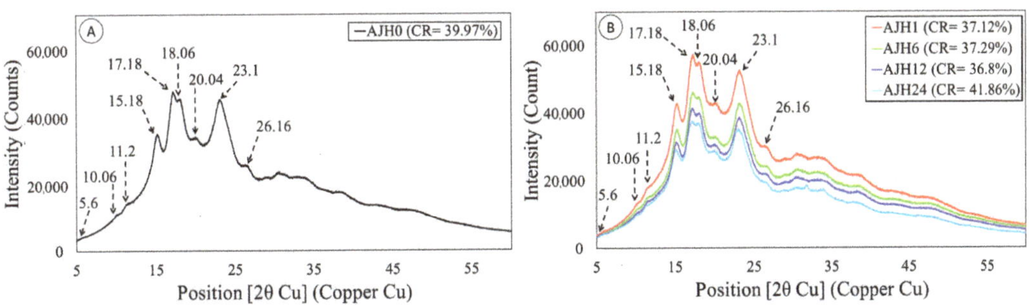

Figure 6. Diffraction patterns of jicama starch. (**A**) native starch (AJH 0) (left); (**B**) acid hydrolyzed modified starch 1h (AJH 1), 6h (AJH 6), 12h (AJH 12), and 24 h (AJH 24), and their respective behavior

at peaks 5.6° (1), 10.6° (2), 11.2° (3), 15.18° (4), 23.1° (7), 26.16° (8), and the duplex at peaks 17.18° and 18.06° (5). The relative crystallinity percentage is shown with the symbol CR.

3.8. Viscoamylographic Profile Changes in Jicama Starch by Acid Hydrolysis

The native jicama starch presents a typical behavior of starches, forming a curve that starts with the minimum temperature required for granule swelling, reaching a pasting temperature (PT) of 76 ± 1 °C. Following, the maximum viscosity observed at the equilibrium point between swelling and leaching of the polymer is a peak viscosity (PV) of 2137.4 ± 98.4 mPa.s, which corresponds to the peak temperature (PkT) of 94.6 ± 0.1 °C. Collectively, these parameters signify the water-binding capacity of the starch [41,42]. Then, the minimum viscosity obtained after the peak viscosity during the cooking process, the trough viscosity (TV), was 1003.1 ± 43.2 mPa.s. It is associated with the total rupture of the granule during gelatinization, and attainment a breakdown viscosity (BDV) of 1134.3 ± 58.9 mPa.s. The native jicama starch also showed a setback viscosity (SBV), relative to retrogradation process, of 658.4 ± 61.4 mPa.s and a final viscosity (FV) of 1661.5 ± 102.8 mPa.s (Table 5). The values obtained are very similar to those reported by Contreras-Jiménez et al. (2019) [12], and it is important to note that viscosity is concentration-dependent for starches and dextrins [61,62].

This curve provides insight into the behavior of jicama starch during the gelatinization process where, due to the action of moisture and temperature, the granule structure collapses, the crystallites melt, the double helixes unwind and hydrogen bonds break [58], and then by association and formation of bonds between chains, a continuous network or gel is created [61]. This process is shown to be different for hydrolyzed jicama starch in that pasting temperature (PT) remains unchanged up to 12 h (74.8 ± 4.5 °C) and decreases significantly ($p \leq 0.05$) at 24 h (62.8 ± 4.2 °C). In turn, PV, TV, BDV, SBV, and FV decrease by 97.7 ± 0.4% during the first hour of acid treatment related to native starch (0 h), and remain without significant differences afterwards, regardless of the degree of hydrolysis achieved in the time studied. On the other hand, PkT, which together with PV indicates the water holding capacity of starch before granule rupture during swelling, is around 82.7 ± 1.7 °C for jicama starch hydrolyzed between 0 and 12 h of hydrolysis and decreases to 69.5 ± 1.2 °C at 24 h. These results are in agreement with those reported by Chen et al. (2023) [62], finding that acid hydrolysis strongly impacts starch swelling capacity and viscosity development, which can be associated with the shortening of amylose and amylopectin chains and modification of crystallinity during hydrolysis [62].

3.9. Effect of Jicama Starch Hydrolysis Time and Starch Concentration on EHDA Solution Physical Properties

3.9.1. Apparent Viscosity (η), and Rate Index (n)

Figure 7 presents the surface response graphics of apparent viscosity (Figure 7A) and rate index (Figure 7B). These values (η, n) were obtained from the flow curves of the native and hydrolyzed jicama starch solutions, which were fitted to the Herschel–Bulkley model, in which the sample presents an initial yield stress, followed by a shear thinning behavior. The specific data results of this test are shown in Table S1 of the supplementary material.

The graphs (Figure 7A,B) show that both the hydrolysis time (directly related to the hydrolysis degree) and the concentration (w/v) at which the EHDA solutions were prepared had a direct impact on the value of the apparent viscosity (Figure 7A) while only the hydrolysis time determined the behavior of the rate index (Figure 7B). According to Kanyuck et al. (2019) [61], the formation of the maltodextrin gel network during starch acid hydrolysis occurs in two stages: initially through the helical association of two chains and subsequently, with aggregation between the double helices into crystallized regions and longer chains connecting the aggregates of helices into a gel. In the present work, acid hydrolysis affected the swelling capacity and viscosity development of starch. Concentration also plays an important role in the rheology of jicama starch gels (based on native and hydrolyzed), as expected [63].

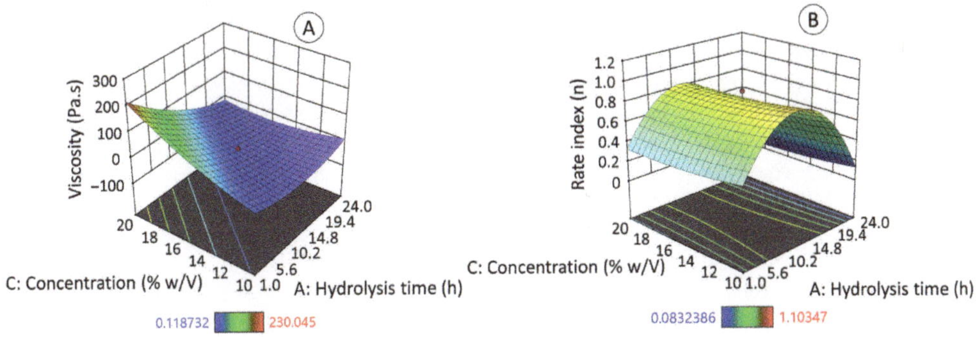

Figure 7. Effect of acid hydrolysis degree of jicama starch on apparent viscosity (**A**) and rate index (**B**) as a function of hydrolysis time and starch concentration.

The apparent viscosity (η) and the rated index (n) equations follow a quadratic model (Equation (2), $R^2_{ADJ} = 0.8951$ and Equation (3), $R^2_{ADJ} = 0.7387$); both mathematical models are statistically significant ($p \leq 0.05$).

$$\eta = 0.28A^2 + 0.89C^2 - 57.23AC + 2.96A - 5.06C + 41.41 \quad (2)$$

$$n = -0.63A^2 - 0.14A + 0.83 \quad (3)$$

where A refers to hydrolysis time (hours), B to voltage (kV), and C (% *w*/*v*) to starch concentration.

3.9.2. Surface Tension (γ)

Figure 8 presents the behavior of the surface tension (γ) of native and hydrolyzed starch EHDA solutions, when varying hydrolysis time and hydrolyzed starch concentration. In this case, data was adjusted to a linear model (Equation (4), $p \leq 0.05$) with an adjusted determination coefficient (R^2_{ADJ}) of 0.7714.

$$\gamma = -6.10A - 0.4750AC - 0.5072C + 58.81 \quad (4)$$

where A refers to hydrolysis time (hours), B to voltage (kV), and C (% *w*/*v*) to starch concentration.

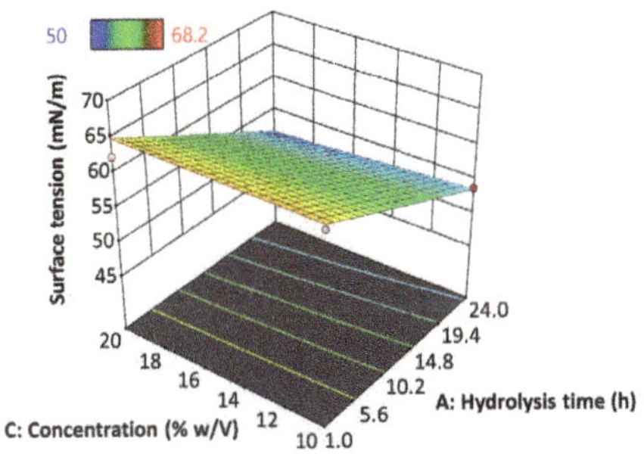

Figure 8. Effect of hydrolysis time of jicama starch on the surface tension of an EHDA solution.

The results show that the impact of hydrolysis time on surface tension is predominant over concentration, such that γ tends to decrease as the degree of starch hydrolysis increases. The specific data results of this test are shown in Table S1 of the supplementary material.

3.9.3. Electrical Conductivity

Electrical conductivity is a pivotal parameter in electrohydrodynamic atomization, since the driving force is the charge exerted on the solutions for polymer elongation as well as for jet breakdown. The presence of charge carriers, such as ions or polarized molecules, becomes a relevant factor in dielectric materials, for example, starch [64]. This property is often influenced by the concentration and type of polymer, solvent, and the availability of ionizable compounds [26]. However, Figure 9 shows that the electrical conductivity is directly influenced by the degree of hydrolysis of the starch, regardless of the starch concentration used in the solution. In this sense, the conductivity tends to increase as the hydrolysis progresses. Its highest point was found at 24 h of hydrolysis. On the other hand, the best-like microsphere particle formation was obtained at a low hydrolysis degree when low electrical conductivity is present. This contrasts with studies based on potato starch solubilized in formic acid, where an increase in exposure time led to a decrease in electrical conductivity [65]. These authors also reported that electrical conductivity does not affect the morphology and diameter of the fibers synthesized by EHDA [26]. The specific data results of this test are shown in Table S1 of the Supplementary Material.

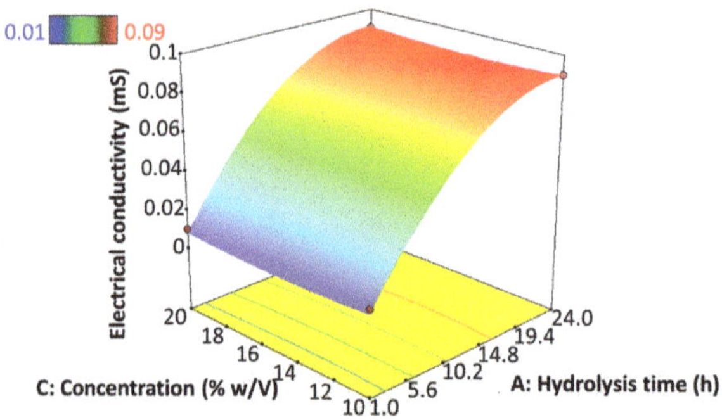

Figure 9. Effect of hydrolysis time of jicama starch on the electrical conductivity of an EHDA solution.

3.10. Morphometric Characteristics of the Electrosprayed Jicama Starch Particles

The effect of hydrolysis time on the Feret diameter (Fe) of the particles generated by EHDA (Figure 8) fits a quadratic model (Equation (5), $R^2_{ADJ} = 0.8698$, $p \leq 0.05$):

$$Fe = 602.4A^2 + 376.4C^2 - 884.5AC - 969.7A + 713.7C - 53.3 \quad (5)$$

where A refers to hydrolysis time (hours), B to voltage (kV), and C (% w/v) to starch concentration.

Figure 10 shows that particle diameter is significantly influenced by concentration and hydrolysis time ($p \leq 0.05$). Concentration and Feret diameter have a directly proportional relationship; however, this is true up to 6 h. In the modified starch for 12 (DE = 19.3) and 24 h (DE = 42.7), particle formation was not verified independently of solids concentration. It is known that the concentration of native starch influences the particle shape in electrospraying, in which high concentrations yield fibers, while low concentrations produce beads [9,66,67].

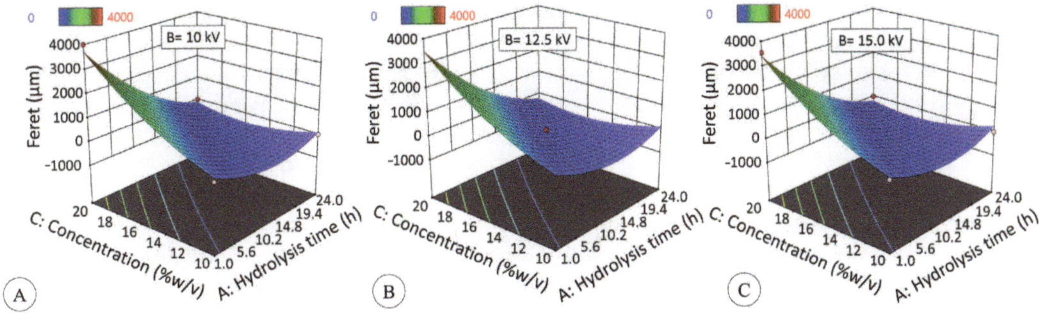

Figure 10. Effect of hydrolysis time (h) and starch concentration (% *w*/*v*) at 10 kV (**A**), 12.5 kV (**B**) and 15 kV (**C**) on the Feret diameter (μm) of modified starch particles generated by electrospraying. Data are fitted to quadratic model ($p \leq 0.05$, R^2ADJ = 0.8698) described by Equation (5).

Regarding roundness (RD, Figure 11), both hydrolysis time and solids concentration had a significant effect ($p \leq 0.05$), adjusting data to a quadratic model (R^2_{ADJ} = 0.9396, $p \leq 0.05$):

$$RD = -0.39A^2 - 0.05C^2 + 0.12AC - 0.31A - 0.1C + 0.7 \quad (6)$$

where A refers to hydrolysis time (hours), B to voltage (kV), and C (% *w*/*v*) to starch concentration. The specific data results of these both analyses, Feret and roundness, are shown in Table S1 of the Supplementary Material, and the bright field micrographs corresponding to the experimental design in Table 1 are shown in Figures S1 and S2.

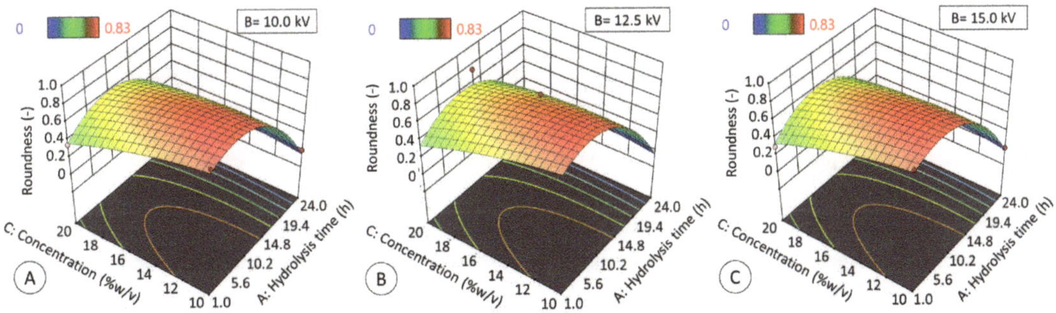

Figure 11. Effect of hydrolysis time (h) and starch concentration (% *w*/*v*) at 10 kV (**A**), 12.5 kV (**B**), and 15 kV (**C**) on the roundness of modified starch particles generated by electrospraying. The data obtained fit the quadratic model (R^2ADJ = 0.9396, $p \leq 0.05$) described by Equation (6).

On the other hand, although voltage did not have a significant effect on Feret diameter and roundness under the experimental conditions ($p > 0.05$), it is noteworthy that the application of voltage resulted in a modification of these parameters in comparison to the absence of voltage in the synthesis process. However, the effect of hydrolysis is more significant than the impact of voltage or solid concentration, as it modifies the structure of the starch and the properties of the gels. Consequently, the ability of jicama starch to be used as a coating material in the electrospray synthesis of particles or microspheres appears to be determined by the degree of hydrolysis.

4. Conclusions

Studies using starch as the only polymer are limited and require further investigation in electrohydrodynamic atomization and the food field [26]. In this regard, jicama could be a potential source of starch; however, this has not yet been tested in this technology [9]. Concerning their structural characteristics, starches encounter difficulties when employed as a sole wall material in EHDA applications. In an attempt to resolve this issue, studies have been conducted on the use of modified starches such as debranched starch, resistant maltodextrins, or solubilized in organic acids, among others. We hypothesize that the properties of jicama (*Pachyrhizus erosus*) starch can be modified by acid hydrolysis with sulfuric acid to improve its application in the EHDA process. We found that acid hydrolysis of jicama starch produced dextrins within 1 to 12 h with a degree of hydrolysis (expressed as dextrose equivalent (DE)) of less than 20%, and syrup solids at 24 h (DE of 40%) [45]. The FTIR results indicate only a polymer chain cut, and no functionalization was detected. The hydrolysis process mainly affected the water retention capacity of the starch, as well as the viscosity, surface tension, and electrical conductivity of the gels. In this regard, it is known that in EHDA, the polymer structure and solution properties impact the shape and size of the electrospray products [8]. In the present study, it was found that the surface tension and viscosity of jicama starch gels decrease as the degree of hydrolysis increases, although the electrical conductivity increases at longer times of acid hydrolysis, ranging from 0.01 to 0.09 mS. This may facilitate the separation of droplets during the electrospraying process, enabling the synthesis of particles from hydrolyzed jicama starch with a dextrose equivalent between 0.4 and 6.3. Additionally, the size and morphology were influenced more by the degree of hydrolysis than by the concentration of solids. However, with hydrolysis times of 12 and 24 h, it was not possible to form like-microspheres particles by electrospray under the conditions of the study. This indicates that jicama starch dextrins with a degree of hydrolysis below 6.3% can be used as the sole wall material to form particles by electrospraying.

Supplementary Materials: The following supporting information can be downloaded at: https://www.mdpi.com/article/10.3390/polym17152069/s1, Table S1: Experimental design results from Table 1; Table S2: Bright field micrographs corresponding to the experimental design in Table 1; Figure S1: Jicama starch hydrolyzed-based solutions (24 h, 20% *p*/*v*) without voltage (a, b) and after electrospraying (c, d).

Author Contributions: Conceptualization, F.S.S.-V. and G.C.-D.; methodology, F.S.S.-V., J.A.T.-M., V.C.-S. and R.R.F.-R.; software, F.S.S.-V., J.A.T.-M., V.C.-S. and G.C.-D.; validation, E.M.-S. and G.C.-D.; formal analysis, F.S.S.-V. investigation, F.S.S.-V., E.M.-S. and G.C.-D.; resources, G.C.-D. and E.M.-S.; writing—original draft preparation, F.S.S.-V.; writing—review and editing, F.S.S.-V., G.C.-D. and E.M.-S.; visualization, G.C.-D., E.M.-S. and R.R.F.-R.; supervision, G.C.-D. and E.M.-S.; project administration, G.C.-D.; funding acquisition, G.C.-D., E.M.-S. and R.R.F.-R. All authors have read and agreed to the published version of the manuscript.

Funding: This research was funded by Instituto Politécnico Nacional, through proyects: SIP 20221471, 20230985, 20240510, and 20251117.

Institutional Review Board Statement: Not applicable.

Data Availability Statement: The original contributions presented in the study are included in the article, further inquiries can be directed to the corresponding author.

Acknowledgments: Fatima Sarahi Serrano-Villa, thanks to Escuela Nacional de Ciencias Biológicas, Centro de Investigación en Ciencia Aplicada y Tecnología Avanzada Querétaro, both from the Instituto Politécnico Nacional, for the experimental support to synthesis and analysis of electrosprayed starch particles. And Fatima Sarahi Serrano-Villa thanks CONAHCYT and BEIFI-IPN for the scholarships provided.

Conflicts of Interest: The authors declare no conflicts of interest.

References

1. Dhiman, A.; Suhag, R.; Singh, A.; Prabhakar, P.K. Mechanistic understanding and potential application of electrospraying in food processing: A review. *Crit. Rev. Food Sci. Nutr.* **2022**, *62*, 8288–8306. [CrossRef]
2. Yang, H.; Yang, Y.; Li, B.Z.; Adhikari, B.; Wang, Y.; Huang, H.L.; Chen, D. Production of protein-loaded starch microspheres using water-in-water emulsion method. *Carbohydr. Polym.* **2020**, *231*, 115692. [CrossRef]
3. Gülmez, H.B.; Topuz, A. Electrospraying of phytosterols and their controlled release characteristics under in vitro digestive conditions. *Food Chem.* **2024**, *435*, 137549. [CrossRef]
4. Tian, S.; Xue, X.A.; Wang, X.; Chen, Z. Preparation of starch-based functional food nano-microcapsule delivery system and its controlled release characteristics. *Abbreviated Front. Nutr.* **2022**, *9*, 982370. [CrossRef] [PubMed]
5. Shan, Y.; Li, J.; Nie, M.; Li, D.; Zhang, Y.; Li, Y.; Wang, L.; Liu, L.; Wang, F.; Tong, L.T. A comprehensive review of starch-based technology for encapsulation of flavor: From methods, materials, and release mechanism to applications. *Carbohydr. Polym.* **2024**, *348*, 122816. [CrossRef]
6. Rostamabadi, H.; Falsafi, S.R.; Rostamabadi, M.M.; Assadpour, E.; Jafari, S.M. Electrospraying as a novel process for the synthesis of particles/nanoparticles loaded with poorly water-soluble bioactive molecules. *Adv. Colloid. Interface Sci.* **2021**, *290*, 102384. [CrossRef]
7. Castro-Coelho, S.; Nogueiro-Estevinho, B.N.; Rocha, F. Encapsulation in food industry with emerging electrohydrodynamic techniques: Electrospinning and electrospraying—A review. *Food Chem.* **2021**, *339*, 127850. [CrossRef]
8. Liu, L.; Tao, L.; Chen, J.; Zhang, T.; Xu, J.; Ding, M.; Wang, X.; Zhong, J. Fish oil-gelatin core-shell electrospun nanofibrous membranes as promising edible films for the encapsulation of hydrophobic and hydrophilic nutrients. *LWT* **2021**, *146*, 111500. [CrossRef]
9. Cuellar-Sánchez, V.; González-Vázquez, M.G.; García-Hernández, A.B.; Serrano-Villa, F.S.; de la Paz Salgado Cruz, M.; García-Bórquez, A.; Morales-Sánchez, E.; Farrera-Rebollo, R.R.; Calderón-Domínguez, G. Starch Biodegradable Films Produced by Electrospraying. In *Starch—Evolution and Recent Advances*, 1st ed.; Martins, O.E., Ed.; ItechOpen: London, UK, 2022; ISSN 2632-0983.
10. Lim, L.T. Electrospinning and electrospraying technologies for food and packaging applications. In *Electrospun Polymers and Composites*; Woodhead Publishing: Cambridge, UK, 2021; Volume 3, pp. 217–259.
11. Gürtler, A.L.; Linseisen, I.; Grohganz, H.; Heinz, A. Coaxial electrospinning of polycaprolactone—A design of experiments approach. *Eur. Polym. J.* **2024**, *208*, 112886. [CrossRef]
12. Contreras-Jiménez, B.; Vázquez-Contreras, G.; de los Ángeles Cornejo-Villegas, M.; del Real-López, A.; Rodríguez-García, M.E. Structural, morphological, chemical, vibrational, pasting, rheological, and thermal characterization of isolated jicama (*Pachyrhizus* spp.) starch and jicama starch added with $Ca(OH)_2$. *Food Chem.* **2019**, *283*, 83–91. [CrossRef] [PubMed]
13. Chakraborty, I.; Pooja, N.; Mal, S.S.; Paul, U.C.; Rahman, M.H.; Mazumder, N. An insight into the gelatinization properties influencing the modified starches used in food industry: A review. *Food Bioprocess Technol.* **2022**, *15*, 1195–1223. [CrossRef]
14. Sánchez-de la Concha, B.B.; Agama-Acevedo, E.; Nuñez-Santiago, M.C.; Bello-Perez, L.A.; Garcia, H.S.; Alvarez-Ramirez, J. Acid hydrolysis of waxy starches with different granule size for nanocrystal production. *J. Cereal Sci.* **2018**, *79*, 193–200. [CrossRef]
15. Chiu, C.; Solarek, D. Modification of Starches. In *Starch: Chemistry and Technology*, 3rd ed.; Academic Press: Cambridge, MA, USA, 2009; pp. 629–655.
16. Kong, X.; Kasapis, S.; Bao, J.; Corke, H. Influence of acid hydrolysis on thermal and rheological properties of amaranth starches varying in amylose content. *JSFA* **2012**, *92*, 1800–1807. [CrossRef]
17. Pratiwi, M.; Faridah, D.N.; Lioe, H.N. Structural changes to starch after acid hydrolysis, debranching, autoclaving-cooling cycles, and heat moisture treatment (HMT): A review. *Starch-Stärke* **2018**, *70*, 1700028. [CrossRef]
18. Rostamabadi, H.; Falsafi, S.R.; Jafari, S.M. Nanostructures of starch for encapsulation of food ingredients. In *Biopolymer Nanostructures for Food Encapsulation Purposes*; Academic Press: Cambridge, MA, USA, 2019; pp. 419–462.
19. de Siqueira, G.L.D.A.; Hornung, P.S.; da Silveira, A.C.; da Silveira Lazzarotto, S.R.; do Prado Cordoba, L.; Schnitzler, E.; Lazzarotto, M. Impact of treatment with HCL/alcoholic in the modifica tion of corn starch. *JTAC* **2017**, *129*, 1705–1713.
20. Li, C.; Hu, Y. Effects of acid hydrolysis on the evolution of starch fine molecular structures and gelatinization properties. *Food Chem.* **2021**, *353*, 129449. [CrossRef] [PubMed]
21. BeMiller, J.N. Dextrins. In *Encyclopedia of Food Sciences and Nutrition*, 2nd ed.; Academic Press: Cambridge, MA, USA, 2003; pp. 1773–1775.
22. Singh, G.P.; Bangar, S.P.; Aayush, K.; Yang, T.; Verma, R.; Kuca, K.; Kumar, D.; Phimolsiripol, Y. Value addition of mango kernel for development and characterization of starch with starch nanoparticles for packaging applications. *Int. J. Biol. Macromol.* **2024**, *274*, 133185. [CrossRef]
23. Kishore, A.; Patil, R.J.; Singh, A.; Pati, K. Jicama (*Pachyrhizus* spp.) a nonconventional starch: A review on isolation, composition, structure, properties, modifications and its application. *Int. J. Biol. Macromol.* **2024**, *258*, 129095. [CrossRef]

24. Ramírez-Miranda, M.; Ribotta, P.D.; Silva-González, A.Z.Z.; Salgado-Cruz, M.D.L.P.; Andraca-Adame, J.A.; Chanona-Pérez, J.J.; Calderón-Domínguez, G. Morphometric and crystallinity changes on jicama starch (*Pachyrizus erosus*) during gelatinization and their relation with in vitro glycemic index. *Starch-Stärke* **2017**, *69*, 1600281. [CrossRef]
25. González-Lemus, L.B.; Calderón-Domínguez, G.; Salgado-Cruz, M.D.L.P.; Díaz-Ramírez, M.; Ramírez-Miranda, M.; Chanona-Pérez, J.J.; Güemes-Vera, N.; Farrera-Rebollo, R.R. Ultrasound-assisted extraction of starch from frozen jicama (*P. erosus*) roots: Effect on yield, structural characteristics and thermal properties. *CYTA—J. Food* **2018**, *10*, 1146–1158. [CrossRef]
26. Fonseca, L.M.; da Cruz, E.P.; Crizel, R.L.; Jansen-Alves, C.; Dias, A.R.G.; da Rosa Zavareze, E. New advances of electrospun starch fibers, encapsulation, and food applications: A review. *Trends Food Sci. Technol.* **2024**, *147*, 104467. [CrossRef]
27. Pareta, R.; Edirisinghe, M.J. A novel method for the preparation of biodegradable microspheres for protein drug delivery. *J. R. Soc. Interface* **2006**, *3*, 573–582. [CrossRef]
28. Ghaeb, M.; Tavanai, H.; Kadivar, M. Electrosprayed maize starch and its constituents (amylose and amylopectin) nanoparticles. *Polym. Adv. Technol.* **2015**, *26*, 917–923. [CrossRef]
29. Zaeim, D.; Sarabi-Jamab, M.; Ghorani, B.; Kadkhodaee, R. Double layer co-encapsulation of probiotics and prebiotics by electro-hydrodynamic atomization. *LWT* **2019**, *110*, 102–109. [CrossRef]
30. Lin, Q.; Liu, Y.; Zhou, L.; Ji, N.; Xiong, L.; Sun, Q. Green preparation of debranched starch nanoparticles with different crystalline structures by electrostatic spraying. *Food Hydr.* **2022**, *127*, 107513. [CrossRef]
31. Pérez-Masiá, R.; Lagaron, J.M.; López-Rubio, A. Surfactant-aided electrospraying of low molecular weight carbohydrate polymers from aqueous solutions. *Carbohydr. Polym.* **2014**, *101*, 249–255. [CrossRef]
32. Aguirre-Güitrón, L.; Calderón-Santoyo, M.; Lagarón, J.M.; Prieto, C.; Ragazzo-Sánchez, J.A. Formulation of the biological control yeast Meyerozyma caribbica by electrospraying process: Effect on postharvest control of anthracnose in mango (*Mangifera indica* L.) and papaya (*Carica papaya* L.). *JSFA.* **2021**, *102*, 696–706. [CrossRef]
33. Biduski, B.; Kringel, D.H.; Colussi, R.; dos Santos Hackbart, H.C.; Lim, L.T.; Dias, A.R.G.; da Rosa Zavareze, E. Electrosprayed octenyl succinic anhydride starch capsules for rosemary essential oil encapsulation. *Int. J. Biol. Macromol.* **2019**, *132*, 300–307. [CrossRef]
34. Fonseca, L.M.; dos Santos Cruxen, C.E.; Bruni, G.P.; Fiorentini, Â.M.; da Rosa Zavareze, E.; Lim, L.T.; Dias, A.R.G. Development of antimicrobial and antioxidant electrospun soluble potato starch nanofibers loaded with carvacrol. *Int. J. Biol. Macromol.* **2019**, *139*, 1182–1190. [CrossRef] [PubMed]
35. Pacheco-da Cruz, E.; Martins-Fonseca, L.; Radünz, M.; Tavares-da Silva, F.; Avila-Gandra, E.; da Rosa-Zavareze, E.; Dellinghausen-Borges, C. Pinhão coat extract encapsulated in starch ultrafine fibers: Thermal, antioxidant, and antimicrobial properties and in vitro biological digestion. *J. Food Sci.* **2021**, *86*, 2886–2897. [CrossRef] [PubMed]
36. López-Hernández, R.E.; García-Solís, S.E.; Monroy-Rodríguez, I.; Cornejo-Mazón, M.; Calderón-Domínguez, G.; Alamilla-Beltrán, L.; Hernández-Sánchez, H.; Gutiérrez-López, G.F. Preparation and characterization of canola oil-in-water Pickering emulsions stabilized by barley starch nanocrystals. *J. Food Eng.* **2022**, *326*, 111037. [CrossRef]
37. González-Vázquez, M.; Calderón-Domínguez, G.; Mora-Escobedo, R.; Salgado-Cruz, M.P.; Arreguín-Centeno, J.H.; Monterrubio-López, R. Polysaccharides of nutritional interest in jicama (*Pachyrhizus erosus*) during root development. *Food Sci. Nutr.* **2022**, *10*, 1146–1158. [CrossRef]
38. Yun, S.H.; Matheson, N.K. Estimation of amylose content of starches after precipitation of amylopectin by concanavalin-A. *Starch-Stärke* **1990**, *42*, 302–305. [CrossRef]
39. Lu, X.; Xiao, J.; Huang, Q. Pickering emulsions stabilized by media-milled starch particles. *Food Res. Int.* **2018**, *105*, 140–149. [CrossRef]
40. Babu, A.S.; Parimalavalli, R.; Rudra, S.G. Effect of citric acid concentration and hydrolysis time on physicochemical properties of sweet potato starches. *Int. J. Biol. Macromol.* **2015**, *80*, 557–565. [CrossRef]
41. Shoukat, R.; Cappai, M.; Pilia, L.; Pia, G. Rice Starch Chemistry, Functional Properties, and Industrial Applications: A Review. *Polymers* **2025**, *17*, 110. [CrossRef] [PubMed]
42. Mauro, R.R.; Vela, A.J.; Ronda, F. Impact of starch concentration on the pasting and rheological properties of gluten-free gels. Effects of amylose content and thermal and hydration properties. *Foods* **2023**, *12*, 2281. [CrossRef]
43. Li, S.; Kong, L.; Ziegler, G.R. Electrospinning of octenylsuccinylated starch-pullulan nanofibers from aqueous dispersions. *Carbohydr. Polym.* **2021**, *258*, 116933. [CrossRef]
44. Rentería-Ortega, M.; Salgado-Cruz, M.D.L.P.; Morales-Sánchez, E.; Alamilla-Beltrán, L.; Farrera-Rebollo, R.R.; Valdespino León, M.; Calderón-Domínguez, G. Effect of electrohydrodynamic atomization conditions on morphometric characteristics and mechanical resistance of chia mucilage-alginate particles. *CYTA—J. Food.* **2020**, *18*, 461–471. [CrossRef]
45. Kong, L.; Ziegler, G.R. Quantitative relationship between electrospinning parameters and starch fiber diameter. *Carbohydr. Polym.* **2013**, *92*, 1416–1422. [CrossRef]
46. Wang, S.; Chao, C.; Huang, S.; Yu, J. Phase Transitions of Starch and Molecular Mechanisms. In *Starch Structure, Functionality and Application in Foods*; Wang, S., Ed.; Springer: Singapore, 2020; pp. 9–27.

47. Stevenson, D.G.; Jane, J.L.; Inglett, G.E. Characterization of Jicama (Mexican Potato) (*Pachyrhizus erosus* L. Urban) Starch from taproots Grown in USA and Mexico. *Starch/Stärke* **2007**, *59*, 132–140. [CrossRef]
48. Amaya-Llano, S.L.; Martínez-Bustos, F.; Martínez Alegría, A.L.; de Jesús Zazueta-Morales, J. Comparative studies on some physico-chemical, thermal, morphological, and pasting properties of acid-thinned jicama and maize starches. *Food Bioprocess Technol.* **2011**, *4*, 48–60. [CrossRef]
49. Kim, H.Y.; Lee, J.H.; Kim, J.Y.; Lim, W.J.; Lim, S.T. Characterization of nanoparticles prepared by acid hydrolysis of various starches. *Starch-Stärke* **2012**, *64*, 367–373. [CrossRef]
50. Martínez-Bustos, F.; Amaya-Llano, S.L.; Carbajal-Arteaga, J.A.; Chang, Y.K.; Zazueta-Morales, J.D.J. Physicochemical properties of cassava, potato and jicama starches oxidised with organic acids. *J. Sci. Food Agric.* **2007**, *87*, 1207–1214. [CrossRef]
51. Fatokun, O.T. Micrometrics and morphological properties of starch. In *Chemical Properties of Starch*, 1st ed.; Martins, E., Ed.; IntechOpen: London, UK, 2019; p. 143.
52. Guo, C.; Geng, S.; Shi, Y.; Yuan, C.; Liu, B. Effect of sulfuric acid hydrolysis on the structure and Pickering emulsifying capacity of acorn starch. *Food Chem. X* **2024**, *22*, 101277. [CrossRef]
53. Vidal, B.C., Jr.; Rausch, K.D.; Tumbleson, M.E.; Singh, V. Kinetics of Granular Starch Hydrolysis in Corn Dry-Grind Process. *Starch-Stärke* **2009**, *61*, 448–456. [CrossRef]
54. Nijman, M. Thermal Analysis. In *Practice: Tips and Hints*; Mettler Toledo: Greifensee, Switzerland, 2018; Volume 2.
55. Mason, W.R. Starch use in foods. In *Starch*; Academic Press: Cambridge, MA, USA, 2009; pp. 745–795.
56. Wang, S.; Copeland, L. Effect of acid hydrolysis on starch structure and functionality: A review. *CRFSN* **2015**, *55*, 1081–1097. [CrossRef]
57. Shi, X.D.; Huang, J.J.; Han, J.Z.; Wang, S.Y. Physicochemical and functional properties of starches from pachyrhizus erosus with low digestibility. *eFood* **2021**, *2*, 154–161. [CrossRef]
58. Wigati, L.P.; Wardana, A.A.; Tanaka, F.; Tanaka, F. Edible film of native jicama starch, agarwood Aetoxylon Bouya essential oil and calcium propionate: Processing, mechanical, thermal properties and structure. *Int. J. Biol. Macromol.* **2022**, *209*, 597–607. [CrossRef]
59. Marta, H.; Hasya, H.N.L.; Lestari, Z.I.; Cahyana, Y.; Arifin, H.R.; Nurhasanah, S. Study of changes in crystallinity and functional properties of modified sago starch (*Metroxylon* sp.) using physical and chemical treatment. *Polymers* **2022**, *14*, 4845. [CrossRef]
60. Zhao, X.; Xing, J.J.; An, N.N.; Li, D.; Wang, L.J.; Wang, Y. Succeeded high-temperature acid hydrolysis of granular maize starch by introducing heat-moisture pre-treatment. *Int. J. Biol. Macromol.* **2022**, *222*, 2868–2877. [CrossRef]
61. Kanyuck, K.M.; Mills, T.B.; Norton, I.T.; Norton-Welch, A.B. Temperature influences on network formation of low DE maltodextrin gels. *Carbohydr. Polym.* **2019**, *218*, 170–178. [CrossRef]
62. Chen, X.; Hou, Y.; Wang, Z.; Liao, A.; Pan, L.; Zhang, M.; Xue, Y.; Wang, J.; Huang, J. A comparative study of resistant dextrins and resistant maltodextrins from different tuber crop starches. *Polymers* **2023**, *15*, 4545. [CrossRef] [PubMed]
63. Van der Sman, R.G.M.; Ubbink, J.; Dupas-Langlet, M.; Kristiawan, M.; Siemons, I. Scaling relations in rheology of concentrated starches and maltodextrins. *Food Hydrocoll.* **2022**, *124*, 107306. [CrossRef]
64. Iranshahi, K.; Defraeye, T.; Rossi, R.M.; Müller, U.C. Electrohydrodynamics and its applications: Recent advances and future perspectives. *Int. J. Heat. Mass. Transf.* **2024**, *232*, 125895. [CrossRef]
65. Pires, J.B.; Fonseca, L.M.; Siebeneichler, T.J.; Crizel, R.L.; Santos, F.N.; Hackbart, H.C.S.H.; Kringel, D.H.; Meinhart, A.D.; Zavareze, E.R.; Dias, A.R.G. Curcumin encapsulation in capsules and fibers of potato starch by electrospraying and electrospinning: Thermal resistance and antioxidant activity. *Food Res. Int.* **2024**, *162*, 112111. [CrossRef]
66. Stijnman, A.C.; Bodnar, I.; Tromp, R.H. Electrospinning of food-grade polysaccharides. *Food Hydrocoll.* **2011**, *25*, 1393–1398. [CrossRef]
67. Cai, J.; Zhang, D.; Zhou, R.; Zhu, R.; Fei, P.; Zhu, Z.-Z.; Cheng, S.Y.; Ding, W. Hydrophobic interface starch nanofibrous film for food packaging: From bioinspired design to self-cleaning action. *J. Agric. Food Chem.* **2021**, *69*, 5067–5075. [CrossRef]

polymers

MDPI

Article

An Assessment of Anion Exchange Membranes for CO_2 Capture Processes: A Focus on Fumasep® and Sustainion®

Kseniya Papchenko [1], Sandra Kentish [2,*] and Maria Grazia De Angelis [1,*]

[1] School of Engineering, Institute of Materials and Processes, University of Edinburgh, Edinburgh EH9 3FB, UK; kpapchen@ed.ac.uk
[2] Department of Chemical Engineering, The University of Melbourne, Parkville, VIC 3010, Australia
* Correspondence: sandraek@unimelb.edu.au (S.K.); grazia.deangelis@ed.ac.uk (M.G.D.A.)

Abstract: Anion exchange membranes are utilised in cutting-edge energy technologies including electrolysers and fuel cells. Recently, these membranes have also emerged as a promising tool in CO_2 capture techniques, such as moisture-driven direct air capture and the separation of CO_2 from other gases, leveraging the moisture-induced sorption/desorption and diffusion of CO_2 in its ionic forms. In this study, we examine the absorption and permeation of CO_2 and CH_4 in two commercially available anion exchange membranes, Fumasep® and Sustainion®, under dry conditions. With the exception of CO_2 sorption in Fumasep®, these measurements have not been previously reported. These new data points are crucial for evaluating the fundamental separation capabilities of these materials and for devising innovative CO_2 capture strategies, as well as for the simulation of novel combined processes. In a dry state, both materials demonstrate similar CO_2 absorption levels, with a higher value for Sustainion®. The CO_2 solubility coefficient decreases with pressure, as is typical for glassy polymers. Fumasep® exhibits higher CO_2/CH_4 ideal solubility selectivity, equal to ~10 at sub-ambient pressures, and higher diffusivity. The CO_2 diffusion coefficient increases with the CO_2 concentration in both membranes due to swelling of the matrix, varying between 0.7 and 2.2×10^{-8} cm^2/s for Fumasep® and between 1.6 and 9.0×10^{-9} cm^2/s for Sustainion®. CO_2 permeability exhibits a minimum at a pressure of approximately 2–3 bar. The CO_2 permeability in the dry state is higher in Fumasep® than in Sustainion®: 3.43 and 0.72 Barrer at a 2-bar transmembrane pressure, respectively. The estimated perm-selectivity was found to reach values of up to 40 at sub-ambient pressures. The CO_2 permeability and CO_2/CH_4 estimated perm-selectivity in both polymers are of a similar order of magnitude to those measured in fluorinated ion exchange membranes such as Nafion®.

Keywords: CO_2 capture; anion exchange membranes; moisture swing capture; diffusion; sorption

1. Introduction

Escalating CO_2 emissions from a broad range of industrial sectors have disrupted Earth's natural carbon cycle, exacerbated the greenhouse effect, and led to climate change. In response to this pressing environmental challenge, CO_2 capture and utilisation strategies have garnered significant interest due to their potential to mitigate anthropogenic greenhouse gas emissions and convert CO_2 into high-value products. Membrane-based technologies have gained traction in recent decades due to their energy efficiency, modularity, and cost-effectiveness. Their application in carbon capture from various sources has been envisioned, offering potential for both point-source decarbonisation [1,2] and direct carbon sequestration from the air [3,4], when used either independently or integrated with other technologies [5,6].

Within the spectrum of novel energy and decarbonisation technologies, ion exchange materials have emerged as crucial components. Such membranes are used in devices such as electrolysers and fuel cells, which will be important players in the energy transition in place in the industrial and energy sectors. Furthermore, they have been evaluated in recent years as materials for assisting CO_2 capture processes. Their high perm-selectivity is particularly beneficial in striving toward net-zero greenhouse gas emission goals. Such membranes have been evaluated mostly in the context of supporting electrochemical CO_2 abatement processes such as electrodialysis (ED) and the electrochemical reduction of CO_2, which are instrumental in the capture and utilisation of CO_2 [7–11]. In these applications, the ion exchange membrane, characterised by a high charge density and high ion perm-selectivity, facilitates the directional migration, separation, and categorisation of ionic substances under a direct current field. In CO_2 reduction, it has been observed that anion exchange membranes (AEMs) may perform better than proton exchange membranes [7,9].

More recently, the use of anion exchange membranes or resin beads has been proposed in the process of moisture-driven carbon capture from the air or other dilute sources [12–14]. In such a non-electrochemical process, the driving force is represented by humidity switches or gradients [15–17]. These processes exploit the fact that the CO_2 uptake into an AEM appropriately ion-exchanged into bicarbonate or carbonate form changes as a function of the degree of the membrane's hydration. If a moisture gradient is applied across the membrane, CO_2 is sorbed on the low-humidity side, reacts to form HCO_3^-, diffuses in this ionic form through the membrane, and converts back into CO_2 on the humid side. This process is a cost-effective approach for CO_2 capture: the energy required to concentrate CO_2 is provided by the evaporation of water, significantly reducing the energy consumption compared to that in thermal and vacuum regeneration methods. The design of the appropriate direct air capture schemes within a membrane or cyclic adsorption/desorption layout requires the estimation of several parameters, including the CO_2 sorption and diffusion rates inside the membranes, at different humidity levels.

In this paper, we provide a direct estimation of the CO_2 sorption and diffusion coefficients in the dry state in two AEMs inspected at 30 °C. Although these kinds of gas transport data are rather rare in the literature on ion exchange membranes, they are important for assessing and decoupling the effects of the membrane's chemical structure and free volume on the gas transport properties. Such factors are best observed and determined in the dry state, when the solution–diffusion mechanism holds, rather than during humid transport, as this process is driven by the membrane charges. Furthermore, knowing the dry CO_2 transport parameters is crucial for assessing membranes for the moisture-driven air capture process. Indeed, in such a process, an opposing CO_2 pressure gradient may develop within the membrane due to the enrichment in CO_2 on the downstream side. Thus, membranes demonstrating a lower dry CO_2 permeability are preferred, as they minimise the undesirable CO_2 back-diffusion phenomenon [14].

To assess the usability of such membranes in other relevant CO_2 capture processes, e.g., in the removal of CO_2 from biogas, we also performed CH_4 sorption tests. The gas sorption in both membranes was measured as a function of pressure in the range between 0 and 8 bar. These data, obtained in the dry state, set baseline values for CO_2 permeability and CO_2/CH_4 perm-selectivity, along with their changes with pressure due to polymer swelling. If extrapolated to higher pressures, this information could also be useful for the natural gas treatment processes. The CO_2 permeability and perm-selectivity values recorded in the dry state are regarded as baseline measurements since they are anticipated to rise under humid conditions. This increase is expected because of the emergence of a facilitated transport mechanism, which enhances the mobility of CO_2 in its ionic forms relative to that of CH_4. However, this hypothesis has not been confirmed, as tests were not conducted to verify it.

Furthermore, we have interpolated the experimental data using a correlative model, the Dual-Mode Sorption model, which allowed us to extrapolate the CO_2 uptake outside the experimentally probed region. In addition, we have assessed the sorption of CH_4 under the same pressure conditions. The results allow us to evaluate the gas sorption and sorption-based selectivity values in a range of pressures of industrial interest, such as those encountered in biogas upgrading.

Using CO_2 diffusivity data and empirical correlations for CH_4 diffusivity, we were also able to assess the membranes' diffusion selectivity and perm-selectivity for the CO_2/CH_4 couple and compare them to the available data for other ion exchange membranes.

2. Materials and Methods

2.1. Materials

Two common anion exchange membrane materials with covalently bonded nitrogen-based cationic functional groups were analysed in the present work. Fumasep® FAA-3-50 was purchased from the Fuel Cell Store, Bryan, TX, USA; Sustainion® X37-50 Grade RT was purchased from Dioxide Materials, Boca Raton, FL, USA. Fumasep® is based on a proprietary polymer backbone with covalently bonded quaternary ammonium functionality. The chemical structure of Sustainion® is shown in Figure 1: it contains the imidazolium group as the cation instead of the usual quaternary ammonium group.

Both membranes were ion-exchanged into OH form by soaking them in 1 M KOH solution for 24 h at room temperature. The membranes were then rinsed with deionised water and dried overnight at 105 °C under a vacuum for the complete removal of water prior to their further characterisation.

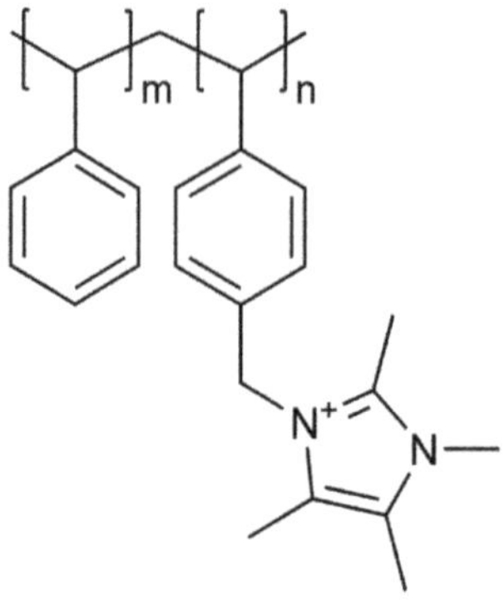

Figure 1. Chemical formula of Sustainion® [18,19].

The thickness was measured to be 50.5 ± 1.4 μm for Fumasep® and 50.0 ± 1.5 μm for Sustainion®, consistent with the declared commercial value of 50 μm of the dry membranes. The density values of the dry membranes were, according to the literature, 1.3 g/cm^3 for Fumasep® [14,20] and 0.9 g/cm^3 for Sustainion® [18,21]. Different sources from the literature show rather different density values for these materials, depending on the humidity conditions in which they were measured and on the counterion. The density value does not affect the estimate of diffusivity derived in this work and only affects the solubility value, expressed per unit mass of the polymer, to a limited extent, estimated to be within 5% using the range of densities reported in the literature.

CO_2 and CH_4, used for the gas sorption tests, were purchased from BOC UK as specialty-grade gas cylinders (Research Grade) with guaranteed minimum purities $\geq$ 99.995%.

2.2. Gas Sorption

The solubility of CO_2 and CH_4 in the two anion exchange membranes in the dry state was determined at 30 °C at pressures of up to 8 bar in a manometric apparatus, built in-house according to ASTM D1434 [22]. The measurements were repeated at least twice for each gas.

The full equipment setup is reported in Figure 2a. A known mass of the polymer sample is placed into a sample cell, and the sample is conditioned at the test temperature under a dynamic vacuum overnight prior to the experiment. The sample chamber is then isolated by closing the interconnecting valve, and the pre-chamber is filled with the desired amount of gas and left until equilibration. The interconnecting valve is then opened, allowing the gas to expand into the sample cell. The mass uptake by the sample is evaluated by measuring the pressure decrease in the closed and calibrated volume over time until a constant value of pressure is reached. The amount of gas absorbed inside the polymer at such a pressure is calculated through a mass balance given the knowledge of the amount of gas in the sample cell prior to expansion, the amount of gas loaded into the pre-chamber, and the residual amount of gas in the gaseous phase in the sample cell after reaching equilibrium. To perform the next incremental step of the isotherm, the cell is once again isolated, the pre-chamber is filled to a higher pressure, and the gas is again left to expand into the sample cell. The procedure is repeated in this stepwise manner until the desired equilibrium pressure is reached, and the whole sorption isotherm is obtained. The system is placed in a thermostatic chamber with forced air convection (Binder KT series), where the temperature is fixed to within ±0.1 °C. The pressure is monitored using two pressure sensors (Druck UNIK 5000, Baker Hughes, Aberdeen, UK) with full scales of 11 and 36 bar and an accuracy of ±0.04% of the full-scale value. The two sensors can be connected separately to the sample cell to allow for an accurate reading and high sensitivity across different pressure intervals. The pressure data are collected using a Eurotherm Nanodac™ recorder (Watlow, West Sussex, UK), controlled via Eurotherm iTools proprietary software (version 9.87).

The solubility coefficient, S, at each equilibrium pressure, p, is determined as the ratio between the concentration of the gas at equilibrium in the sample, c, and the equilibrium pressure, as follows:

$$S = \frac{c}{p} \tag{1}$$

The dependence of the gas concentration on pressure can then be described using several models, based on the nature of the penetrant and the state of the polymer [23].

Sorption isotherms which are concave to the pressure axis, typical for more condensable gases absorbed in glassy polymers, can be described by applying the Dual-Mode Sorption (DMS) model. The DMS model assumes two populations of gas in equilibrium with one another absorbed into the polymer: one absorbed into the dense equilibrium matrix and described by Henry's law, and one absorbed into the non-equilibrium excess volume associated with the glassy state and described by the Langmuir isotherm. The variation in the gas concentration with pressure is then described as the sum of these two contributions:

$$c = k_D p + \frac{C'_H b p}{1 + b p} \tag{2}$$

where k_D is the Henry's law constant, C'_H is the Langmuir capacity constant, and b is the Langmuir affinity parameter. For light gases, linear sorption isotherms can be described by Henry's law as $c = k_D p$. The parameters for the DMS model and Henry's law were obtained from the best fit to the experimental data using Weighted Least Squares (WLS) minimisation, to account for uncertainty in the experimental values.

The penetrant diffusivity in the film, D, can be evaluated from the sorption kinetics at each sorption step in the isothermal run by assuming Fickian diffusion and accounting for the variation in the penetrant's concentration at the film interface [24].

If a dense homogeneous membrane is exposed to a penetrant in a limited volume, as in a pressure-decay experiment, the penetrant's concentration in the bulk, and consequently at the film's interface, will decrease as the penetrant is absorbed into the membrane. The expression for the mass uptake as a function of time, M_t, in this case is given by [24]:

$$\frac{M_t - M_0}{M_\infty - M_0} = 1 - \sum_{n=1}^{\infty} \frac{2\alpha(1+\alpha)}{1+\alpha+\alpha^2 q_n^2} \exp\left(-D\frac{q_n^2 t}{l^2}\right) \quad (3)$$

where M_0 and M_∞ are the initial and final mass uptake, respectively; α is the ratio between the volume of solution and that of the membrane, corrected for the partition coefficient of the penetrant between the gaseous phase and the polymer; and l is the semithickness of the membrane, while the q_n variables are the positive, nonzero solutions of the equation $tg(q_n) = -\alpha\, q_n$. By fitting the experimental data for the mass uptake versus time to the aforementioned equation, using Least Squares (LS) minimisation, we obtain the average diffusivity value within the concentration interval inspected in the differential sorption step. An example of the data output from a sorption step and the relative fitting of the data to Equation (3) to obtain D are reported in Figure 2b. Figure S1 shows the kinetic CO_2 uptake in Sustainion®.

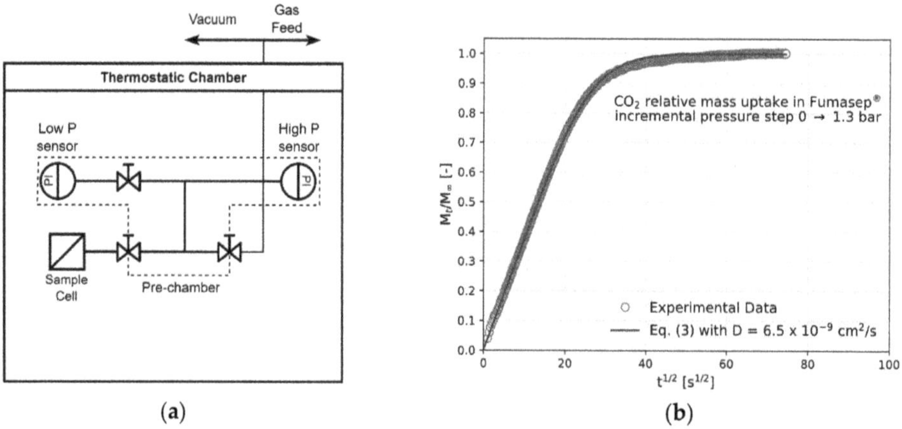

Figure 2. (**a**) A schematic of the pressure-decay apparatus for gas sorption. (**b**) The sorption kinetics of CO_2 in Fumasep® at 30 °C and the fitting of D with Equation (3). The step's initial pressure was 0, and the final pressure was 1.3 bar.

When swelling is associated with the sorption of the penetrant into the polymer, the diffusion coefficient can increase with the concentration. Oftentimes, this dependence can be described through an empirical exponential correlation [25,26]:

$$D = D_0 \exp(\beta c_{av}) \quad (4)$$

where D_0 is the diffusion coefficient at the limit of a penetrant concentration of zero, β is the exponential factor that accounts for the plasticising ability of the penetrant, and c_{av} is the average concentration of the penetrant in the membrane during the sorption step. The parameters for Equation (4) were obtained through best fit to the experimental data, again using the LS minimisation method.

The CO_2 diffusion coefficients were determined for the anion exchange membranes in the present work and correlated to the CO_2 concentration. Due to the low sorption levels, it was not possible to determine the diffusivity values for CH_4 with reasonable accuracy.

Given the dense and homogeneous nature of the investigated materials, the CO_2 permeability coefficient, P, was estimated as a product of its average solubility coefficient and diffusivity coefficient in each step, under the solution–diffusion framework [27,28]:

$$P = \overline{S} \cdot D \text{ with } \overline{S} = \frac{C_{i+1} - C_i}{p_{i+1} - p_i} \quad (5)$$

Gas permeability is typically measured in [mol m m^{-2} s^{-1} Pa^{-1}] in an SI system. A widely adopted unit in the membrane community is [Barrer] = [10^{-10} cm^3(STP) cm cm^{-2} s^{-1} $cmHg^{-1}$]. The conversion between these two units is 1 mol m m^{-2} s^{-1} Pa^{-1} = 2.99 × 10^{15} Barrer.

The ideal perm-selectivity, α_{ij}, between gas i and j is, under negligible downstream pressure conditions, the ratio between the pure gas permeabilities and can be seen as the product of the solubility selectivity, α_{ij}^S, and the diffusivity selectivity, α_{ij}^D:

$$\alpha_{ij} = \frac{P_i}{P_j} = \frac{S_i}{S_j} \cdot \frac{D_i}{D_j} = \alpha_{ij}^S \cdot \alpha_{ij}^D \quad (6)$$

The ideal perm-selectivity is generally used as a first estimate of the separation performance of new materials and a comparison with existing data from the literature. In the present work, we were able to determine the CO_2/CH_4 solubility selectivity for the two anion exchange membranes as function of pressure, which gave an initial indication of the separation capability of these materials.

3. Results

3.1. The CO_2 and CH_4 Sorption Isotherms

The amount of CO_2 and CH_4 absorbed at the end of an isothermal pressure step is reported in Figure 3a and Table S1, in terms of mmol/g of the dry polymer.

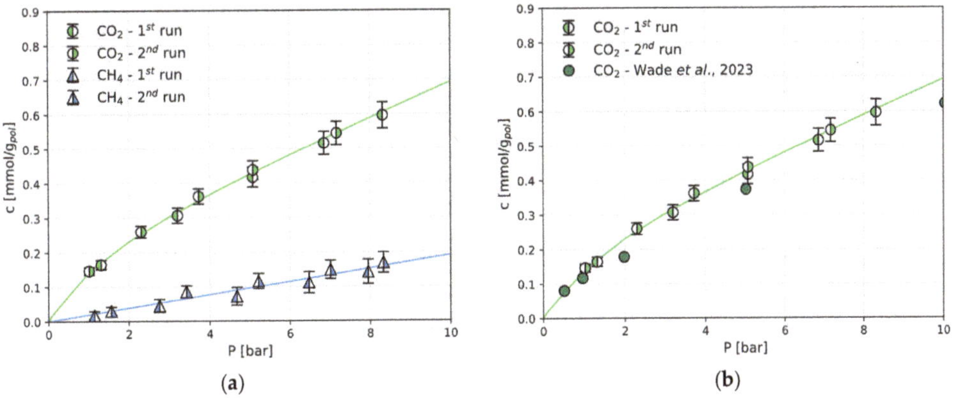

Figure 3. The sorption isotherms in Fumasep® at 30 °C: (**a**) CO_2 and CH_4 sorption isotherms; (**b**) CO_2 sorption in comparison with the literature [14]. The continuous lines are obtained by fitting using the DMS model (Equation (2)) or the linear Henry's law model.

CH_4 sorption follows the linear behaviour of Henry's law in the Fumasep® membrane, while the CO_2 sorption in the same material follows a convex shape (Figure 3a). For the

CO_2 sorption in dry Fumasep®, experimental data are available from Wade et al. [14] under the same conditions used here and are shown in Figure 3b for comparison. It is clear that the data obtained in this work are in excellent agreement with those obtained by independent researchers who have reported the CO_2 sorption in Fumasep®.

The CO_2 and CH_4 sorption isotherms measured under the same conditions in Sustainion® are reported in Figure 4. For this polymer, these are the first experimental data of this kind, and there are no data from the literature to use as benchmark.

Sustainion® is characterised by larger CO_2 sorption levels than those in Fumasep® (Figure 5a), while the same qualitative behaviour is followed. The CH_4 sorption follows a linear trend, as is the case for Fumasep®. The DMS parameters for both gases in both polymers are reported in Table 1: linear isotherms are represented simply by Henry's law via the coefficient k_D. CH_4 uptake is slightly lower in Fumasep® than that in Sustainion®.

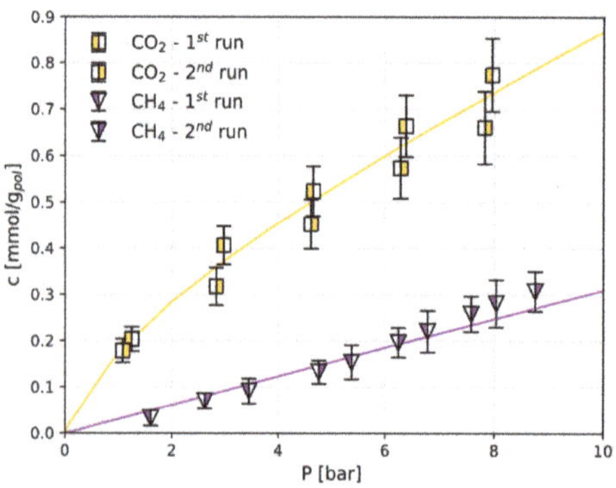

Figure 4. The CO_2 and CH_4 sorption isotherms in Sustainion® at 30 °C in the dry state. The solid lines are obtained by fitting the DMS model (Equation (2)) or a linear model.

Due to the shape of the solubility isotherm, the solubility coefficient ($S = c/p$) of CO_2 in both materials decreases rapidly with pressure and thus the ideal solubility selectivity, $\alpha^S_{CO2/CH4}$, reported in Figure 5b, decreases from values of approximately 10 in Fumasep® and 8 in Sustainion® to above 2, leading to a 64% change in solubility in both materials in the range expected. Interestingly, the best separation performance occurs at a low partial pressure, which is a value of interest for several CO_2 capture processes.

The relatively large uncertainties in the gas solubility values are translated into uncertainties in the model parameters used to describe them, as reported in Table 1. From these, the error in the solubility selectivity can be estimated and is here reported as the shaded areas in Figure 5b. The solubility selectivity at 1 bar is affected by a relative error of 10% in Fumasep® and is equal to 7.3 ± 0.7, while an uncertainty of 20% is estimated for Sustainion®, where $\alpha^S_{CO2/CH4}$ is equal to 5.6 ± 1.1.

Overall, the differences between the two materials are limited and quantitative, rather than qualitative: Fumasep® sorbs slightly less CO_2 but is more selective with respect to CH_4 than Sustainion® in pure gas conditions. The presence of humidity upstream would reduce the gaseous CO_2 uptake [17,29,30]. The total carbon uptake, however, is expected to increase considerably at low humidity levels, as gaseous CO_2 is converted into bicarbonate (HCO_3^-) and carbonate ($CO_3{}^{2-}$) ions. The concentration of these species would then decrease as full-saturation conditions were approached. This peculiar behaviour is the basis of moisture-swing direct carbon capture process, here considered as one potential application. A full scheme of the reactions in the humidified membranes, which is outside of the scope of the present work that deals only with dry membranes, can be found elsewhere [14].

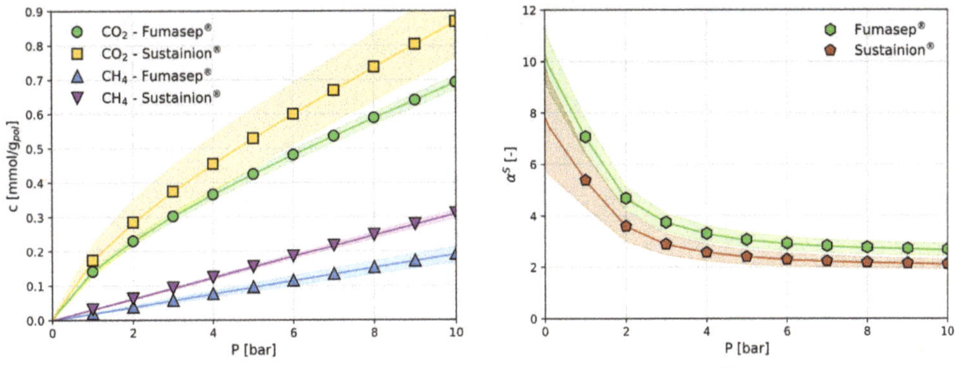

Figure 5. A comparison of the CO_2 and CH_4 sorption in Fumasep® and Sustainion®: (a) the CO_2 and CH_4 sorption isotherms as determined by the DMS (Equation (2)) and linear models; (b) CO_2/CH_4 solubility selectivity $\alpha^S_{CO2/CH4}$ as function of pressure, evaluated using DSM models for each gas. The errors associated with the model parameters are reported as shaded areas.

Table 1. The DMS model (Equation (2)) correlation parameters for the CO_2 and CH_4 sorption isotherms in the anion exchange membranes investigated, together with the coefficient of determination.

Material	CO_2				CH_4	
	$k_D \times 10^2$ mmol g^{-1} bar^{-1}	$C'_H \times 10^1$ mmol g^{-1}	b bar^{-1}	R^2	$k_D \times 10^2$ mmol g^{-1} bar^{-1}	R^2
Fumasep®	4.84 ± 0.04	2.42 ± 0.18	0.62 ± 0.01	0.991	1.92 ± 0.24	0.841
Sustainion®	6.19 ± 0.24	2.90 ± 0.86	0.63 ± 0.11	0.969	3.10 ± 0.12	0.963

3.2. CO_2 and CH_4 Diffusion

The diffusion coefficients for CO_2 were estimated from the transient mass uptake in the polymers, as described in the Methods section, and are shown in Figure 6 and reported in Table S1. These are the first experimentally obtained CO_2 diffusion values for these polymers: a value was previously given in the literature for Fumasep® (3.8×10^{-9} cm^2/s [14]), but this was estimated only as the ratio between independent permeability and solubility data. In our case, the diffusion coefficient was obtained directly by fitting the transient sorption data using Equation (3). Interestingly, the CO_2 diffusion coefficients in both materials follow an exponential dependence on concentration, described by Equation (4) and with relevant parameters reported in Table 2. The CO_2 diffusivity in Sustainion® increases from 1.6 to 9.0×10^{-9} cm^2/s within the range examined, while the diffusion in Fumasep® is significantly faster and changes from 0.7 to 2.2×10^{-8} cm^2/s. These differences could be due to a smaller initial free volume of Sustainion® compared to that in Fumasep®. However, as is typical of membranes characterised by a smaller initial free volume, the dependence on concentration represented by the parameter β is higher for the Sustainion® membrane than that for Fumasep®. The slope is significant and allows the diffusion coefficients to be increased by one order of magnitude by increasing the CO_2 pressure from 0 to 8 bar, indicating an appreciable effect of CO_2-induced swelling on the transport behaviour, which should be taken into account when designing the separation process.

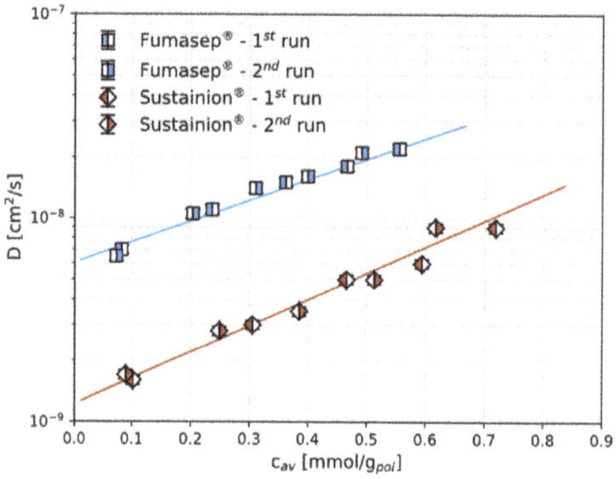

Figure 6. The CO_2 diffusion coefficients in Fumasep® and Sustainion®, determined from the sorption transient at 30 °C in the dry state, as a function of average CO_2 concentration. The continuous lines are obtained from a best fit to the exponential correlation (Equation (4)). In some cases, the error bars are comparable or smaller than the symbol size.

Table 2. The exponential fitting (Equation (4)) correlation parameters for CO_2 diffusion at 30 °C in the AEM investigated, together with the coefficient of determination.

Material	$D_0 \times 10^9$ $cm^2\ s^{-1}$	β $mmol\ g^{-1}$	R^2
Fumasep®	6.07 ± 0.19	2.34 ± 0.15	0.977
Sustainion®	1.22 ± 0.11	2.96 ± 0.31	0.921

The low level of sorption of CH_4 into these polymers did not allow us to obtain reliable sorption transient data and estimate the diffusivity values. Nevertheless, an estimation of the ideal diffusivity selectivity, $\alpha^D_{CO2/CH4}$, can be performed based on the empirical correlation with the kinetic diameters of the penetrants [31,32]. The small-molecule diffusion in dense polymers, where the interchain distance is comparable to or smaller than the penetrant size, is controlled by the free volume elements (FVEs) formed by the thermally activated motion of the polymer chain segments. The pre-exponential term of the diffusion coefficient from Equation (4), $D_{0,i}$, can then be empirically correlated with the activation energy of diffusion, $E_{D,i}$:

$$\ln D_{0,i} = a\frac{E_{D,i}}{RT} - b \tag{7}$$

Here, parameters a and b are independent of the nature of the penetrant, as observed empirically, with a also being independent of the polymer type and exhibiting the universal value of 0.64 [33]. The parameter b is equal to 9.2 and 11.5 in rubbery and glassy polymers, respectively [32,34]. Such a correlation stems from the free volume theory of diffusion, where the migration of the penetrant molecule between neighbouring FVEs depends on their size being large enough to accommodate it.

Within this theory, the activation energy of diffusion, $E_{D,i}$, should be proportional to the volume of the FVE where the molecule jumps and thus be dependent on the penetrant's characteristic dimension, usually taken to be equal to its kinetic diameter, d_i [23,34]:

$$E_{D,i} = cd_i^2 - f \tag{8}$$

Here, c and f are empirically derived polymer-dependent parameters, with c representing the jump length and $\sqrt{f/c}$ representing the interchain distance. These parameters account for the polymeric matrix's flexibility, as a lower energy is required to create an FVE of a required size when the chain can move freely.

By combining Equations (7) and (8), the following equation for the diffusion coefficient of a small molecule i in the polymer, D_i, can be written:

$$\ln D_i = -\left(\frac{1-a}{RT}\right)cd_i^2 + f\left(\frac{1-a}{RT}\right) - b \quad (9)$$

and the following relation can be obtained for the CO_2/CH_4 diffusivity selectivity:

$$\ln \frac{D_{CO_2}}{D_{CH_4}} = \left(\frac{1-a}{RT}\right)c\left(d_{CH_4}^2 - d_{CO_2}^2\right) \quad (10)$$

The values of the polymer-dependent adjustable parameter c can range between 250 cal mol^{-1} Å^{-2} for flexible polymers and 1100 cal mol^{-1} Å^{-2} for stiff-chain polymers [32,33].

Considering d_{CO_2} = 3.3 Å and d_{CH_4} = 3.8 Å (cit); c in the range of 250 to 1100 cal mol^{-1} Å^{-2}; and T = 30 °C, a value for $\alpha^D_{CO2/CH4}$ in the range of 1.7 to 10.3 is obtained, suggesting a positive contribution of the diffusion selectivity to the overall separation capability of Fumasep® and Sustainion®. Assuming medium flexibility for these materials, i.e., c = 650 cal mol^{-1} Å^{-2}, a value of ~4 can be taken as indicative of the diffusivity selectivity in these polymers.

It is expected that, in the presence of humidity, i.e., during a moisture-driven type of transport, CO_2 transport will be facilitated, as it will diffuse as bicarbonate or carbonate anions, resulting in a higher effective carbon diffusion coefficient. CH_4, on the other hand, will continue to diffuse as an uncharged species, with increased transport selectivity as a direct outcome.

3.3. CO_2 and CH_4 Permeation

Finally, the permeability coefficients were evaluated as the product between the diffusivity and solubility for CO_2 in the two materials (Equation (5)) and reported in Figure 7. Due to the combination of solubility and diffusivity effects, Fumasep® shows a higher CO_2 permeability than that of Sustainion®. The permeability varies from 3.4 to 4.7 Barrer and from 0.7 to 1.7 for Sustainion. Also, as S decreases sharply in the low-pressure range while D continues to increase exponentially as S flattens out at medium pressures, the permeability shows a minimum (3.3 Barrer for Fumasep® at about 3 bar and 0.7 Barrer for Sustainion® at 2 bar). This threshold is sometimes indicated as the plasticisation pressure in the literature on membranes and is typically observed in the CO_2 permeation through glassy polymers.

Finally, we report the CO_2 dry permeability values obtained in the literature for other ion exchange membranes, namely two fluorinated matrices which go under the trade names of Aquivion® [35,36] and Nafion® [37,38], which are classified as cation exchange membranes (CEMs). In particular, Nafion NRE 211 was tested after being dried at 80 °C for 48 h [37] at 35 °C and a 2-bar upstream pressure, while Nafion 115 was dried under a vacuum at 35 °C and tested at the same temperature [38]. Aquivion E87-12S was either pre-dried at 50 °C for 24 h and tested at 35 °C and 1 bar [35] or pre-dried at 100 °C for 24 h and tested at 35 °C and 2 bar [36]. The permeability of Nafion® and Aquivion®, shown in Figure 7, lies in between the values obtained in this work for Sustainion® and Fumasep®, respectively.

Table 3 reports the permeability values for CO_2 and CH_4 in the four materials, together with the solubility, diffusivity, and corresponding selectivity values where available. The diffusivity selectivity, estimated for Fumasep® and Sustainion® based on the ratio of the kinetic diameters of the two penetrants, as described in Section 3.2, allows for an estimate of the ideal CO_2/CH_4 selectivity at different pressures. At low pressure, α_{CO_2/CH_4} reaches values of 40 and 31 for Fumasep® and Sustainion®, respectively, showing their strong potential for applications such as biogas upgrading. While the CO_2 permeability and CO_2/CH_4 estimated selectivity of Fumasep® and Sustainion® are comparable with those of other commercially available ion exchange membranes, both materials investigated here offer a 10 times higher CO_2 solubility but a comparable solubility selectivity with respect to these values for Nafion®.

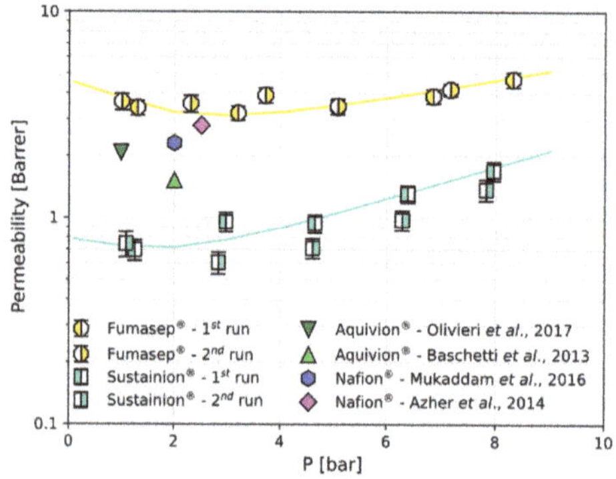

Figure 7. The CO_2 permeability coefficients in Fumasep® and Sustainion® at 30 °C in the dry state, evaluated as the product of the solubility and diffusivity coefficients. The continuous lines are reported as a guide for the eye. The CO_2 permeability in Aquivion® [35,36] and Nafion® [37,38] at 35 °C is reported for comparison. In some cases, the error bars are comparable to or smaller than the symbol size.

Table 3. Permeability, solubility, diffusivity, and ideal selectivity values for materials investigated in the present work and the literature. Values marked with (*) are estimated according to Equation (10).

Material	T °C	P bar	P_{CO_2} Barrer	$S_{CO_2} \times 10^2$ mmol g^{-1} bar^{-1}	$D_{CO_2} \times 10^9$ cm^2 s^{-1}	α_{CO_2/CH_4}	$\alpha^S_{CO_2/CH_4}$	$\alpha^D_{CO_2/CH_4}$	Ref.
Fumasep®	30	2	3.43	9.24	9.55	19.3 *	4.82	4.00 *	This work
Sustainion®	30	2	0.72	11.1	2.40	14.4 *	3.59	4.00 *	This work
Aquivion®	35	1	2.07	-	-	23.8	-	-	[35]
Nafion®	35	2	2.3	1.46	27.0	27.7	4.58	6.00	[37]

The dry-state data reported here are directly applicable to carbon capture design processes involving these materials. One such application is moisture-driven carbon capture from air or other dilute sources. Here, a continuous CO_2 removal process can be sustained upon the provision of an appropriate humidity gradient, with an upstream low-humidity side and a downstream high-humidity side. The design of such a process requires knowledge of the CO_2 sorption and diffusion levels inside the membrane at different humidity levels, with the dry-state parameters serving as reference values. Wade et al. [14] used the pure CO_2 permeability and solubility values in dry Fumasep®, coupled with humidity uptake and transport measurements, to assess its suitability for the moisture-driven separation process. Carbon moves across the AEM as $CO_3{}^{2-}$ and HCO_3—when hydrated, and the transport rate of the ions is determined by the CO_2-H_2O reaction rate and the anion exchange capacity (AEC) of the materials. Wade et al. [14] found that a low molecular CO_2 permeability guarantees higher CO_2 capture rates when such a humidity gradient is applied by limiting the gaseous CO_2 back-diffusion in response to an opposing CO_2 partial pressure gradient. Based on the results presented here, Sustainion® seems to be promising for moisture-swing air capture applications given that its CO_2 permeability is nearly five times lower than that of Fumasep®.

It is important to underline the fundamental difference in the carbon transport in humidified conditions across AEMs such as Fumasep® and Sustainion® and CEMs such as Nafion® and Aquivion®: the CO_2 transport in AEMs occurs primarily in ionic form and is more efficient compared to that in CEMs, where carbon transport occurs indirectly through the hydrophilic channels formed in hydrated conditions [14,36,38–40]. Thus, while the net CO_2 permeability increases in both types of materials when uniformly exposed to a higher relative humidity, the CO_2 uptake in an AEM decreases with an increasing moisture uptake, allowing for the moisture-swing mechanism which is absent in CEMs [29,30,41].

4. Conclusions

In this study, we conducted a detailed analysis of the dry gas transport properties of two commercially available anion exchange membranes, Fumasep® and Sustainion®. Both membranes feature covalently attached nitrogen-based cationic functional groups and show potential for use in various clean energy and carbon-neutral applications. Their increasing application in fuel cells, electrolysis, and notably in CO_2 capture and utilisation processes makes understanding their basic transport properties such as their sorption and diffusion coefficients vital for process design and simulation.

We carried out dry CO_2 and CH_4 sorption tests at 30 °C and pressures of up to 8 bar. These tests revealed that CO_2 exhibits a convex-shaped sorption isotherm in both types of membranes, with Sustainion® displaying a marginally higher sorption capacity than that of Fumasep®. The sorption isotherms for CO_2 can be described using the DSM model. In contrast, CH_4 displayed a linear sorption isotherm, consistent with Henry's law, and showed a lower mass uptake compared to that in CO_2. From these findings, we calculated the ideal solubility selectivity for CO_2 over CH_4 in each membrane. Fumasep® demonstrated a higher selectivity, which decreased with an increasing pressure from 10 to 3 over the pressure range examined.

The diffusion coefficients for CO_2 were determined from the transient phase of the sorption tests at varying concentrations. In both membranes, CO_2 diffusivity followed an exponential relationship with concentration, suggesting significant swelling effects within the membrane structure induced by CO_2. Notably, Fumasep® exhibited a considerably higher CO_2 diffusivity compared to that of Sustainion®, likely due to a greater free volume. This suggests that Sustainion® may be promising for moisture-swing air capture applications, where a low molecular CO_2 diffusion coefficient should guarantee lower back-diffusion of molecular CO_2 when a humidity gradient is applied. The diffusivity selectivity was estimated as equal to ~4 in these materials on the basis of the ratio of the kinetic diameters of the penetrants, suggesting a positive contribution of the diffusion mechanism to the overall separation capabilities of these materials.

The permeability curves, calculated as a function of pressure using the solution–diffusion model, showed that the dry CO_2 permeabilities of both membranes had a comparable order of magnitude to that observed in fluorinated ion exchange polymers like Nafion® and Aquivion®. The permeability decreased initially with an increasing pressure, reaching a minimum at around 2–3 bar, and then increased exponentially. This pattern is typical for glassy polymers, highlighting the complex interplay of solubility and diffusivity under varying pressures. The perm-selectivity was estimated to reach values up to 40 in Fumasep® at low pressures, with the values at higher pressures comparable to those for the other ion exchange polymers.

In conclusion, the results of this study establish a solid groundwork for the development of advanced CO_2 capture and utilisation technologies utilising the selective properties and high charge density of anion exchange membranes. This includes promising applications such as moisture-driven direct air capture, biogas purification, and other innovative processes.

Supplementary Materials: The following supporting information can be downloaded at https://www.mdpi.com/article/10.3390/polym17111581/s1. Table S1. CO_2 and CH_4 sorption isotherms and diffusivity coefficients in Fumasep® and Sustainion®, measured at 30 °C; Figure S1. Sorption kinetics of CO_2 in Sustainion® at 30 °C and fitting of D using Equation (3).

Author Contributions: Conceptualisation: M.G.D.A. Methodology: S.K. and K.P. Software: K.P. Investigation: K.P. Data Curation, K.P. Writing—Original Draft Preparation: M.G.D.A. Writing—Review and Editing: M.G.D.A., K.P. and S.K. Supervision: M.G.D.A. All authors have read and agreed to the published version of the manuscript.

Funding: Prof. Sandra Kentish would like to acknowledge the Australian Research Council Discovery Project DP240101405 for the financial support.

Institutional Review Board Statement: Not applicable.

Data Availability Statement: The data can be made available by the authors upon request.

Conflicts of Interest: The authors declare no conflicts of interest.

References

1. Merkel, T.C.; Lin, H.; Wei, X.; Baker, R. Power Plant Post-Combustion Carbon Dioxide Capture: An Opportunity for Membranes. *J. Membr. Sci.* **2010**, *359*, 126–139. [CrossRef]
2. Ferrari, M.-C.; Amelio, A.; Nardelli, G.M.; Costi, R. Assessment on the Application of Facilitated Transport Membranes in Cement Plants for CO_2 Capture. *Energies* **2021**, *14*, 4772. [CrossRef]
3. Fujikawa, S.; Selyanchyn, R.; Kunitake, T. A New Strategy for Membrane-Based Direct Air Capture. *Polym. J.* **2021**, *53*, 111–119. [CrossRef]
4. Castel, C.; Bounaceur, R.; Favre, E. Membrane Processes for Direct Carbon Dioxide Capture From Air: Possibilities and Limitations. *Front. Chem. Eng.* **2021**, *3*, 668867. [CrossRef]
5. Luis, P.; Van Gerven, T.; Van Der Bruggen, B. Recent Developments in Membrane-Based Technologies for CO_2 Capture. *Prog. Energy Combust. Sci.* **2012**, *38*, 419–448. [CrossRef]
6. Drioli, E.; Stankiewicz, A.I.; Macedonio, F. Membrane Engineering in Process Intensification—An Overview. *J. Membr. Sci.* **2011**, *380*, 1–8. [CrossRef]
7. Wang, H.; Yan, J.; Song, W.; Jiang, C.; Wang, Y.; Xu, T. Ion Exchange Membrane Related Processes towards Carbon Capture, Utilization and Storage: Current Trends and Perspectives. *Sep. Purif. Technol.* **2022**, *296*, 121390. [CrossRef]
8. Li, M.; Yang, K.; Abdinejad, M.; Zhao, C.; Burdyny, T. Advancing Integrated CO_2 Electrochemical Conversion with Amine-Based CO_2 Capture: A Review. *Nanoscale* **2022**, *14*, 11892–11908. [CrossRef]
9. Rehberger, H.; Rezaei, M.; Aljabour, A. Challenges and Opportunities of Choosing a Membrane for Electrochemical CO_2 Reduction. *Membranes* **2025**, *15*, 55. [CrossRef]
10. Pimlott, D.J.D.; Kim, Y.; Berlinguette, C.P. Reactive Carbon Capture Enables CO_2 Electrolysis with Liquid Feedstocks. *Acc. Chem. Res.* **2024**, *57*, 1007–1018. [CrossRef]
11. Li, M.; Irtem, E.; Iglesias Van Montfort, H.-P.; Abdinejad, M.; Burdyny, T. Energy Comparison of Sequential and Integrated CO_2 Capture and Electrochemical Conversion. *Nat Commun* **2022**, *13*, 5398. [CrossRef]
12. Wang, T.; Lackner, K.S.; Wright, A. Moisture Swing Sorbent for Carbon Dioxide Capture from Ambient Air. *Environ. Sci. Technol.* **2011**, *45*, 6670–6675. [CrossRef]
13. Prajapati, A.; Sartape, R.; Rojas, T.; Dandu, N.K.; Dhakal, P.; Thorat, A.S.; Xie, J.; Bessa, I.; Galante, M.T.; Andrade, M.H.S.; et al. Migration-Assisted, Moisture Gradient Process for Ultrafast, Continuous CO_2 Capture from Dilute Sources at Ambient Conditions. *Energy Environ. Sci.* **2022**, *15*, 680–692. [CrossRef]
14. Wade, J.L.; Lopez Marques, H.; Wang, W.; Flory, J.; Freeman, B. Moisture-Driven CO_2 Pump for Direct Air Capture. *J. Membr. Sci.* **2023**, *685*, 121954. [CrossRef]
15. Liu, S.; Hu, J.; Zhang, F.; Zhu, J.; Shi, X.; Wang, L. Robust Enhancement of Direct Air Capture of CO2 Efficiency Using Micro-Sized Anion Exchange Resin Particles. *Sustainability* **2024**, *16*, 3601. [CrossRef]
16. Kaneko, Y.; Lackner, K.S. Kinetic Model for Moisture-Controlled CO_2 Sorption. *Phys. Chem. Chem. Phys.* **2022**, *24*, 21061–21077. [CrossRef]

17. Wang, T.; Lackner, K.S.; Wright, A.B. Moisture-Swing Sorption for Carbon Dioxide Capture from Ambient Air: A Thermodynamic Analysis. *Phys. Chem. Chem. Phys.* **2013**, *15*, 504–514. [CrossRef]
18. Chakraborti, T.; Sharma, R.; Krishnamoorthy, A.N.; Chaudhari, H.; Mamtani, K.; Singh, J.K. Unravelling the Effect of Molecular Interactions on Macroscale Properties in Sustainion Anion Exchange Membrane (AEM) under Hydrated Conditions Using MD Simulations. *J. Membr. Sci.* **2024**, *705*, 122887. [CrossRef]
19. Liu, Z.; Yang, H.; Kutz, R.; Masel, R.I. CO_2 Electrolysis to CO and O_2 at High Selectivity, Stability and Efficiency Using Sustainion Membranes. *J. Electrochem. Soc.* **2018**, *165*, J3371–J3377. [CrossRef]
20. Jang, E.-S.; Kamcev, J.; Kobayashi, K.; Yan, N.; Sujanani, R.; Talley, S.J.; Moore, R.B.; Paul, D.R.; Freeman, B.D. Effect of Water Content on Sodium Chloride Sorption in Cross-Linked Cation Exchange Membranes. *Macromolecules* **2019**, *52*, 2569–2579. [CrossRef]
21. Kubannek, F.; Zhegur-Khais, A.; Li, S.; Dekel, D.R.; Krewer, U. Model-Based Insights into the Decarbonation Dynamics of Anion-Exchange Membranes. *Chem. Eng. J.* **2023**, *459*, 141534. [CrossRef]
22. *ASTM D1434-82(2015)e1*; Standard Test Method for Determining Gas Permeability Characteristics of Plastic Film and Sheeting. F02 Committee ASTM International: West Conshohocken, PA, USA, 2015. [CrossRef]
23. Ricci, E.; Minelli, M.; De Angelis, M.G. Modelling Sorption and Transport of Gases in Polymeric Membranes across Different Scales: A Review. *Membranes* **2022**, *12*, 857. [CrossRef] [PubMed]
24. Crank, J. *The Mathematics of Diffusion*, 2nd ed.; Oxford Science Publications; Clarendon Press: London, UK; Oxford University Press: Oxford, UK, 1975; ISBN 978-0-19-853411-2.
25. Duthie, X.; Kentish, S.; Powell, C.; Nagai, K.; Qiao, G.; Stevens, G. Operating Temperature Effects on the Plasticization of Polyimide Gas Separation Membranes. *J. Membr. Sci.* **2007**, *294*, 40–49. [CrossRef]
26. Minelli, M. Modeling CO_2 Solubility and Transport in Poly(Ethylene Terephthalate) above and below the Glass Transition. *J. Membr. Sci.* **2014**, *451*, 305–311. [CrossRef]
27. Wijmans, J.G.; Baker, R.W. The Solution-Diffusion Model: A Review. *J. Membr. Sci.* **1995**, *107*, 1–21. [CrossRef]
28. Wijmans, J.G.H.; Baker, R.W. The Solution–Diffusion Model: A Unified Approach to Membrane Permeation. In *Materials Science of Membranes for Gas and Vapor Separation*; Yampolskii, Y., Pinnau, I., Freeman, B., Eds.; Wiley: Hoboken, NJ, USA, 2006; pp. 159–189. ISBN 978-0-470-85345-0.
29. Shi, X.; Xiao, H.; Liao, X.; Armstrong, M.; Chen, X.; Lackner, K.S. Humidity Effect on Ion Behaviors of Moisture-Driven CO_2 Sorbents. *J. Chem. Phys.* **2018**, *149*, 164708. [CrossRef]
30. Shi, X.; Xiao, H.; Kanamori, K.; Yonezu, A.; Lackner, K.S.; Chen, X. Moisture-Driven CO_2 Sorbents. *Joule* **2020**, *4*, 1823–1837. [CrossRef]
31. Freeman, B.D. Basis of Permeability/Selectivity Tradeoff Relations in Polymeric Gas Separation Membranes. *Macromolecules* **1999**, *32*, 375–380. [CrossRef]
32. Checchetto, R.; De Angelis, M.G.; Minelli, M. Exploring the Membrane-Based Separation of CO_2/CO Mixtures for CO_2 Capture and Utilisation Processes: Challenges and Opportunities. *Sep. Purif. Technol.* **2024**, *346*, 127401. [CrossRef]
33. Van Krevelen, D.W.; te Nijenhuis, K. *Properties of Polymers: Their Correlation with Chemical Structure: Their Numerical Estimation and Prediction from Additive Group Contributions*, 4th ed.; Elsevier: Amsterdam, The Netherlands, 2009; ISBN 978-0-08-054819-7.
34. Meares, P. The Diffusion of Gases Through Polyvinyl Acetate[1]. *J. Am. Chem. Soc.* **1954**, *76*, 3415–3422. [CrossRef]
35. Olivieri, L.; Aboukeila, H.; Giacinti Baschetti, M.; Pizzi, D.; Merlo, L.; Sarti, G.C. Humid Permeation of CO_2 and Hydrocarbons in Aquivion® Perfluorosulfonic Acid Ionomer Membranes, Experimental and Modeling. *J. Membr. Sci.* **2017**, *542*, 367–377. [CrossRef]
36. Giacinti Baschetti, M.; Minelli, M.; Catalano, J.; Sarti, G.C. Gas Permeation in Perflurosulfonated Membranes: Influence of Temperature and Relative Humidity. *Int. J. Hydrogen Energy* **2013**, *38*, 11973–11982. [CrossRef]
37. Mukaddam, M.; Litwiller, E.; Pinnau, I. Gas Sorption, Diffusion, and Permeation in Nafion. *Macromolecules* **2016**, *49*, 280–286. [CrossRef]
38. Azher, H.; Scholes, C.A.; Stevens, G.W.; Kentish, S.E. Water Permeation and Sorption Properties of Nafion 115 at Elevated Temperatures. *J. Membr. Sci.* **2014**, *459*, 104–113. [CrossRef]
39. Ma, S.; Skou, E. CO_2 Permeability in Nafion® EW1100 at Elevated Temperature. *Solid State Ion.* **2007**, *178*, 615–619. [CrossRef]
40. Ziv, N.; Mondal, A.N.; Weissbach, T.; Holdcroft, S.; Dekel, D.R. Effect of CO_2 on the Properties of Anion Exchange Membranes for Fuel Cell Applications. *J. Membr. Sci.* **2019**, *586*, 140–150. [CrossRef]
41. Wang, T.; Wang, X.; Hou, C.; Liu, J. Quaternary Functionalized Mesoporous Adsorbents for Ultra-High Kinetics of CO_2 Capture from Air. *Sci. Rep.* **2020**, *10*, 21429. [CrossRef]

polymers

MDPI

Article

Effect of Side Substituent on Comb-like Polysiloxane Membrane Pervaporation Properties During Recovery of Alcohols C2-C4 from Water

Evgenia Grushevenko [1], Islam Chechenov [1], Tatyana Rokhmanka [1], Tatiana Anokhina [1], Stepan Bazhenov [1] and Ilya Borisov [1,2,*]

[1] A.V. Topchiev Institute of Petrochemical Synthesis, Russian Academy of Sciences, Leninsky Prospect 29, 119991 Moscow, Russia; evgrushevenko@ips.ac.ru (E.G.); chechenov@ips.ac.ru (I.C.); rokhmankatn@ips.ac.ru (T.R.); tsanokhina@ips.ac.ru (T.A.); sbazhenov@ips.ac.ru (S.B.)

[2] Federal Research Center Kazan Scientific Center of Russian Academy of Sciences, Lobachevskogo St. 2/31, 420111 Kazan, Tatarstan, Russia

* Correspondence: boril@ips.ac.ru

Abstract: The pervaporation properties of membranes based on comb-like polysiloxanes when C_2-C_4 alcohols are removed from water were studied for the first time. It was established that membranes based on comb-like polysiloxanes with linear aliphatic and organosilicon substituents have increased permeability selectivity for C_{3+} alcohols. The obtained results were interpreted from the point of view of the solubility of the components of the separated mixture in polysiloxanes. It was shown that membranes based on polysiloxanes with linear substituents have increased butanol/water permeability selectivity (2.5–3.7). The achieved selectivity values correspond to the level of highly selective zeolite membranes, which allows for a reduction in energy consumption for the pervaporation removal of butanol by more than two times.

Keywords: comb-like polysiloxane; pervaporation; alcohol recovery; water treatment; polymeric membrane

1. Introduction

The wastewater treatment of various chemical industries is an important environmental issue. One of the types of wastewater pollutants are oxygenates (ethers and lower alcohols). These substances are present in the effluents of GTL processes, namely petrochemical production (for example, effluents of the isomerization process).

Alcohols are an important intermediate, final, and by-product of industrial organic synthesis. For instance, the effluents of GTL processes contain a great number of lower alcohols, C_1-C_5 (22.8 g/L) [1]. These substances are not concentrated as a separate value-added product but are processed together with the total water effluent in the flotation and floaculation unit to remove suspended solids, followed by processing in a bioreactor, in which organic matter is disposed of to the level that allows wastewater to be discharged into reservoirs [2].

In the case of the targeted production of alcohols, they often need to be concentrated from aqueous media. According to various data, up to 50% of C_2-C_5 alcohols are formed in the aqueous phase during the synthesis of alcohols from CO and hydrogen, while the content of C_2-C_4 alcohols reaches 60–70% [3–5]. To obtain a marketable product, the alcohols obtained must be concentrated.

Pervaporation, or evaporation through a membrane, is one of the most promising methods of recovery of bio-alcohols from water [6]. It has advantages such as the absence of reagents, recovery of products at low temperatures, and modularity of the separation unit [7]. The main advantage of pervaporation over biopurification is the possibility of recovery and further use of alcohols.

There is a large number of works on the recovery of oxygen-containing organic compounds from aqueous media by pervaporation [6–17]. However, the number of membrane materials used for the recovery of oxygenates from aqueous media is small, comprising hydrophobic silicalites [8,17], silicone rubbers (for example, PDMS, POMS) [9–11], and highly permeable polymer glasses (for example, PTMSP, PIM-1) [12,13].

First of all, polydimethylsiloxane (PDMS) is used as a material of pervaporation membranes due to its high permeability, stable transport properties, and chemical and thermal resistance. The high flexibility of the PDMS chain, as well as weak intermolecular interactions, provide increased values of the free volume fraction and segmental mobility of the polymer, which determines its high gas and liquid permeability. The transport of gasses and vapors through nonporous membranes is determined by the "solution-diffusion" mechanism. This polymer has a record low diffusion selectivity [18]. Together with high solubility selectivity, it provides the increased selectivity of the permeability of PDMS membranes for components with increased solubility coefficients, and the chemical cross-linking of the polymer ensures the stability of the membranes in the media being separated. Polysiloxane membranes are widely used to purify liquid and gas streams from volatile organic compounds (VOCs). PDMS membranes have high permeability and stable transport properties, but their selectivity is insufficient to effectively solve many urgent separation problems. In this regard, the development of methods to increase the selectivity of polysiloxane membranes is a major challenge in this field of membrane science and technology.

Membranes based on polysiloxanes represent a subject of significant scientific and practical importance, particularly in the context of hydrocarbon capture and hydrophobic pervaporation processes [4], owing to their exceptional permeability and thermal stability [19]. Moreover, their cross-linked structure confers them with remarkable chemical stability in a wide range of solvents [20,21]. The separation capabilities of polydimethylsiloxane (PDMS) and polyoctylmethylsiloxane (POMS) membranes have been extensively explored [22–25]. Researchers have explored the potential of chemically modifying polysiloxane chains to enhance membrane separation performance. Of particular interest is the process of hydrophobizing polysiloxanes, which involves introducing alkyl side substituents with lengths ranging from 7 to 10 atoms into the polysiloxane backbone. This modification significantly improves the separation efficiency, as demonstrated by an increase in the selectivity factor for the separation of 1% weight fraction of methyl-tret-butyl ether from water, ranging from 111 for the methyl group to 161, 169, and 180, respectively. The incorporation of 10% diacetate groups into the polysiloxane backbone resulted in a 15% enhancement of the phenol flux from an aqueous solution compared to polydimethylsiloxane (PDMS), resulting in an increase in the phenol–water partition coefficient to 21.5. When separating a mixture of methyl isobutyl ketone and water, it was demonstrated that the addition of long alkyl substituents, such as octyl and tridecyl, significantly enhanced the separation efficiency, with partition coefficients increasing from 705 in PDMS to 1030 for octyl and 1200 for tridecyl.

Membranes based on polysiloxanes are of great scientific and practical interest, in particular for hydrocarbon capture and hydrophobic pervaporation [26], due to their high permeability and thermal stability [27]. In addition, they are chemically stable in a large number of solvents due to their cross-linked structure [28,29]. The separation ability of polydimethylsiloxane (PDMS) and polyoctylmethylsiloxane (POMS) membranes has been extensively studied [30–33]. Chemical modification of the polysiloxane chain can improve the separation characteristics of membranes. In particular, the hydrophobization of polysiloxanes is of great interest to researchers. Thus, the introduction of alkyl side substituents of lengths 7, 8, and 10 into the polysiloxane chain allowed for an increase in the separation factor of 1 wt.% methyl-tretbutyl ether in water from 111 (for the methyl group) to 161, 169, and 180, respectively [34]. The introduction of 10% diacetate groups into the polysiloxane chain allowed for an increase in the phenol flow from an aqueous solution by 15% compared with polydimethylsiloxane, which led to an increase in the phenol/water separation factor to 21.5 [35]. When separating a mixture of methylisobutyl ketone–water, it was shown that the introduction of long alkyl substituents (octyl, tridecyl) led to a significant increase in separation efficiency; the separation factor increased from 705 for PDMS to 1030 (octyl) and 1200 (tridecyl) [35].

Numerous studies have shown that polysiloxanes containing long linear aliphatic substituents have increased selectivity in the recovery of VOCs from water compared with other polysiloxanes [26]. The vast majority of works on the study of gas transport and pervaporation properties of polymers of this type are devoted to polyoctylmethylsiloxane (POMS) [10]. It was demonstrated in [32] that the introduction of a long hydrocarbon substituent C_8 into the side chain of polysiloxane leads to an increase in the hydrophobicity of the membrane material (water flow for a PDMS membrane is 117 $g \cdot m^{-2} \cdot h^{-1}$, for a POMS membrane—85 $g \cdot m^{-2} \cdot h^{-1}$). Due to this, the selectivity of the recovery of non-polar aromatic substances in relation to water increases. However, it is unclear from the open sources what caused the choice of this polymer, since work on the directional design of comb-like polysiloxanes for the recovery of VOCs from water has not yet been carried out. This is largely due to the complexity of the synthesis of the membrane materials of this type. There are several methods for the synthesis of comb-like polysiloxanes and membranes on their basis [26]. All of them are multi-stage and require the synthesis of expensive monomers, the recovery of the polymer from the reaction mass, and the use of various catalysts for the synthesis of the polymer and its cross-linking [26]. The creation of an effective and simple method for the synthesis and cross-linking of polysiloxanes with different chemical structures and side chain length made it possible to synthesize a representative number of comb-like polysiloxanes and conduct systematic studies of the fundamental relationship between the structure and transport properties of this important group of membrane materials for hydrocarbons [36]. Thus, the successful solution of the problem of developing new and simple methods for the molecular design of polysiloxanes opens up opportunities for creating materials with increased selectivity during the pervaporation recovery of VOCs from water.

This work is aimed at studying the influence of the geometry of the side substituent of polymethylsiloxanes on their pervaporation properties during the recovery of lower alcohols from aqueous media. The novelty of the work lies in the establishment of a fundamental relationship between the chemical structure of side chains and the transport properties of membrane materials based on comb-shaped polysiloxanes for obtaining highly selective membranes in the process of the pervaporation separation of alcohols from water.

2. Materials and Methods

2.1. Materials

For the synthesis of polymethylorganosiloxanes, polymethylhydrosiloxane (PMHS) was used with a number average molecular weight, Mn, of 1700–3200 g/mol (Sigma-Aldrich, St. Louis, MO, USA), Karstedt's catalyst (platinum complex of 1,3-divinyl-1,1,3,3-tetramethyldisiloxane in xylene, Sigma-Aldrich, St. Louis, MO, USA), 1-hexene (97%, Sigma-Aldrich, St. Louis, MO, USA), 1-heptene (97%, Sigma-Aldrich, St. Louis, MO, USA), 1-octene (98%, Sigma-Aldrich, St. Louis, MO, USA), 1-decene (94%, Sigma-Aldrich, St. Louis, MO, USA), 1-dodecene (95%, Sigma-Aldrich, St. Louis, MO, USA), 1-tetradecene (92%, Sigma-Aldrich, St. Louis, MO, USA), 3,3-dimethylbutene-1 (95%, ABCR, Karlsruhe, Germany), 4,4-dimethylpentene-1 (99%, Sigma-Aldrich, St. Louis, MO, USA), vinyltrimethylsilane (97%, ABCR, Karlsruhe, Germany), allyltrimethylsilane (98%, ABCR, Karlsruhe, Germany), 1,7-octadiene (98%, Sigma-Aldrich, St. Louis, MO, USA), n-hexane (99%, Chimmed, Podolsk, Russia) without further purification. To obtain polydimethylsiloxane, a vinyl-terminated polydimethylsiloxane (PDMS) was used at Mn 25,000 g/mol (Sigma-Aldrich, St. Louis, MO, USA), polymethylhydrosiloxane (PMHS) at Mn 1700–3200 g/mol (Sigma-Aldrich, St. Louis, MO, USA), Karstedt catalyst (platinum complex of 1,3-divinyl-1,1,3,3-tetramethyldisiloxane in xylene, Sigma-Aldrich, St. Louis, MO, USA), and toluene (99.8%, Chimmed, Podolsk, Russia) without further purification.

Ethanol (96.3%, Ferein, Moscow, Russia), propanol-1 (Chimmed, Podolsk, Russia), n-butanol (Chimmed, Podolsk, Russia), and distilled water were used to study the pervaporation membrane properties.

2.2. Synthesis and Production of Membranes

A molding solution and polymethylsiloxane membranes with various side groups were obtained using a recently proposed method [37]. This method consists of the in situ modification and cross-linking of polymethylhydrosiloxane (PMHS) with α-olefin (allyltrimethylsilane, 3,3-dimethylbutene-1, 1-hesene, 1-heptene (1-octene, 1-decene, 1-dodecene, 1-tetradecene) and diene hydrocarbon (1,7-octadiene) in the presence of a Karstedt catalyst by a hydrosilylation reaction. The reaction scheme is shown in Figure 1.

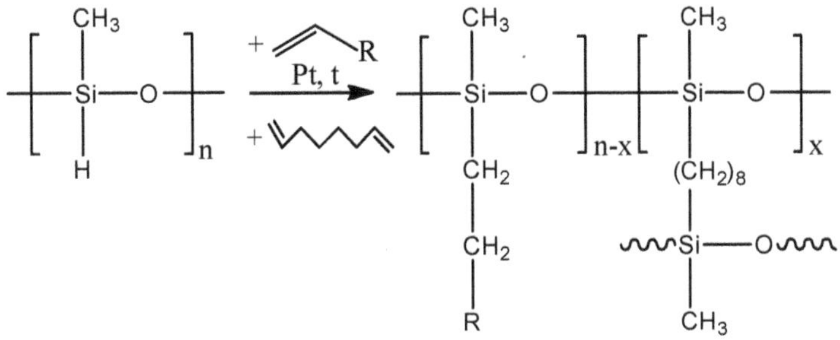

Figure 1. The scheme of the reaction for obtaining polymethylsiloxanes substituted by a side group. n—polymerization degree of initial PMHS, x—monomer units entering into a cross-linking reaction.

According to the method, the reaction was carried out in an n-hexane medium. For this purpose, a 3 wt.% solution of PMHS was pre-prepared in it. The reaction was carried out for 3 h while the solution was continuously stirred at 60 °C under a reflux condenser. The membranes were obtained by watering the reaction mixture on a stainless-steel mesh (cell size 40 μm) fixed on a Teflon surface, followed by drying to a constant weight at 60 °C for 24 h. The thickness of the films varied in the range of 40–50 μm. A polydimethylsiloxane (PDMS) membrane was used as an object of comparison: vinyl-terminated PDMS was cross-linked with PMHS by a hydrosilylation reaction in the presence of a Karstedt catalyst. The ratio of PDMS:PMHS:catalyst was 10:1:0.01. The films on the mesh were obtained by the method described above. The conventional symbols of the membranes obtained are shown in Table 1.

Table 1. Conventional symbols of the samples of membranes under study.

Symbol of Membrane	Polymer
PHexMS	Polyhexylmethylsiloxane
PHepMS	Polyheptylmethylsiloxane
POMS	Polyoctylmethylsiloxane
PDecMS	Polydecylmethylsiloxane
PDDMS	Polydodecylmethylsiloxane
PTDMS	Polytetradecylmethylsiloxane
PVTBMS	Poly-3,3-dimethylbutylmethylsiloxane
PATBMS	Poly-4,4-dimethylpentene-1methylsiloxane
PVTMSMS	Poly-3,3-dimethylsilbutylmethylsiloxane
PATMSMS	Poly-4,4-dimethylsilpentylmethylsiloxane
PDMS	Polydimethylsiloxane

2.3. Vacuum Pervaporation

The pervaporation experiments were carried out on the installation shown in Figure 2 and described in detail in [19].

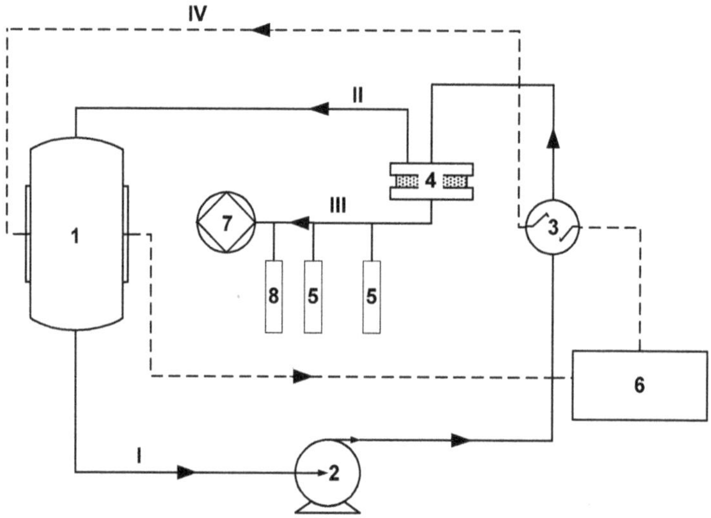

Figure 2. Layout of the vacuum pervaporation installation [19].

The initial feed mixture was poured into a 1 L container (1) and pumped in a circulating mode using an Ismatec gear pump (2). The volumetric flow rate of the feed mixture varied from 20 to 220 mL/min. The mixture heated in the heat exchanger (3) was fed into the membrane module (4). The effective membrane area was 13.85 cm^2. Permeate vapors were condensed in glass traps placed in Dewar vessels with liquid nitrogen (−196 °C) (5). The continuous operation of the installation throughout the experiment was ensured by the presence of two parallel traps. To prevent permeate vapor from entering the vacuum pump, a safety trap was used (8). The temperature of the feed mixture was maintained with an accuracy of ±0.1 °C using a LOIP LT-100 liquid thermostat (6). To create the driving force of the mass transfer process in the submembrane space, a pressure of ~0.05 mbar was maintained with an Ebara PDV-250 vacuum pump (7).

The pervaporation was carried out at a temperature of the feed mixture of 30 (±0.1) °C. A solution of n-butanol, n-propanol, and ethanol in water with an organic component content of 1.0, 1.0, and 3.0 wt.%, respectively, was used as a feed mixture. The solution used was prepared from an organic solvent (purity grade–reagent grade) by the gravimetric method.

The concentration of the initial mixture and permeate was determined by gas chromatography on a Crystallux-4000M chromatograph equipped with a thermal conductivity detector.

The process of vacuum pervaporation was characterized using the following parameters: permeate flow, separation factor, and the index of pervaporation separation.

Total permeate flux (J, kg/m^2 h) was calculated as

$$J = \frac{m}{S \cdot t}, \quad (1)$$

where m is the weight of the permeate (kg) penetrated through the membrane with the area S (m^2) for a known period of time t (h).

The separation factor (α) was determined by the following formula:

$$\alpha = \frac{y_o \cdot x_w}{y_w \cdot x_o}, \quad (2)$$

where x_o and x_w are the mass fractions of the organic component and water in the separated mixture and y_o and y_w are the mass fractions of the organic component and water in the permeate.

It is worth noting that the parameters listed above depend on the experiment conditions, involving temperature, composition of the feed mixture, pressure in the submembrane space, and the type of membrane [20]. Therefore, it is preferable to compare the transport properties of membranes using such process parameters as permeability (P/l) and selectivity, which was determined from the ratio of the permeability of the *i*-th component to the permeability of water.

The permeability coefficient (*P*, Barrer) for component *i* was calculated according to the following equation:

$$P = \frac{J_i l}{\left(P_i^f - P_i^p\right)}, \quad (3)$$

where J_i is the component *i* flow (cm^3/(cm·m^2·s)), p_i^f and p_i^p are the vapor pressure of component *i* in the initial mixture and permeate (cm hg), respectively, and *l* is the thickness of the membrane (m). To determine the vapor pressure of permeate and the initial mixture in the case of the binary mixtures of n-propanol–water and ethanol–water, activity coefficients were calculated using the Aspen Plus 8.6 software package using the NRTL (non-random two-liquid) model [21]. As it was shown in [22], in the case of the binary mixture of n-butanol–water, it is preferable to use the approximated 4-parametric Margules equation to calculate the activity coefficients.

$$lgy_b = \left(1 - x_b\right)^2 \left[A + 2(B - A - D)x_b + Dx_b^2\right], \quad (4)$$

$$lgy_w = x_b \left[B + 2(A - B - D)(1 - x_b) + 3D\left(1 - x_b\right)^2, \quad (5)$$

2.4. Determination of Sorption and Solubility Coefficients

The solubility coefficients of butanol in the polymer were determined under isothermal conditions according to the method developed by us. Sorption was carried out in a solution of 1 wt.% of 1-butanol in water. A sample of polysiloxane with mass (m_{pol}) and density (ρ_{pol}) is placed in a closed space filled by half with a model mixture (V_0). Next, the sample was thermostated (LOIP LT-108a thermostat, Saint-Petersburg, Russia) at 30 °C for 24 h. It was found out that sorption equilibrium is established during this time, since the concentration of butanol in the solution does not change in the subsequent time. The amount of the sorbed butanol substance was calculated as the difference between the initial amount of the butanol substance and the amount of butanol substance after 24 h of sorption in the solution. The amounts of the substance were calculated based on the mass concentrations of butanol in the solution, which were determined by gas chromatography. The concentration of butanol was determined by analyzing the equilibrium concentrations of the vapor phase in order to increase the accuracy of the analysis. The vapor phase analysis was performed using a headspace sampler (DRP-10, NPF Meta-Chrome LLC, Yoshkar-Ola, Russia), the main feature of which is the use of a thermostatically controlled gas syringe for sampling. This approach makes it possible to increase the measurement accuracy by increasing the area of peaks observed on the chromatogram. The initial mass concentration w_0 in the solution and the mass concentration obtained after 24 h of sorption w_{24} were determined using a Crystallux-4000M gas chromatograph (NPF Meta-Chrome LLC, Yoshkar-Ola, Russia). The chromatographic analysis parameters were as follows: evaporator temperature—230 °C; column temperature—180 °C; and detector temperature—230 °C. The analyses were carried out using a 1 m long packed column filled with Porapak Q sorbent. The concentration obtained from the chromatograms w_{24} corresponds to the concentration of a sorbate in the solution after 24 h of contact with the polymer.

The amount of the substance (mol) of butanol was found by the following formula:

$$n^b = \frac{w^b \cdot m_s}{M^b}, \tag{6}$$

where w^b and M^b are a mass fraction and the molecular weight (g/mol) of butanol and m_s is the mass of the solution (g).

The solubility coefficient (cm^3/(cm^3 cmHg) was calculated using the following formula:

$$S_i = \frac{n_s \cdot R \cdot T}{V_{pol} \cdot p_{sorb}}, \tag{7}$$

where p_{sorb} and T are the sorbate vapor pressure (cmHg) and temperature (K).

Polymer volume (cm^3), as the ratio of polymer mass and density:

$$V_{pol} = m_{pol} / \rho_{pol}, \tag{8}$$

where m_{pol} (g) and ρ_{pol} (g/cm^3) are the polymer mass and density.

The amount of the substance sorbed in the polymer (mol) was determined as

$$n_s = n_0 - n_{24}, \tag{9}$$

The solubility of water in the polymer was determined gravimetrically. The polymer was weighed in dry air and saturated water vapor at 30 °C on a Sartorius ENTRIS124-1S scale. The amount of the substance of the sorbed water was determined by the following formula:

$$n^w = \frac{m_{pol}^{24} - m_{pol}^{0}}{M^w}, \tag{10}$$

where m_{pol}^0 and m_{pol}^{24} are the mass of the polymer in the dry state and in water vapor (g) and M^w is the molecular weight of water (g/mol).

2.5. Scanning Electron Microscopy (SEM)

SEM was used to characterize the structure and morphology of the membranes. SEM was carried out on a Thermo Fisher Phenom XL G2 Desktop SEM (Thermo Scientific, Waltham, MA, USA). Cross-sections of the membranes were obtained by fracturing in liquid nitrogen after a preliminary impregnation of the specimens in isopropanol. A thin (5–10 nm) gold layer was deposited on the prepared samples in a vacuum chamber (~0.01 mbar) using a desktop magnetron sputter "Cressington 108 auto Sputter Coater" (Cressington Scientific Instruments Ltd., Watford, UK). The accelerating voltage during image acquisition was 15 kV.

3. Results and Dicussion

3.1. Synthesis of Membranes from Comb-like Polysiloxane

The proposed one-step method for the synthesis of film and composite membranes based on comb-like polysiloxanes consists of in situ modification and cross-linking, excluding the stages of the separation and purification of the polymer between the stages of modification and cross-linking (Figure 1). A distinctive feature of the proposed method is that the modification and cross-linking reaction is carried out in the same reaction medium in the presence of a single Pt-containing catalyst (Karstedt catalyst). An important simplification of this approach to the synthesis of polysiloxane membranes is the use of cheap and commercially available starting compounds, such as polymethylhydrosiloxane (a product of organosilicon production), α-olefins, and dienes (products of basic organic synthesis). The main advantages of the new approach in comparison with traditional methods for obtaining comb-like polysiloxanes are as follows:

- The use of the reaction mixture as a molding solution;
- The use of a single catalyst for the modification and cross-linking of the membrane material;
- The absence of stages of polymer recovery and purification (low solvent consumption);
- The simplicity of the modification of the material chemical structure (variability in the use of modifying agents with terminal double bonds);
- Little time spent on membrane production.

Using a new method, a series of 10 membranes based on comb-like polysiloxanes with linear and branched structures, as well as silicon-containing side substituents, were synthesized and studied in the process of pervaporation (Figure 1). Typical SEM images of obtained membranes are represented in Figure 3.

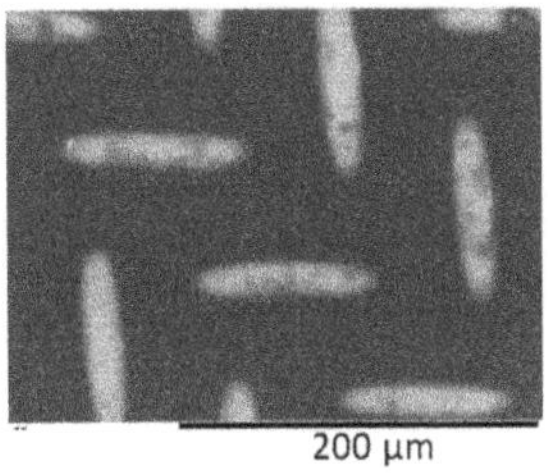

(a)

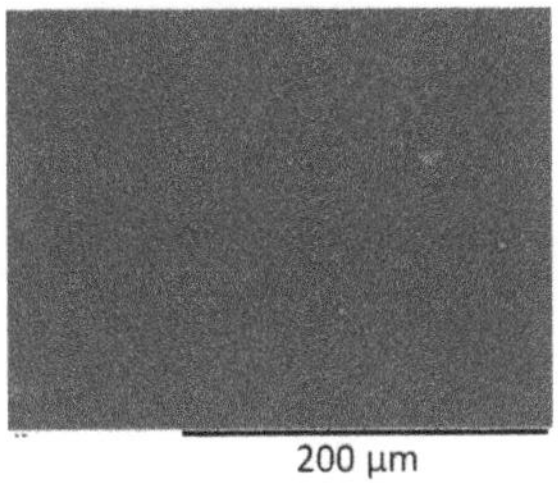

(b)

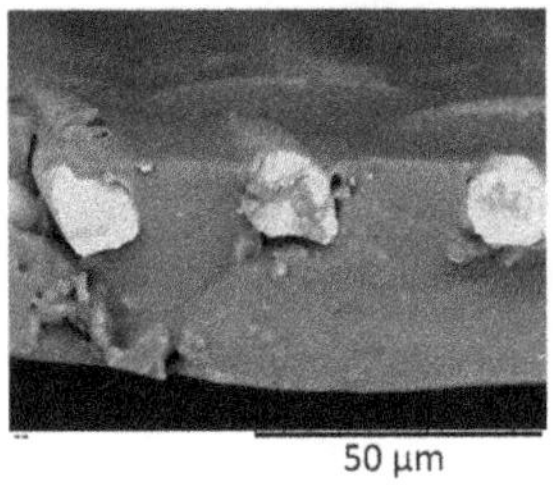

(c)

Figure 3. SEM images of a polysiloxane membrane on a stainless-steel mesh: (**a**) top view, (**b**) bottom view, (**c**) cross-section.

3.2. Transport Properties of the Obtained Membranes Based on Comb-like Polysiloxane

The transport properties of the obtained membranes were studied during the pervaporation separation of a model mixture of n-butanol, n-propanol, and ethanol in water with an organic component content of 1.0, 1.0, and 3.0 wt.%. The experimental results are shown in Figure 4 and Table 2.

Table 2. Partial fluxes for 1-butanol, 1-propanol, ethanol, and water and alcohol–water separation factors for studied polysiloxane membranes.

Membrane	Partial Flux, g/(m^2 h)				Separation Factor (X/Water)		
	BuOH	PrOH	EtOH	Water	BuOH	PrOH	EtOH
PDMS	18.7	7.4	10.5	84.1	42	17.2	8.1
PHexMS	7.6	2.0	2.4	17.8	80.4	22.4	8.7
PHepMS	6.7	1.8	1.7	12.9	97.5	28.0	8.5
POMS	5.4	1.6	1.4	11.0	92.3	27.5	8.1
PDecMS	4.4	1.4	1.1	10.0	83.9	26.2	7.1
PDDMS	2.4	0.7	0.5	6.9	65.0	20.1	5.0
PTDMS	1.8	0.6	0.4	5.7	58.1	19.1	4.7
PATMSMS	3.6	1.3	1.2	13. 6	49.5	18.3	5.5
PVTMSMS	2.6	0.8	0.8	10.7	46.5	14.9	4.5
PATBMS	2.2	0.8	0.8	11.6	36.1	13.5	4.4
PVTBMS	0.8	0.3	0.3	5.5	27.5	10.7	3.5

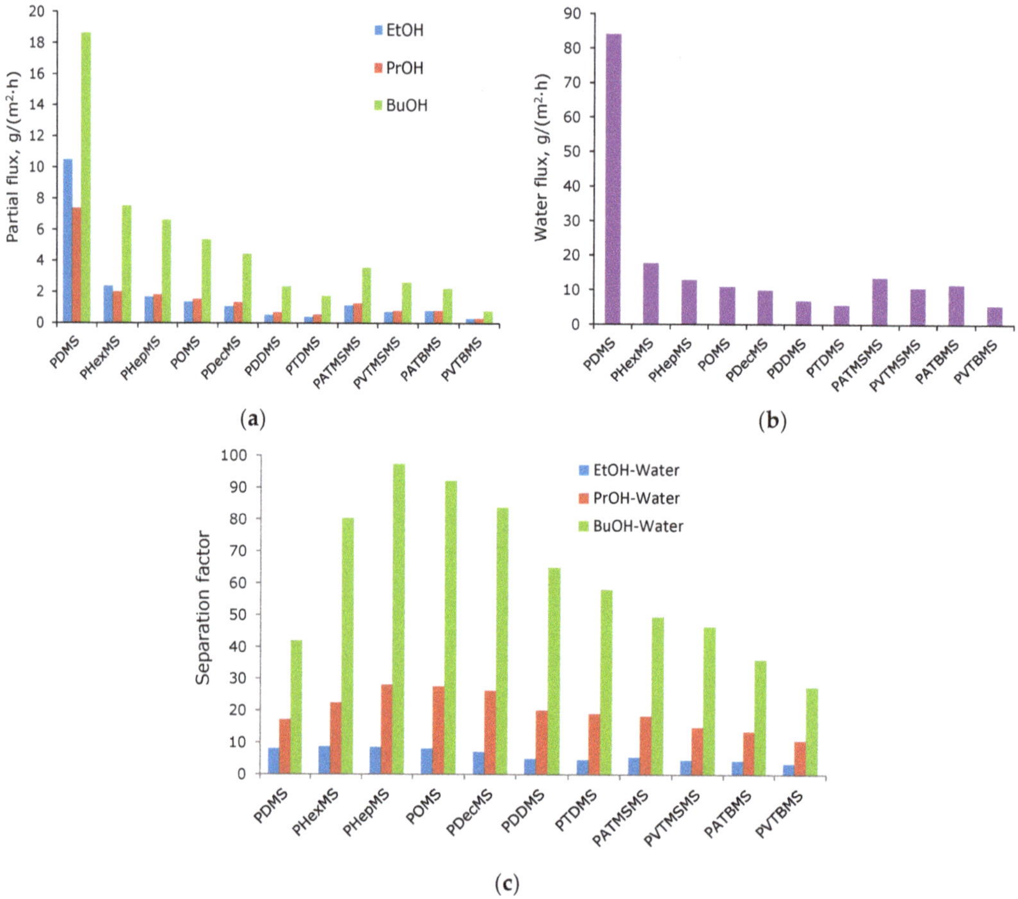

Figure 4. (**a**) The dependence of the component fluxes on the type of side substituent (BuOH—1-butanol; PrOH—1-propanol; EtOH—ethanol). (**b**) the dependence of the water flux on the type of side substituent. (**c**) The dependence of the alcohol–water separation factor on the type of side substituent (BuOH—1-butanol; PrOH—1-propanol; EtOH—ethanol; W—water).

In Figure 4a,b, there is a tendency to decrease the partial flows of components with an increase in the size of the side substituent. In the case of polysiloxanes with linear side chain substituents, the flow decreases monotonously with increasing side chain length. For polysiloxanes with branched substituents, the flow increases with increasing length of the flexible aliphatic junction between the bulk group (tertbutyl or trimethylsilyl) and the main chain of two to three methylene fragments. This behavior is associated with a decrease in conformational rotation when the bulk group is removed from the main chain. Moreover, the fluxes for polysiloxanes containing a trimethylsilyl fragment in the side chain are higher than for those containing a tertbutyl fragment, which may be a consequence of the high mobility of the trimethylsilyl group [23,24]. The PDMS membrane (the most permeable in the series being studied) is characterized by a high water flux value, which exceeds the value of the 1-butanol flux by more than 2.5 times, and in the case of propanol and ethanol, by 5 and 10 times. In the case of comb-like polysiloxanes, the ratio of water and alcohol fluxes decreases sharply compared to PDMS, which leads to a proportional increase in the alcohol–water separation factor (Figure 4c). This can be explained by an increase in the hydrophobicity of the material with the introduction of large hydrocarbon side substituents [32].

Thus, PDMS is characterized by relatively low values of the separation factor (BuOH-W—42; PrOH-W—17; EtOH-W—8.5). For linear polyalkylsiloxanes, the C3-C4 alcohol separation factors are significantly higher (BuOH-W—58–97; PrOH-W—19–28 (Table 2)). The EtOH-W separation factor varies slightly in the series of polysiloxanes being considered (7.0–8.5). The maximum value of separation factors was demonstrated by the PHepMS membrane (BuOH-W—97.5; PrOH-W—29; EtOH-W—8.5 (Table 2)). However, for branched polyalkylsiloxanes, the separation factors are significantly lower. For example, PVTBMS demonstrates minimum separation factor values, even lower than PDMS (BuOH-W—27.5; PrOH-W—10.7; EtOH-W—3.5). Flux and the separation factor are the parameters that depend on the magnitude of the driving force of the process and the thickness of the membrane. To explain these results, it is necessary to involve studies of the structure of the material, as well as the analysis of the results in terms of the permeability coefficients of polymers and their components (diffusion and solubility coefficients) [20].

It can be seen from Figure 5 and Table 3 that the dependencies of the values of the permeability and selectivity coefficients correlate well with the data on the values of the fluxes of components and alcohol–water separation factors. Among the studied polysiloxanes, the most permeable membrane material is PDMS (P (BuOH) = 38,700 Barrer; P (PrOH) = 29,500 Barrer; P (EtOH) = 19,300 Barrer; P (H_2O) 23,900 Barrer). As the length of the linear alkyl side substituent increases, the permeability decreases for all components (Figure 5a). However, a more favorable situation is observed for alcohols C_3 and C_4. For example, the selectivity for butanol is in the range of 2.5–3.7 for polyalkylsiloxanes with linear C_6-C_{12} substituents, which corresponds to the highly selective zeolite membranes [25]. Such an increase in selectivity leads to a decrease in energy consumption for the pervaporation recovery of butanol by more than two times [25], which indicates the prospects of linear polyalkylsiloxanes for solving the problem of the pervaporation recovery of PrOH and BuOH from aqueous media. At the same time, the maximum selectivity of permeability for alcohols EtOH-W (0.85), PrOH-W (2.0), and BuOH-W (3.7) is demonstrated by the PHepMS membrane. A slight increase in ethanol selectivity is due to its high polarity and hydrophilicity. Indeed, PDMS derivatives have an EtOH/H_2O selectivity either lower or close to one [25].

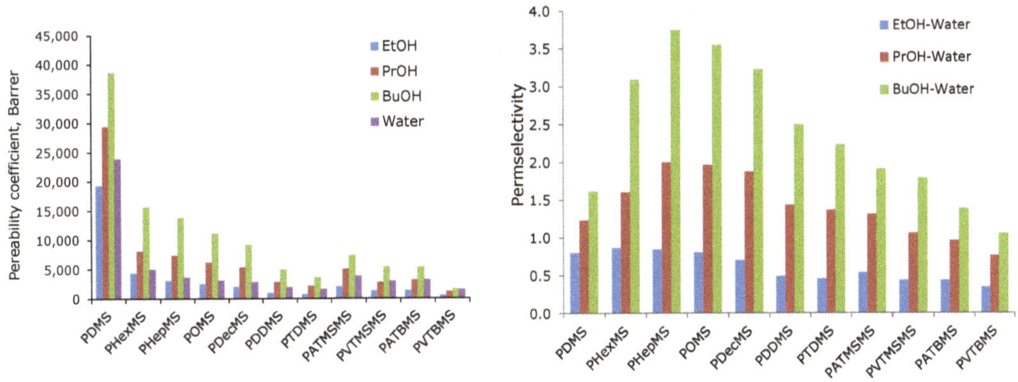

(a) (b)

Figure 5. The dependence of the values of the permeability coefficients on the type of side substituent. (BuOH—1-butanol; PrOH—1-propanol; EtOH—ethanol; W—water (**a**). The dependence of the alcohol–water selectivity values on the type of side substituent. (BuOH—1-butanol; PrOH—1-propanol; EtOH—ethanol; W—water (**b**).

Table 3. Permeability coefficients for 1-butanol, 1-propanol, ethanol, and water and alcohol–water selectivity for studied polysiloxane membranes.

Membrane	Permeability Coefficient, Barrer				Separation Factor (X/Water)		
	BuOH	PrOH	EtOH	Water	BuOH	PrOH	EtOH
PDMS	38,700	29,450	19,330	23,930	1.6	1.2	0.8
PHexMS	15,600	8100	4420	5060	3.1	1.6	0.9
PHepMS	13,800	7360	3130	3680	3.8	2.0	0.9
POMS	11,140	6180	2530	3140	3.5	2.0	0.8
PDecMS	9200	5340	2025	2850	3.2	1.9	0.7
PDDMS	4930	2830	990	1970	2.5	1.4	0.5
PTDMS	3630	2220	760	1620	2.2	1.4	0.5
PATMSMS	7360	5060	2120	3870	1.9	1.3	0.5
PVTMSMS	5430	2760	1380	3040	1.8	1.1	0.5
PATBMS	5340	3190	1470	3250	1.4	1.0	0.4
PVTBMS	1660	1200	550	1570	1.1	0.8	0.4

The following question arises: what is the reason for the increased selectivity of linear polyalkylsiloxanes, and why is the selectivity of polysiloxanes with bulk substituents significantly lower than that of the former? The fact is that the selectivity of permeability is the product of the selectivity of diffusion and solubility. The preferred permeability of alcohols C3 and C4 compared with water in the polysiloxanes being studied is determined by significantly lower water solubility coefficients compared with butanol. As shown in Figure 5, the introduction of nonpolar side substituents reduces the water solubility coefficients in polysiloxanes compared to PDMS, whereas the solubility of butanol decreases not so significantly (Figure 6a,b). As a result, the selectivity of the solubility αs for PDMS and PHepMS differ 2.7 times (48 and 128, respectively) (Figure 6b). As can be seen from Figure 5, it is the selectivity of solubility that makes the main contribution to the selectivity of the alcohol/water permeability of polysiloxane membranes.

In each of the studied homologous series of polymers (with linear alkyl, trimethylsilyl, tertbutyl, and cyclohexyl substituents), the selectivity of solubility increases with an increase in the hydrocarbon fragment (Figure 6b). This is due to the fact that n-butanol is soluble in aliphatic hydrocarbons and the solubility of water in alkanes is close to zero. As was shown in our previous work [36], ordered structures are formed in polysiloxanes with linear substituents, which are nanoscale layers formed by parallel-oriented side chains. The proportion of such ordered areas in the polymer was estimated based on X-ray phase analysis data obtained at liquid nitrogen temperature. Such membrane materials are heterogeneous, and their transport properties are determined by the supramolecular structure of the comb-like polyalkylsiloxane. An increase in the proportion of the ordered domains in the polymer leads to a decrease in the selectivity of diffusion for small molecules and an increase in the selectivity of permeability for larger organic components.

Based on the results obtained, it is possible to assume the reason for the increased selectivity of alcohol/water polysiloxanes with linear substituents. As shown in [36], the selectivity of diffusion for small molecules of polymers with linear substituents is lower than for PDMS, and for comb-like polysiloxanes with bulk substituents, it is at the level of or significantly higher than PDMS. The low selectivity of water/alcohol diffusion is one of the factors ensuring the high selectivity of alcohol/water permeability of comb-like polysiloxanes with linear side substituents capable of forming a hexagonal packing of side chains. This can be explained by the fact that due to plasticization, the organic component is embedded in a hexagonal structure formed by side substituents, forming additional transport channels for the diffusion of molecules. Such a diffusion mechanism is unlikely for water, since its solubility in hydrocarbons is negligible. In its turn, such hexagonal packing is not implemented for side substituents of a nonlinear structure. Therefore, the selectivity of diffusion for polymers with bulk substituents is higher compared to linear polyalkylsiloxanes. As a result, the permeability selectivity of polymers with bulk substituents is lower than with linear ones. In addition, for polymers, the tertbutyl side groups are even lower than for PDMS (Figure 6).

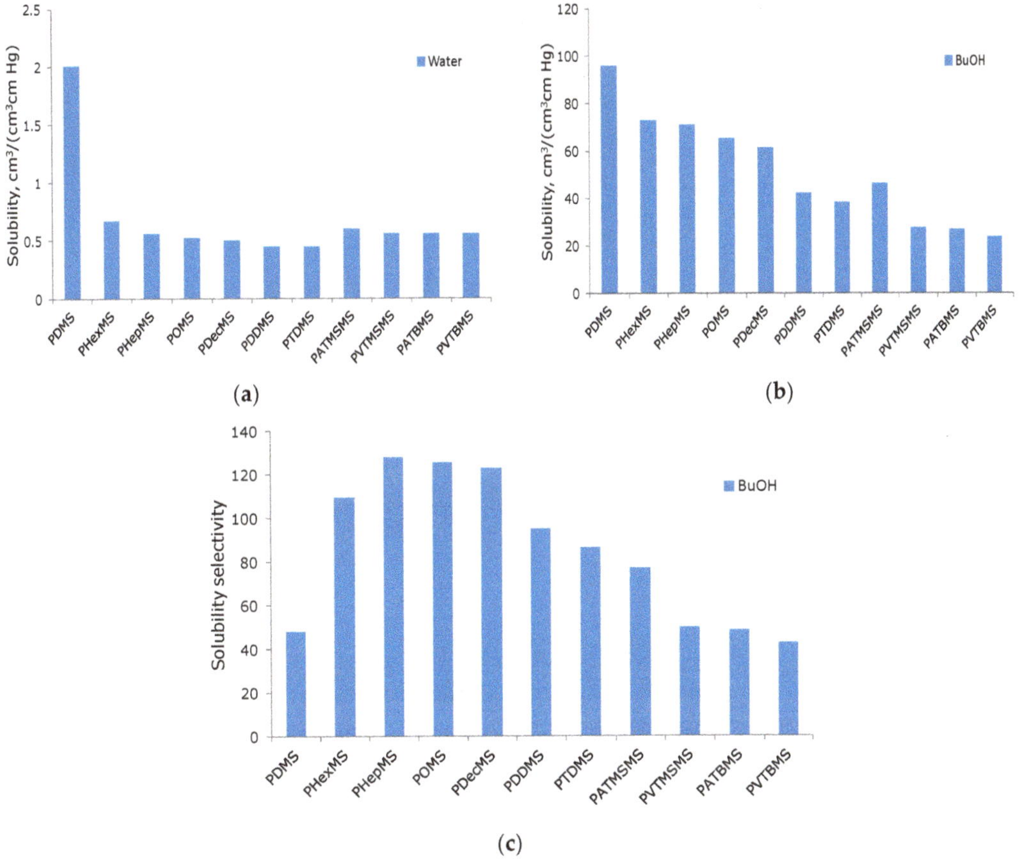

Figure 6. The dependence of the values of the water and butanol solubility on the type of side substituent (**a**,**b**). The dependence of the butanol–water solubility selectivity values on the type of side substituent (c).

4. Conclusions

The pervaporation properties of membranes based on comb-like polysiloxanes were studied during the separation of a four-component mixture of BuOH-PrOH-EtOH-H_2O for the first time. It has been established that membranes based on comb-like polysiloxanes with linear aliphatic and organosilicon substituents have an increased selectivity of permeability to alcohols C_{3+}, which can be more than twice as high as the selectivity of membranes based on polydimethylsiloxane due to an increase in the selectivity of alcohol/water solubility. The membrane based on polyheptylmethylsiloxane demonstrates the greatest selectivity (separation factors 97, 29, and 9 for 1-butanol, 1-propanol, and ethanol, respectively), which is consistent with the high solubility coefficients of alcohols in this material. It has been shown that membranes based on polysiloxanes with linear substituents have an increased butanol/water permeability selectivity (2.5–3.7) relative to membranes made of polymers with bulk substituents (1.0–1.9) due to the low selectivity of water/alcohol diffusion. The achieved selectivity values correspond to the level of highly selective zeolite membranes, which allows for a reduction in energy consumption for the pervaporation recovery of butanol by more than two times [25]. Thus, it can be argued that membranes based on polyalkylsiloxanes with linear C_6-C_{12} substituents are promising for solving the problem of the pervaporation recovery of C_{3+} alcohols from aqueous media.

Author Contributions: Conceptualization, I.B.; data curation, T.R.; formal analysis, E.G. and T.A.; funding acquisition, S.B.; investigation, E.G. and I.C.; methodology, E.G. and I.B.; project administration, S.B.; resources, T.A.; supervision, I.B.; validation, T.R.; writing—original draft, T.A. and I.B.; writing—review and editing, E.G., T.R. and S.B. All authors have read and agreed to the published version of the manuscript.

Funding: The work was carried out with the financial support of the Ministry of Science and Higher Education of the Russian Federation using funds from the Agreement on the provision of a grant from the federal budget for the implementation of major scientific projects in priority areas of scientific and technological development No. 075-15-2024-646 dated 12 July 2024.

Institutional Review Board Statement: Not applicable.

Data Availability Statement: The original contributions presented in this study are included in the article. Further inquiries can be directed to the corresponding author.

Acknowledgments: The authors acknowledge Podtynnikov I. for their assistance in performing pervaporation experiments.

Conflicts of Interest: The authors declare no conflicts of interest.

Abbreviations

y_i, x_i, w_i	the component mass fraction $[g \cdot g^{-1}]$
J_i	component mass permeate flux $[kg \cdot m^{-2} \cdot h^{-1}]$
l	membrane selective layer thickness [cm]
M_i	molar mass of the component $[kg \cdot mol^{-1}]$
P_i	permeability coefficient [Barrer]
S_i	solubility coefficient $[cm^3 \cdot cm^{-3} \cdot cm\ Hg^{-1})]$
$\alpha_{i/j}$	separation factor alcohol/water [-]
n_i	amount of substance [mol]
m_i	mass of component [g]
V_i	volume of component $[cm^3]$
ρ_i	density of component $[g/m^3]$
BuOH	n-butanol
PrOH	n-propanol
EtOH	ethanol
W	water

References

1. Majone, M.; Aulenta, F.; Dionisi, D.; D'Addario, E.N.; Sbardellati, R.; Bolzonella, D.; Beccari, M. High-rate anaerobic treatment of Fischer–Tropsch wastewater in a packed-bed biofilm reactor. *Water. Res.* **2010**, *44*, 2745–2752. [CrossRef] [PubMed]
2. De Klerk, A. *Fischer–Tropsch Refining*; John Wiley & Sons: Hoboken, NJ, USA, 2012; 619p.
3. Krylova, A.Y.; Kryazhev, Y.G.; Kulikova, M.V.; Kurkin, V.I.; Lyadov, A.S.; Sagitov, S.A. Synthesis of Alcohols from CO and H2 on Iron Catalysts Containing Carbon Fiber. *Solid Fuel Chem.* **2011**, *45*, 322–326. [CrossRef]
4. Krylova, A.Y.; Kurkin, V.I.; Kulikova, M.V.; Lyadov, A.S.; Sagitov, S.A. Synthesis of Monohydric Alcohols from CO and H2 on Fe/Sibunit Catalysts. *Solid Fuel Chem.* **2011**, *45*, 281–285. [CrossRef]
5. Kölbel, H.; Ralek, M. The Fischer–Tropsch Synthesis in the Liquid Phase. *Cat. Rev* **1980**, *21*, 225–274. [CrossRef]
6. Ezeji, T.C.; Qureshi, N.; Blaschek, H.P. Butanol fermentation research: Upstream and downstream manipulations. *Chem. Rec.* **2004**, *4*, 305–314. [CrossRef]
7. Liu, G.; Wei, W.; Jin, W. Pervaporation Membranes for Biobutanol Production. *ACS Sustain. Chem. Eng.* **2013**, *2*, 546–560. [CrossRef]
8. Ikegami, T.; Yanagishita, H.; Kitamoto, D.; Haraya, K.; Nakane, T.; Matsuda, H.; Koura, N.; Sano, T. Production of highly concentrated ethanol in a coupled fermentation/pervaporation process using silicalite membranes. *Biotechnol. Tech.* **1997**, *11*, 921–924. [CrossRef]

9. Rozicka, A.; Niemisto, J.; Keiski, R.L.; Kujawski, W. Apparent and intrinsic properties of commercial PDMS based membranes in pervaporative removal of acetone, butanol and ethanol from binary aqueous mixtures. *J. Membr. Sci.* **2014**, *453*, 108–118. [CrossRef]
10. Rom, A.; Friedl, A. Investigation of pervaporation performance of POMS membrane during separation of butanol from water and the effect of added acetone and ethanol. *Sep. Purif. Technol.* **2016**, *170*, 40–48. [CrossRef]
11. Borisov, I.L.; Ushakov, N.V.; Volkov, V.V.; Finkel'shtein, E.S. Polydimethylsilalkylene-Dimethylsiloxanes as Advanced Membrane Materials for Thermopervaporative Recovery of Oxygenates from Aqueous Reaction Media. *Petr. Chem.* **2016**, *56*, 798–804. [CrossRef]
12. Zák, M.; Klepic, M.; Štastná, L.C.; Sedláková, Z.; Vychodilová, H.; Hovorka, S.; Friess, K.; Randová, A.; Brozová, L.; Jansen, J.C.; et al. Selective removal of butanol from aqueous solution by pervaporation with a PIM-1 membrane and membrane aging. *Sep. Purif. Technol.* **2015**, *151*, 108–114. [CrossRef]
13. Borisov, I.L.; Malakhov, A.O.; Khotimsky, V.S.; Litvinova, E.G.; Finkelshtein, E.S.; Ushakov, N.V.; Volkov, V.V. Novel PTMSP-based membranes containing elastomeric fillers: Enhanced 1-butanol/water pervaporation selectivity and permeability. *J. Membr. Sci.* **2014**, *466*, 322–330. [CrossRef]
14. Kim, H.J.; Jang, K.S.; Galebach, P.; Gilbert, C.; Tompsett, G.; Conner, W.C.; Jones, C.W.; Nair, S. Seeded growth, silylation, and organic/water separation properties of MCM-48 membranes. *J. Membr. Sci.* **2013**, *427*, 293–302. [CrossRef]
15. Bakhtin, D.S.; Kulikov, L.A.; Maksimov, A.L.; Volkov, A.V. Stabilization of gas transport properties of composite membranes based on poly [1-trimethylsilyl-1-propine] by introducing porous aromatic frameworks nanoparticles obtained by the Friedel-Crafts reaction. *Russ. J. Appl. Chem.* **2020**, *93*, 257–263. [CrossRef]
16. Bakhtin, D.S.; Kulikov, L.A.; Legkov, S.A.; Khotimskiy, V.S.; Levin, I.S.; Borisov, I.L.; Maksimov, A.L.; Volkov, V.V.; Karakhanov, E.A.; Volkov, A.V. Aging of thin-film composite membranes based on PTMSP loaded with porous aromatic frameworks. *J. Membr. Sci.* **2018**, *554*, 211–220. [CrossRef]
17. Kanemoto, M.; Negishi, H.; Sakaki, K.; Ikegami, T.; Chohnan, S.; Nitta, Y.; Ohta, H. Efficient butanol recovery from acetone-butanol-ethanol fermentation cultures grown on sweet sorghum juice by pervaporation using silicalite-1 membrane. *J. Biosci. Bioeng.* **2016**, *121*, 697–700. [CrossRef]
18. Zhmakin, V.V.; Teplyakov, V.V. The evaluation of the C1–C4 hydrocarbon permeability parameters in the thin film composite membranes. *Sep. Purif. Technol.* **2017**, *186*, 145–155. [CrossRef]
19. Borisov, I.L.; Grushevenko, E.A.; Podtynnikov, I.A.; Bakhtin, D.S.; Bondarenko, G.N. Novel Membrane Material Based on Polybutadiene and Polydimethylsiloxane for Gas Separation and Hydrophobic Pervaporation. *Petr. Chem.* **2018**, *58*, 1113–1122. [CrossRef]
20. Baker, R.W.; Wijmans, J.G.; Huang, Y. Permeability, permeance and selectivity: A preferred way of reporting pervaporation performance data. *J. Membr. Sci.* **2010**, *348*, 346–352. [CrossRef]
21. Kujawska, A.; Knozowska, K.; Kujawa, J.; Kujawski, W. Influence of downstream pressure on pervaporation properties of PDMS and POMS based membranes. *Sep. Purif. Technol.* **2016**, *159*, 68–80. [CrossRef]
22. Borisov, I.L.; Golubev, G.S.; Vasilevsky, V.P.; Volkov, A.V.; Volkov, V.V. Novel hybrid process for bio-butanol recovery: Thermopervaporation with porous condenser assisted by phase separation. *J. Membr. Sci.* **2017**, *523*, 291–300. [CrossRef]
23. Finkelshtein, E.; Makovetskii, K.; Gringolts, M.; Rogan, Y.; Golenko, T.; Starannikova, L.; Yampolskii, Y.; Shantarovich, V.; Suzuki, T. Addition-type polynorbornene with Si(CH3)3 side group: Synthesis, gas permeation and free volume. *Macromolecules* **2006**, *39*, 7022–7029. [CrossRef]
24. Starannikova, L.; Pilipenko, M.; Belov, N.; Yampolskii, Y.; Gringolts, M.; Finkelshtein, E. Addition-type polynorbornene with Si(CH3)3 side groups: Detailed study of gas permeation and thermodynamic properties. *J. Membr. Sci.* **2008**, *323*, 134–143. [CrossRef]
25. Zheng, P.; Li, S.; Wang, N.; Li, J.; An, V. The potential of the first corporation to produce biofuels as a result of fermentation: In terms of energy consumption. *Chin. J. Chem. Eng.* **2019**, *27*, 1296–1306. [CrossRef]
26. Grushevenko, E.A.; Borisov, I.L.; Volkov, A.V. High-Selectivity Polysiloxane Membranes for Gases and Liquids Separation (A Review). *Pet. Chem.* **2021**, *61*, 959–976. [CrossRef]
27. Prabhakar, R.S.; Raharjo, R.; Toy, L.G.; Lin, H.; Freeman, B.D. Self-Consistent Model of Concentration and Temperature Dependence of Permeability in Rubbery Polymers. *Ind. Eng. Chem. Res.* **2005**, *44*, 1547–1556. [CrossRef]
28. Bernardo, P.; Drioli, E.; Golemme, G. Membrane Gas Separation: A Review/State of the Art. *Ind. Eng. Chem. Res.* **2009**, *48*, 4638–4663. [CrossRef]
29. Merkel, T.C.; Gupta, R.P.; Turk, B.S.; Freeman, B.D. Mixed-Gas Permeation of Syngas Components in Poly(Dimethylsiloxane) and Poly(1-Trimethylsilyl-1-Propyne) at Elevated Temperatures. *J. Membr. Sci.* **2001**, *191*, 85–94. [CrossRef]
30. Choi, S.H.; Kim, J.H.; Lee, S.B. Sorption and Permeation Behaviors of a Series of Olefins and Nitrogen through PDMS Membranes. *J. Membr. Sci.* **2007**, *299*, 54–62. [CrossRef]
31. Catarino, M.; Ferreira, A.; Mendes, A. Study and Optimization of Aroma Recovery from Beer by Pervaporation. *J. Membr. Sci.* **2009**, *341*, 51–59. [CrossRef]
32. Börjesson, J.; Karlsson, H.O.E.; Trägårdh, G. Pervaporation of a model apple juice aroma solution: Comparison of membrane performance. *J. Membr. Sci.* **1996**, *119*, 229–239. [CrossRef]

33. Olsson, J.; Trägårdh, G.; Lipnizki, F. The Influence of Permeant and Membrane Properties on Mass Transfer in Pervaporation of Volatile Organic Compounds from Dilute Aqueous Solutions. *Sep. Sci. Technol.* **2002**, *37*, 1199–1223. [CrossRef]
34. Borisov, I.; Podtynnikov, I.; Grushevenko, E.; Scharova, O.; Anokhina, T.; Makaev, S.; Volkov, A.; Volkov, V. High Selective Composite Polyalkylmethylsiloxane Membranes for Pervaporative Removal of MTBE from Water: Effect of Polymer Side-Chain. *Polymers* **2020**, *12*, 1213. [CrossRef]
35. Bennett, M.; Brisdon, B.J.; England, R.; Field, R.W. Performance of PDMS and Organofunctionalised PDMS Membranes for the Pervaporative Recovery of Organics from Aqueous Streams. *J. Membr. Sci.* **1997**, *137*, 63–88. [CrossRef]
36. Borisov, I.L.; Grushevenko, E.A.; Anokhina, T.S.; Bakhtin, D.S.; Levin, I.S.; Bondarenko, G.N.; Volkov, V.V.; Volkov, A.V. Influence of side chains assembly on the structure and transport properties of comb-like polysiloxanes in hydrocarbon separation. *Mater. Today Chem.* **2021**, *22*, 100598. [CrossRef]
37. Grushevenko, E.A.; Borisov, I.L.; Bakhtin, D.S.; Legkov, S.A.; Bondarenko, G.N.; Volkov, A.V. Membrane material based on octyl-substituted polymethylsiloxane for separation of C3/C1 hydrocarbons. *Petr. Chem.* **2017**, *57*, 334–340. [CrossRef]

polymers

MDPI

Article

Plasma Surface Treatment and Application of Polyvinyl Alcohol/Polylactic Acid Electrospun Fibrous Hemostatic Membrane

Xiaotian Ge, Li Zhang, Xuanhe Wei, Xi Long and Yingchao Han *

State Key Laboratory of Advanced Technology for Materials Synthesis and Processing, Biomedical Materials and Engineering Research Center of Hubei Province, Wuhan University of Technology, Wuhan 430070, China; 317173@whut.edu.cn (X.G.); zhanglihyc@whut.edu.cn (L.Z.); 336143@whut.edu.cn (X.W.); longxi528702@163.com (X.L.)

* Correspondence: hanyingchao@whut.edu.cn

Abstract: In this study, an improved PVA/PLA fibrous hemostatic membrane was prepared by electrospinning technology combined with air plasma modification. The plasma treatment was used to modify PLA to enhance the interlayer bonding between the PVA and PLA fibrous membranes first, then modify the PVA to improve the hemostatic capacity. The surfaces of the PLA and PVA were oxidized after air plasma treatment, the fibrous diameter was reduced, and roughness was increased. Plasma treatment enhanced the interfacial bond strength of PLA/PVA composite fibrous membrane, and PLA acted as a good mechanical support. Plasma-treated PVA/PLA composite membranes showed an increasing liquid-enrichment capacity of 350% and shortened the coagulation time to 258 s. The hemostatic model of the liver showed that the hemostatic ability of plasma-treated PVA/PLA composite membranes was enhanced by 79% compared to untreated PVA membranes, with a slight improvement over commercially available collagen. The results showed that the plasma-treated PVA/PLA fibers were able to achieve more effective hemostasis, which provides a new strategy for improving the hemostatic performance of hemostatic materials.

Keywords: atmospheric pressure cold plasma; polyvinyl alcohol; polylactic acid; surface modification; hemostatic material

1. Introduction

Massive bleeding caused by surgery or trauma has a high lethality rate, and the use of hemostatic materials to accelerate hemostasis in the early stage of traumatic bleeding is an effective means of saving lives, so research into hemostatic materials with good fluid absorption capacity and wound sealing properties is necessary [1–3]. Ideal hemostatic materials should have the abilities of achieving rapid hemostasis, good fluid absorption, strong adhesion to tissues and organs, good biocompatibility, and low cost. Altering the wettability and biocompatibility of materials has been a constant endeavor. With the continuous development of hemostatic materials, many new materials have shown good hemostatic ability. To further develop this, there are also requirements for enhancing the enrichment of coagulation factors in materials to achieve methods of hemostasis [4].

In recent years, researchers have developed many composites that enable rapid hemostasis, such as chitosan, hyaluronic acid, starch, collagen, sodium alginate, hydroxyapatite, and other materials with good biocompatibility. Usually, people add additional hemostatic ability or activate coagulation pathways to enhance the hemostatic efficiency of materials. For example, chitosan can accelerate platelet enrichment, and Ca^{2+} can activate coagulation [5–9]. Meanwhile, some high porosity materials with good infiltration ability can also accelerate coagulation by enriching blood, such as electrospun fibers, MOF, zeolites, and aerogels, etc. [10–12]. Polymer materials have shown good performance due to their lower cost and higher biocompatibility. PVA is widely used as a low-cost, non-toxic, easily decomposable material. PVA has excellent hydrophilicity due to the large number of hydroxyl groups, and can be further modified by hydroxyl modification, cross-linking, and oxidization [13]. PVA electrospun fibers in hemostatic applications, due to their limited mechanical properties, have more difficulty facing the complexity of a bleeding situation, particularly in the bleeding of a larger amount, even partially dissolved, so for the PVA fibers membrane, adding a layer of mechanical support is necessary. PLA has good mechanical properties and biocompatibility and is widely used as a low-cost, non-toxic, and easily decomposable material [14]. PVA and PLA can be formed into composite fiber membranes by electrospinning, as an excellent matrix material for tissue repair [15–20]. Based on the high porosity and large specific surface area of electrospun fiber membranes, they also display promising applications in hemostasis

Plasma is considered the fourth state of matter and contains many active species, (neutral particles, excited state radicals, photons, electrons, ions, and so on). Plasma surface modification is an efficient, simple, and green modification technology. Plasma surface modification can effectively change the hydrophilicity of materials by introducing hydrophilic functional groups such as –OH, and –COOH on the material surface, and enhance cell adhesion by increasing the surface roughness [21–24].

In this study, plasma treatment was used to modify PLA to enhance the interlayer bonding between PVA and PLA fibrous membranes first, then modify PVA to improve the liquid-enrichment capacity, resulting in an improved PVA/PLA electrospun fibrous hemostatic membrane. The surface properties (roughness, functional groups, hydrophilicity), tensile modulus, and the hemostatic capacity of fiber membranes were evaluated in a rat liver hemostasis model.

2. Materials and Methods

2.1. Materials

Polyvinyl alcohol (PVA; Aladdin, 1788; average molecular weight: 20,000–150,000; degree of hydrolysis: 87–89%), polylactic acid (PLA; average molecular weight 980,000, Daigang Bio; Jinan, China), dichloromethane (DCM, Sinopharm; Shanghai, China), N,N′-dimethylformamide (DMF, Sinopharm; Shanghai, China), a cell counting kit (CCK8; HYCEZMBIO; Wuhan, China), and Calcein-AM/PI double stain kit (Yeasen Biotechnology; Shanghai, China).

2.2. Preparation of PVA, PLA, and PVA/PLA Fibrous Membranes by Electrospun

The PVA solution was prepared by dissolving 5 g of PVA in 50 mL of deionized water, stirring for 2 h at 80 °C, and then homogenizing for 2 h at room temperature to eliminate air bubbles. The PVA solution was transferred into a 10 mL syringe with a 21 G needle (inner diameter of 0.5 mm). The experiments were carried out at 25 °C and 50% humidity, with an electric field voltage of 11 kV. The pushing rate of the PVA solution was 0.8 ml/h, and the distance from the needle to the receiver was 15 cm. The process lasted for roughly 10 h, to obtain PVA electrospun membranes with a thickness of approximately 0.1 mm.

The PLA solution was prepared by dissolving 0.3 g of PLA in 4 mL of DCM and adding 2 mL of DMF after thorough dissolution. The mass fraction of PLA in DCM was 7.5 wt%. The PLA solution was transferred into a 10 mL syringe with a 20 G needle (inner diameter of 0.6 mm). The experiments were carried out at a temperature of 25 °C and a humidity of 50%, with an electric field of 11 KV. The PLA solution was pushed at a rate of 5 mL/h, and the distance from the needle to the roller receiver was 12 cm, with a roller rotating speed of 1000 r/min. The process lasted for roughly 1 h, to obtain PLA electrospun membranes with a thickness of approximately 0.1 mm.

The PLA electrostatic spinning fibrous membrane was prepared by the above spinning method. Then the PLA fibrous membrane was subjected to plasma treatment for a time set at about 5 min per 2×2 cm^2. After the treatment, the PLA fibrous membrane was re-pasted on tin foil, and the PVA fibers were electrostatically spun on the surface of the PLA fibers by the above-mentioned electrostatic spinning of PVA fibrous membranes. After that, the PVA fiber membrane was plasma treated to obtain the monolayer PVA/PLA composite membrane with a thickness of approximately 0.2 mm.

2.3. Plasma Treatment of Fibrous Membranes

Plasma treatment is carried out at atmospheric pressure for a fixed time every 2×2 cm^2. The plasma (RELYON Plasma Piezo Brush PZ3) was operated at a maximum power of 18 W, with power variation settings of 100% (18 W), 80% (14.4 W), 60% (10.8 W), and 40% (7.2 W). The fibrous membrane was placed 0.5 mm below the plasma generator module and treated at different times. Plasma treatment was performed on a non-conductive base.

2.4. Characterization of Fibrous Membranes

XPS spectra were acquired on a Thermo Scientific K-Alpha instrument (Waltham, MA, USA) and the acquired data were processed by Advantage software. Aluminum targets are used in XPS testing, scanning starts from energy 1361 eV. FTIR spectra were detected using a Nicolet6700, (Thermo Scientific, Waltham, MA, USA) specification instrument in the range of 400–4000 cm^{-1}, with three scans on the same sample. The surface morphology of the fiber membrane was photographed using JSMIT200 (JEOL, Tokyo, Japan), at a voltage of 5 KV, and the fiber diameter was counted using the software ImageJ version 1.51. The process of sample preparation before SEM image shooting included cutting different fiber samples and sticking them onto the sample stage with conductive tape. Finally, after 60 s of gold spraying, the samples were placed in a vacuum chamber for shooting. The number of fiber diameters counted was over 300. AFM images were captured using an atomic force microscope with a Nanoscope IV specification and the average roughness, Ra, was calculated. AFM scanning size of PVA fiber was 5 μm, and the scanning size of PLA fiber was 10 μm, with a tip size of 20 nm. The contact angle was taken using a Dataphysics OCA20. In the contact angle test, deionized water was used with a droplet size of 50 μL to measure its static contact angle. The contact angle on the surface of PVA fibers was the contact angle formed when the droplet contacted the fiber surface instantly, and the contact angle formed between the droplet and the fiber surface after 8 s. PLA was the contact angle formed when the droplet stabilizes on its surface.

2.5. Mechanical Property Test

The mechanical properties of the membranes were tested using an INSTRON 1122 tensiometer. The membranes were cut to a length of 6 cm and a width of 1 cm. The clamping distance was 2 cm, the stretching speed was 10 mm/min, and the stretching was carried out in the length direction. The thickness of the samples was pre-determined using an electronic micrometer and the thickness of the fibrous film was chosen to be around 0.1 mm. The test was repeated five times to calculate the average value and deviation.

2.6. Interlayer Bonding

During the plasma treatment of PLA in the preparation process, the PLA electrostatically spun fiber membrane was divided into the plasma-treated area and the untreated area. After electrospinning PVA fibers on its surface, the fibers were cut to the size of 3×3 cm^2 to obtain two kinds of samples, plasma-treated and untreated. The PVA and PLA surfaces were taped separately, and the other end of the tape was pasted on the fixture of the mechanical testing machine to separate the PVA and PLA layers at a tensile speed of 10 mm/min, and the maximum tensile force required for the separation of the two layers was recorded.

2.7. Water Absorption

Experiments on the water absorption capacity of fiber membranes were carried out with reference to existing standards [25]. Before the experiment, the fiber was cut to the size of 3×3 cm^2 and after drying, when the fiber membrane was of constant weight, the mass of the fiber membrane was recorded as W_0. The fiber membrane was placed in deionized water for 24 h and after removing the excess water droplets from the surface of the fiber membrane, the weight was recorded as W_1 and the water absorption of the fiber membrane was calculated using the following equation:

$$\text{water absorption } (\%) = W_1/W_0 \times 100\% \tag{1}$$

2.8. Whole Blood Coagulation Index

Whole blood coagulation experiments were performed based on previous work [26]. Anticoagulated rat whole blood $CaCl_2$ (0.2 M) was added 200 μL to each sample, and after 5 min of coagulation, uncoagulated erythrocytes were lysed with deionized water to release hemoglobin, and absorbance was measured at 540 nm. The coagulation index was calculated as follows:

$$\text{BCI } (\%) = A_1/A_0 \times 100\% \tag{2}$$

where A_1 and A_0 denote the absorbance of the experimental group and blank control group, respectively.

2.9. Clotting Time

In vitro hemostasis experiments were performed using rat anticoagulated blood, which was used within 24 h of collection. For each hemostasis test, fiber membranes of the same size were added to the glass tube, and then 2 mL of rat anticoagulated blood and 80 uL of $CaCl_2$ (0.2 M) solution were added, then the mixture was placed at 37 °C. The state of the blood was checked at intervals, and the coagulation time was taken as the time when the glass tube was inverted and no blood flowed.

2.10. Hemolysis Assay

The anticoagulated blood containing $CaCl_2$ (0.2 M) was diluted 10-fold with saline and mixed with the material (each ml corresponded to a fibrous membrane area of 3 × 3 cm^2), then the mixed blood was incubated at 37 °C for 4 h. Then, 1 mL of blood was taken and centrifuged at 3000 rpm for 5 min, and the absorbance of 100 μL of supernatant from each sample was measured at 540 nm. The hemolysis rate was calculated as follows:

$$\text{Hemolysis } \% = (A_s - A_n)/(A_p - A_n) \times 100\% \tag{3}$$

A_s is the absorbance of the sample, A_n is the absorbance of the positive control (saline), and A_p is the absorbance of the negative control (deionized water).

2.11. Cytotoxicity

Cytotoxicity was evaluated by means of leachate, which was prepared at 1 mL per 6 × 6 cm^2 surface area of fibrous membrane. The fibrous membranes were immersed in DMEM medium for 3 days and the temperature was maintained at 37 °C to obtain the extract of the fiber sample. Using the fiber extract for cell culture allows the fiber extract to come into contact with cells to evaluate the cytotoxicity of the material. The subgroups were set as the blank group, plasma-treated PLA group, and treated PVA/PLA composite fibrous group. The experiments were all performed using 24-well plates for cell culture, inoculating cells at a cell concentration of approximately 10^4 cell/mL, adding 500 μL of the sample extract per well to the well plates, and co-culturing with L929 cells for 1, 3, and 5 days. The blank group was DMEM complete medium. The cell viability was measured by CCK8 kit, the dehydrogenase in the mitochondria of the surviving cells was reduced by WST-8 in the reagent to an orange-colored product; the more the surviving cells, the deeper the orange-colored product is, and the cell viability of different samples can be reflected by the measurement of the absorbance at 450 nm. When the cell viability is above 95%, it indicates that the material has good biocompatibility. The formula for cell viability is as follows:

$$\text{cell viability} = OD_{sample} - OD_0/OD_{control} - OD_0 \times 100\% \quad (4)$$

OD_{sample}, OD_0, and $OD_{control}$ are the absorbance at 450 nm of the sample, CCK8 solution, and the blank group, respectively.

Cultures of L929 cells were performed in the same way and cells were stained alive and dead with AMPI after 1, 3, and 5 days of culture to determine cell growth.

2.12. Liver Hemostasis Experiment

Sprague–Dawley (SD) rats liver laceration model was conducted according to the standard procedure. The SD rats (weight 200 g–300 g) were randomly divided into five groups, the blank control group, untreated PVA group, treated PVA group, commercially available collagen group, and treated PVA/PLA group. Each rat was anesthetized by intraperitoneal injection of 4% w/v chloral hydrate (0.01 mL/g rat body weight). Rat livers were exposed through an abdominal incision. A bleeding wound (1 cm long and 0.5 cm deep) of the liver was created with a scalpel, and the fibrous membrane was immediately placed on the surface of the wound. Then, the filter paper with adsorbed rat blood was weighed and the amount of bleeding was recorded. All mouse experiments were conducted in accordance with the guidelines of the Laboratory Animal Center of Wuhan University of Technology (Wuhan, China) and approved by the Laboratory Animal Ethics Committee of Wuhan University of Technology. The surviving animals were euthanized at the end of the experiment [1,3]. The percentage reduction in bleeding was calculated as follows:

$$\text{Percentage reduction in bleeding} = (M_0 - M_1)/M_0 \times 100\% \quad (5)$$

M_0 is the blank group and M_1 is the experimental group.

2.13. Statistical Analysis

The results are calculated from three parallel experiments expressed as the mean ± standard deviation. One-way ANOVA was performed to assess significant differences and * $p < 0.05$ was considered statistically significant.

3. Results and Discussion

3.1. Characterizations of Plasma-Treated PLA

The effect of plasma treatment on the chemical state of atoms on the surface of the fibrous membrane was revealed using XPS. As shown in Figure 1b–f and Supporting Information Tables S1 and S2, three peaks (C–C, C–O, and O–C=O) can be observed at binding energies of 284.8 eV, 286 eV, and 288.5 eV split-fit, respectively [27,28]. It can be seen that the relative content of oxygen on the surface of PLA fibers increased from 34% to 47% and the plasma treatment caused the C–C bonds on the surface of PLA fibers to break continuously from a relative content of 47% to 35%, and the oxygen-containing functional groups (O–C=O, C–O) increased continuously. These results indicate that the air plasma treatment is breaking the C–C bond of the PLA molecular chain and introducing some oxygen-containing functional groups [29].

FTIR spectra further proved the changes in the functional groups on the surface of the fibrous membranes. Some of the classical absorption peaks of PLA can be seen from Figure 1a, with the characteristic carbonyl peak near 1753 cm^{-1} and the C–O–C absorption peak at around 1181 cm^{-1}. The weaker hydroxyl absorption peak near 2944 cm^{-1} is due to the fact that the PLA molecular chain has only a small amount of hydroxyl groups at the end [30]. No new functional groups appear in the spectra of the treated fibrous membranes, but their relative contents have changed (Table S3). It was found that the oxygen-containing functional groups on the surface of PLA increased with the treatment time, and the results are generally consistent with XPS.

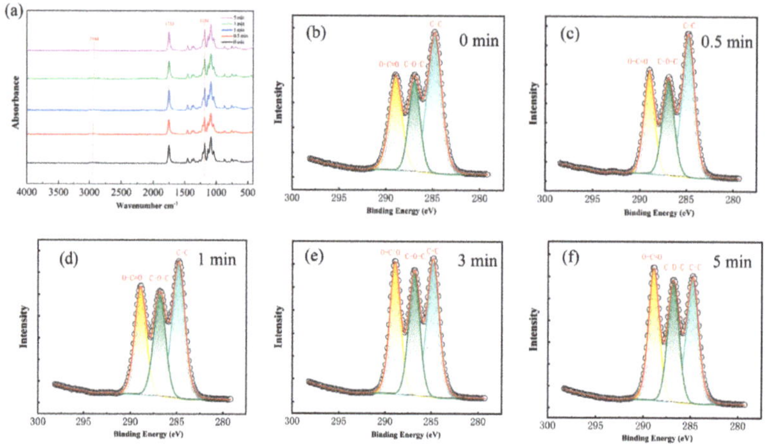

Figure 1. FTIR spectra of PLA with different treatment times (**a**) and C1s fitting with different treatment times of 0 min (**b**), 0.5 min (**c**), 1 min (**d**), 3 min (**e**), 5 min (**f**).

As shown in Figure 2 and Table S4, the diameter of PLA fibers decreased from 2.16 ± 0.58 μm to 1.68 ± 0.62 μm, and the surface roughness gradually increased from 11.27 ± 2.6 nm to 30.73 ± 5.0 nm. The decrease of average diameter and the increase of average roughness can be attributed to the interaction between the plasma and the polymer molecules. The activated high–energy oxygenated plasma strikes the fibers to produce an etching effect [31].

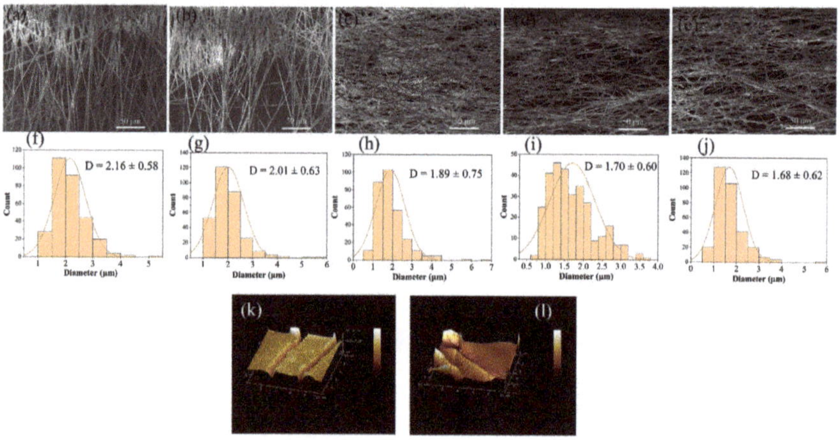

Figure 2. SEM images of PLA electrostatically spun fibrous with different times of plasma treatment and diameter distribution of fibrous (**a**,**f**) 0 min, (**b**,**g**) 0.5 min, (**c**,**h**) 1 min, (**d**,**i**) 3 min, (**e**,**j**) 5 min. (**k**) AFM images of PLA electrostatically spun fibrous membranes without plasma treatment, (**l**) 5 min plasma treatment of the AFM images of PLA electrospun fibrous membranes. The plasma power used was 18 W.

As can be seen in Figure 3, the wettability of PLA changed significantly after plasma treatment, and the contact angle decreased from 122 ± 8.2° to 15 ± 2.1°. There are many factors affecting the wettability of the fiber membrane, and the most important factors affecting PLA should be the increase of polar functional groups on the fiber surface and the microstructure of the fiber surface. From the XPS results, we can see that the increase of oxygen–containing functional groups on the surface of PLA after plasma treatment, which makes the polarity of the fiber surface increase, and the water contact angle on the surface of the fiber with higher polarity is lower. In addition, from the SEM image, we can see that the pore space of the fiber surface has increased, and the liquid droplet infiltrates faster on the surface with higher pore space and rougher surface, which is also one of the reasons for the change in the contact angle of the PLA fibers [28].

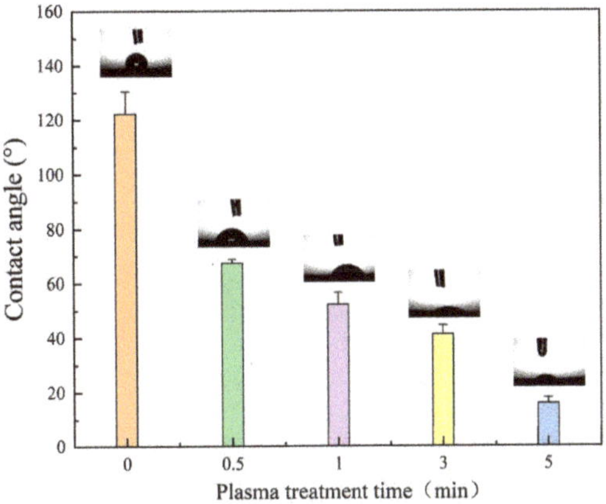

Figure 3. Images of PLA fibrous membranes after different treatment times.

3.2. *Characterizations of Plasma-Treated PVA*

From Figure 4 and Table S5, it can be seen that the oxygen content on the surface of the PVA fibers membrane is slightly increased, along with the processing time from 0 to 5 min. Moreover, the oxygen content on the surface is slightly increased, along with the rise of the plasma power from 7.2 W to 18 W (Figure S1 and Table S6). Therefore, it can be concluded that both increasing power and extending treatment time are beneficial to raising the content of oxygen on the surface of the fiber membrane. As shown in Figure 2, the four peaks of the split-peak fitting can be observed with binding energies of 284.8 eV, 286 eV, 287.5 eV, and 289 eV, corresponding to C–C, C–O, C=O, and O–C=O, respectively [32]. It can be seen that no new functional groups appear and the relative contents of the functional groups change, as shown in Tables S7 and S8. Results show that increasing power and extending treatment time lead to a decrease in C–C bonds and an increase in C=O, C–OOH bonds, which is generally consistent with the results of other studies [33–35]. This implies that plasma treatment generates partial molecular chain breakage and slight carboxylation [36].

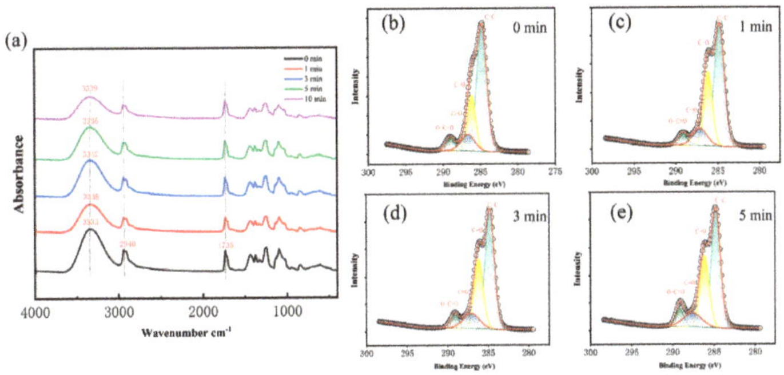

Figure 4. FTIR spectra (**a**) and XPS spectra (**b–e**) of PVA fibrous membrane treated by plasma with different processing time.

As shown in Figure S3, the treated PVA fibrous membranes show typical PVA characteristic peaks of alcohol hydroxyl O–H stretching (3333–3342 cm^{-1}), carbonyl C=O stretching (1735 cm^{-1}), and C–H stretching (2940 cm^{-1}) [37]. The same method was used for the FTIR spectra of PVA, and the analyzed statistics are summarized in Tables S9 and S10. With plasma treatment of 5 min, the increase of power results in the rise of the C=O/C–H ratio, indicating a slight increase in the relative amount of C=O in the molecular chain, which is consistent with the XPS results. Moreover, the O–H/C–H ratio is also decreased, which can be attributed to the oxidation of OH to carboxylate groups. With plasma treatment of 18 W, the increase in treatment time leads to the rise of the C=O/C–H ratio which is consistent with the XPS results. In addition, the ratio of O–H/C–H first increased from 1 min to 3 min, then decreased from 3 min to 5 min. This can be attributed to the interaction between plasma and PVA, involving both the production and oxidation of OH [38]. When the reactive oxygen species generated by plasma treatment are added at different positions on the PVA chain, they correspond to the generation and oxidation of hydroxyl groups. The schematic diagram of possible changes on the PVA chain after plasma treatment is shown in S2. Before 3 min, the production of OH might be the main reaction process and the oxidation reaction process of OH is dominated after 3 min.

As shown in Figure 5, along with the rise of plasma treatment time from 0 min to 5 min, the diameter of PVA fibers gradually decreased from 280 ± 81 nm to 157 ± 53 nm, and the surface roughness of PVA fibers gradually increased from 7.38 ± 0.1 nm to 18.8 ± 0.2 nm (Table S11). Similar to the etching on the surface of PLA fibers, plasma etching causes the breakage of PVA molecular chains, which is macroscopically manifested as a gradual decrease in the diameter of PVA fibers.

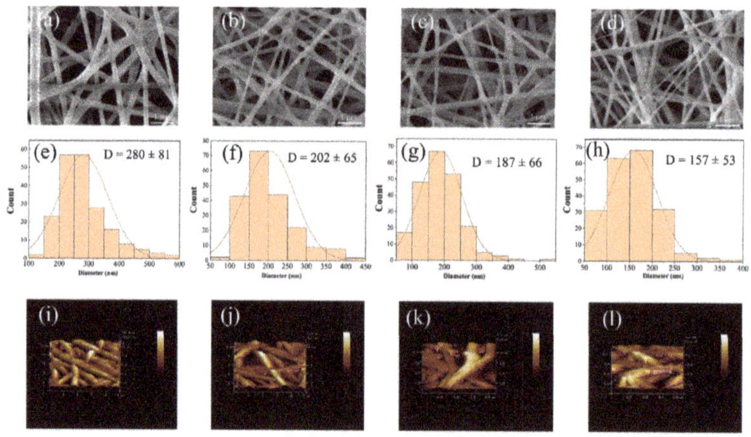

Figure 5. SEM images (**a–d**), fiber diameter distribution (**e–h**), AFM images (**i–l**) of PVA treated by plasma with different times (**a**,**e**,**i**): 0 min, (**b**,**f**,**j**): 1 min, (**c**,**g**,**k**): 3 min, (**d**,**h**,**l**): 5 min.

3.3. *Wettability*

After plasma modification, the wettability of the fibrous membrane is significantly improved. As shown in Figure 6, the contact angle drops from 79 ± 3.5° to 29 ± 0.6°, along with the rise of plasma treatment time from 0 min to 5 min. Moreover, the 5 min-treated PVA fibrous membrane shows much faster spreading of droplets than the untreated fibrous membrane, and droplets on 5 min-treated PVA fibrous membrane spread out in less than 8 s. In addition, the moisture absorption experiment proves that the fiber membrane with plasma treatment of 5 min has a faster moisture absorption speed in the early stages. The mesh structure of the fibrous membrane and the roughness, content of polar functional groups, and fiber fraction of fiber surface are the major influencing factors on fibrous membrane wettability [39–44]. Before plasma treatment, a large contact angle attributed to the meshes maintains the entrapped air between fibers, decreasing the surface free energy of the PVA fiber membrane. After plasma treatment, the increases in surface roughness and polar functional groups of fibers enhance the wettability of fibrous membranes.

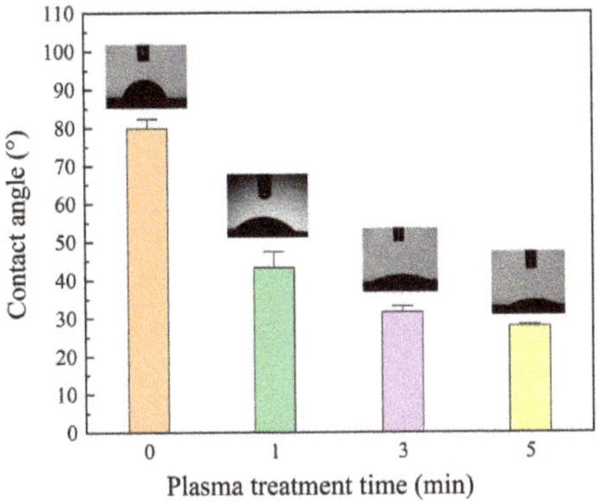

Figure 6. Droplet initial digital photos on PVA fibrous membrane with different treatment time.

3.4. PVA/PLA Tensile Modulus

As shown in Figure 7a, the fracture strength of the plasma-treated PVA/PLA fiber membranes increased by about 1 Mpa, but the fracture elongation decreased to some extent, which indicates that the plasma treatment affected the ductility of the PVA/PLA fiber membranes. The mechanical properties of PVA and PLA at different treatment times were tested separately (Figures S4 and S5). It was found that the tensile strain of PVA and PLA fiber membranes decreased and the tensile modulus increased after plasma treatment. In addition, the fracture strength of the treated PVA increased, and the fracture strength of PLA did not change much. The addition of PLA provided some mechanical support for PVA, while the plasma treatment improved the mechanical properties of the fiber membranes with the loss of part of the ductility.

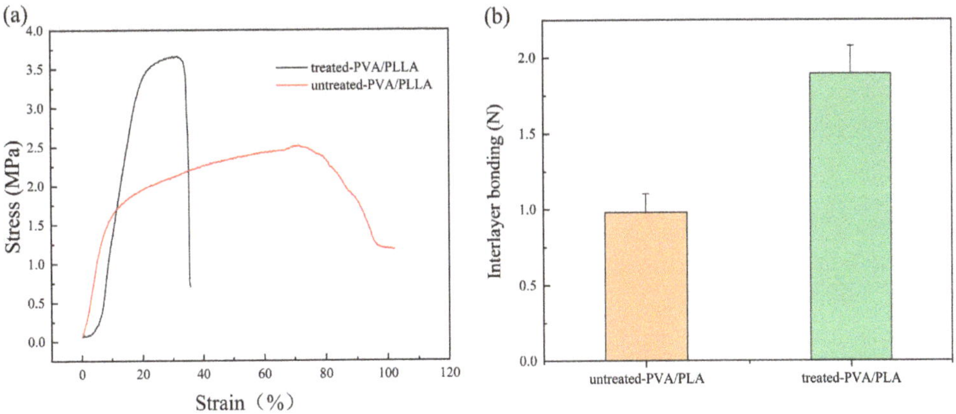

Figure 7. Stress–strain curves (**a**) and interlayer bonding (**b**).

It was reported that air plasma treatment generated the size-related surface effect that the decreasing fiber diameter could increase the tensile modulus dramatically. This is especially evident when the fiber diameter is less than 80 nm. The Young's modulus of fibrous is calculated by the following equation:

$$E = k_{PVA}\, L^3/48\pi ab^3 \quad (6)$$

L is the length of the fibrous in the channel, K_{PVA} is the stiffness of the fibrous; the fibrous is considered as an elliptical cross-section, and a, b are the long and short axes of the ellipse, respectively [45].

In our study, a decrease of fiber diameter of about 40% was observed, so the increase of tensile modulus can be mainly attributed to the decrease in fiber diameter. In addition, other studies showed that plasma treatment could promote the formation of chemical bonds (intermolecular and intramolecular hydrogen) on the surface of the polymer and enhance the cross-linking, leading to an increase in modulus [41].

As shown in Figure 7b, the bonding between the PVA fibers and PLA fibers after plasma treatment increased by about 1 N. The bonding between the fiber membranes before plasma treatment may be due to electrostatic attraction caused by the residual charge of the electrostatic spinning membrane and bonded interlacing between a small number of fibers. The increased probability of bonded interlacing between the two layers of fibers after plasma treatment may be related to the increased roughness of the fiber surface after treatment and the increased void space between the fibers. It has been shown that the stripping energy of PVA fibers from the plasma-treated base increases with the roughness of the base within a certain range [46].

3.5. Water Absorption

As shown in Figure 8, the water absorption rate of PLA increased to 237%, and the water absorption of a single layer of PVA/PLA increases to about 450% after plasma treatment. The large difference in water absorption before and after plasma treatment is mainly due to the fact that PLA electrospun fibers change from a hydrophobic to a hydrophilic state, while untreated PLA electrospun fibers hardly adsorb water.

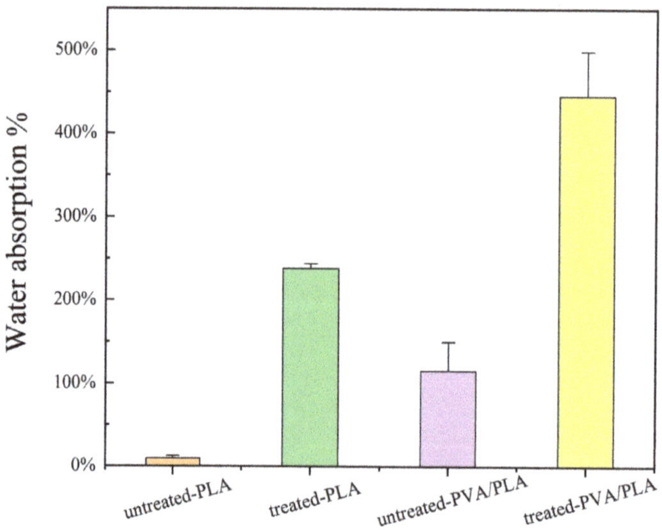

Figure 8. Liquid enrichment capacity of treated and untreated PVA/PLA fiber membranes.

3.6. Hemostatic Performance Test

The biosafety evaluation was performed before the hemostasis test. From Figures S6 and S7, it can be seen that the hemolysis rate of the treated PVA/PLA composite fibers was less than

5%, and the L929 cells showed normal proliferation, which indicates that the fibers have high biosafety [47].

An in vitro coagulation test was used to assess whether the hemostatic capacity of the treated fibrous membranes was improved. As shown in Figure S8, the BCI of treated PVA fibrous membranes was reduced from 47% to 25%, and the BCI of treated PVA/PLA was reduced to 21%, indicating that the treated fibrous membranes had a greater blood enrichment capacity. In addition, the coagulation time of the treated PVA fibrous membrane was reduced to about 438 s, and the coagulation time of the added treated PLA was reduced to about 258 s. The coagulation time of the treated PVA fibrous membrane was reduced to about 438 s, and the coagulation time of the treated PVA/PLA was reduced to about 258 s. The results can be attributed to the increase in the hydrophilicity of the PVA and PLA fiber membranes and the decrease in the diameter of the fibers, allowing the blood to pass through the surface of the fibrous membranes more rapidly to achieve the enrichment of blood between the fibers, which makes the fibrous membranes more structurally conducive to the enrichment of blood [48].

As shown in Figure 9, in the absence of fibrous membrane hemostasis, the rats exhibited rapid hemorrhage, with a total hemorrhage volume of approximately 0.25 ± 0.07 g in three minutes. Hemostasis was achieved in the PVA group, commercially available collagen group, treated PVA group, and treated PVA/PLA group, with a total hemorrhage of 0.14 ± 0.03 g, 0.0418 ± 0.01 g, 0.04 ± 0.02 g, and 0.029 ± 0.03 g in three minutes, respectively. Bleeding was reduced by 70% and 79% for treated PVA fibers and PVA/PLA fibers, respectively, compared to untreated PVA. The addition of PLA mainly provides mechanical support for the composite membrane, and further enriches the blood to accelerate the coagulation of blood to reduce bleeding. Three-dimensional electrospun fiber membranes with a porous structure have a high absorptive capacity and can concentrate blood to promote coagulation, showing a tamponade effect. Therefore, based on the physical barrier effect, PVA/PLA fibrous membranes show a good hemostatic effect [12,49,50]. Plasma treatment generates the enhancements of wetting properties, mechanical performance, and adhesion capacity of the PVA/PLA fiber membrane, enhancing the physical barrier effect for hemostasis.

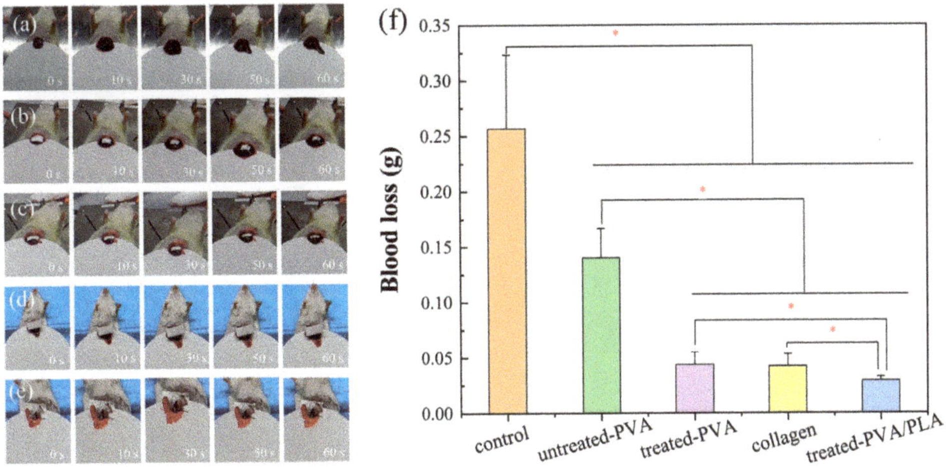

Figure 9. Image of animal experiments with the rat liver hemostasis, (**a**) control, (**b**) PVA, (**c**) plasma, treated PVA, (**d**) commercially available hemostatic collagen, (**e**) treated PVA/PLA, and total blood loss (**f**). * $p < 0.05$ was considered statistically significant.

4. Conclusions

A PVA/PLA hemostatic fiber membrane was prepared using air plasma modification and electrostatic spinning technology. By plasma treatment of PLA surface, the interlayer bonding force between the PVA and PLA fiber membrane increased from 1 N to 2 N. By plasma treatment of PVA, the fracture strength of PVA increased by about 1 MPa, and the amount of liver hemorrhage reduced by about 70%. After the air plasma treatment, the surface oxidation of PLA was somewhat higher than that of PVA, with the oxygen element content elevated by 13% and 2.4%, respectively; the fiber diameters of PLA and PVA were reduced from 2.16 ± 0.58 μm to 1.68 ± 0.62 μm, and 280 ± 81 nm to 157 ± 53 nm, respectively, and the roughness increased from 11.27 ± 2.6 nm to 30.73 ± 5.0 nm and 7.38 ± 0.14 nm to 18.8 ± 0.20 nm. Plasma-treated PVA/PLA composite membranes showed an increasing liquid-enrichment capacity of 350%. The hemostatic model of the liver showed that the hemostatic ability of plasma-treated PVA/PLA composite membranes was enhanced by 79% compared to untreated PVA membranes, and PLA acted as a good mechanical support. PVA/PLA composite membranes therefore have good biocompatibility.

Supplementary Materials: The following supporting information can be downloaded at: https://www.mdpi.com/article/10.3390/polym16121635/s1, Table S1. Plasma treatment of PLA at different times; Table S2. C1s of PLA fibrous membrane with different processing time of plasma treatment (power: 18 W); Table S3. C=O/C–H and C–O/C–H ratios of PLA at different treatment time; Table S4. Surface roughness of PLA electrospun fiber at different treatment times (power: 18 W); Figure S1. XPS spectra of PVA fiber membrane treated by plasma with different powers; Table S5. PVA was treated with plasma at different times with a power of 18 W; Table S6. PVA plasma treatment at different powers for 5 min; Table S7. C1s of PVA fiber membrane with different processing times of plasma treatment (power: 18 W); Table S8. C1s of PVA fibrous membrane with different powers of plasma treatment (time: 5 min); Figure S2. The schematic diagram of possible changes on the PVA chain after plasma treatment; Figure S3. FTIR spectra of PVA fibrous membrane treated by plasma with different powers (a) and different processing times (b); Table S9. Infrared absorbance ratio of O–H/C–H and C=O/C–H at different powers (time: 5 min); Table S10. Infrared absorbance ratio of O–H/C–H and C=O/C–H at different plasma processing times (power: 18 W); Table S11. Surface roughness of fiber at different processing times (power: 18 W); Figure S4. Stress–strain curves (b) and tensile modulus (d) of PLA fibers membrane; Figure S5. Stress–strain curves (b) and tensile modulus (d) of PVA fibers membrane; Figure S6. Hemocompatibility of untreated and treated PVA electrospun fibrous membranes; Figure S7. Cell viability (a) and AMPI staining (b) in blank, treated PLA, and treated PVA/PLA groups for 1, 3, and 5 days; Figure S8. Coagulation index (a), coagulation time (b) of the blank, untreated PVA, treated PVA, and treated PVA/PLA.

Author Contributions: Conceptualization, L.Z.; Methodology, L.Z., X.L. and Y.H.; Software, X.G.; Validation, X.W. and X.L.; Formal analysis, L.Z.; Investigation, X.W.; Data curation, X.G., X.L. and Y.H.; Writing—original draft, X.G.; Writing—review & editing, Y.H. All authors have read and agreed to the published version of the manuscript.

Funding: This work was supported by the Science and Technology Partnership Program, Ministry of Science and Technology of China (KY201602002).

Institutional Review Board Statement: All mouse experiments were conducted in accordance with the guidelines of the Laboratory Animal Center of Wuhan University of Technology (Wuhan, China) and approved by the Laboratory Animal Ethics Committee of Wuhan University of Technology, approval number WHUT2024-037.

Data Availability Statement: Data are contained within the article and Supplementary Materials.

Conflicts of Interest: The authors declare that they have no known competing financial interests or personal relationships that could have appeared to influence the work reported in this article.

References

1. Han, W.; Wang, S. Advances in Hemostatic Hydrogels That Can Adhere to Wet Surfaces. *Gels* **2023**, *9*, 2. [CrossRef]
2. Guo, B.; Dong, R.; Liang, Y.; Li, M. Hemostatic materials for wound healing applications. *Nat. Rev. Chem.* **2021**, *5*, 773–791. [CrossRef]

3. Zheng, Y.; Shariati, K.; Ghovvati, M.; Vo, S.; Origer, N.; Imahori, T.; Kaneko, N.; Annabi, N. Hemostatic patch with ultra-strengthened mechanical properties for efficient adhesion to wet surfaces. *Biomaterials* **2023**, *301*, 122240. [CrossRef]
4. Sedlár, M.; Kacvinská, K.; Fohlerová, Z.; Izsák, D.; Chalupová, M.; Suchy, P.; Dohnalová, M.; Sopuch, T.; Vojtová, L. A synergistic effect of fibrous carboxymethyl cellulose with equine collagen improved the hemostatic properties of freeze-dried wound dressings. *Cellulose* **2023**, *30*, 11113–11131. [CrossRef]
5. Al-Mofty, S.E.; Karaly, A.H.; Sarhan, W.A.; Azzazy, H.M.E. Multifunctional Hemostatic PVA/Chitosan Sponges Loaded with Hydroxyapatite and Ciprofloxacin. *ACS Omega* **2022**, *7*, 13210–13220. [CrossRef]
6. Mao, G.Z.; Wang, Z.; Tian, S.Y.; Li, H.Y.; Yang, W.Z. A novel chitosan-hyaluronic acid-pullulan composite film wound dressing for effectively inhibiting bacteria and accelerating wound healing. *Mater. Today Commun.* **2022**, *33*, 104801. [CrossRef]
7. Hossain, S.; Mohanty, S.; Verma, D. Polyelectrolyte complex based nanofibrous aggregates for fast hemostasis. *Mater. Today Commun.* **2021**, *27*, 102364. [CrossRef]
8. Ko, Y.G.; Kim, B.N.; Kim, E.J.; Chung, H.Y.; Park, S.Y.; Kim, Y.J.; Kwon, O.H. Bioabsorbable Carboxymethyl Starch-Calcium Ionic Assembly Powder as a Hemostatic Agent. *Polymers* **2022**, *14*, 3909. [CrossRef]
9. Lee, M.H.; Lee, D.R.; Chon, J.W.; Chung, D.J. Hemostatic Patches Based on Crosslinked Chitosan Films Applied in Interventional Procedures. *Polymers* **2021**, *13*, 2402. [CrossRef]
10. Pourshahrestani, S.; Zeimaran, E.; Djordjevic, I.; Kadri, N.A.; Towler, M.R. Inorganic hemostats: The state-of-the-art and recent advances. *Mater. Sci. Eng. C-Mater. Biol. Appl.* **2016**, *58*, 1255–1268. [CrossRef]
11. Lv, H.Y.; Zhao, M.; Li, Y.R.; Li, K.; Chen, S.J.; Zhao, W.W.; Wu, S.H.; Han, Y.T. Electrospun Chitosan-Polyvinyl Alcohol Nanofiber Dressings Loaded with Bioactive Ursolic Acid Promoting Diabetic Wound Healing. *Nanomaterials* **2022**, *12*, 2933. [CrossRef]
12. Yang, X.X.; Shi, N.; Liu, J.; Cheng, Q.Q.; Li, G.Y.; Lyu, J.; Ma, F.G.; Zhang, X.T. 3D Printed Hybrid Aerogel Gauzes Enable Highly Efficient Hemostasis. *Adv. Healthc. Mater.* **2023**, *12*, 2201591. [CrossRef]
13. Kumar, A.; Han, S.S. PVA-based hydrogels for tissue engineering: A review. *Int. J. Polym. Mater. Polym. Biomater.* **2017**, *66*, 159–182. [CrossRef]
14. Ilyas, R.A.; Sapuan, S.M.; Harussani, M.M.; Hakimi, M.; Haziq, M.Z.M.; Atikah, M.S.N.; Asyraf, M.R.M.; Ishak, M.R.; Razman, M.R.; Nurazzi, N.M.; et al. Polylactic Acid (PLA) Biocomposite: Processing, Additive Manufacturing and Advanced Applications. *Polymers* **2021**, *13*, 1326. [CrossRef]
15. Agarwal, Y.; Rajinikanth, P.S.; Ranjan, S.; Tiwari, U.; Balasubramnaiam, J.; Pandey, P.; Arya, D.K.; Anand, S.; Deepak, P. Curcumin loaded polycaprolactone-/polyvinyl alcohol-silk fibroin based electrospun nanofibrous mat for rapid healing of diabetic wound: An in-vitro and in-vivo studies. *Int. J. Biol. Macromol.* **2021**, *176*, 376–386. [CrossRef]
16. Lin, M.G.; Shen, J.L.; Qian, Q.A.; Li, T.; Zhang, C.Y.; Qi, H. Fabrication of Poly(Lactic Acid)@TiO_2 Electrospun Membrane Decorated with Metal–Organic Frameworks for Efficient Air Filtration and Bacteriostasis. *Polymers* **2024**, *16*, 889. [CrossRef]
17. Bi, H.Y.; Feng, T.Y.; Li, B.B.; Han, Y.C. In Vitro and In Vivo Comparison Study of Electrospun PLA and PLA/PVA/SA Fiber Membranes for Wound Healing. *Polymers* **2020**, *12*, 839. [CrossRef]
18. Kamoun, E.A.; Loutfy, S.A.; Hussein, Y.; Kenawy, E.S. Recent advances in PVA-polysaccharide based hydrogels and electrospun nanofibers in biomedical applications: A review. *Int. J. Biol. Macromol.* **2021**, *187*, 755–768. [CrossRef]
19. Ji, Y.; Song, W.; Xu, L.; Yu, D.-G.; Bligh, S.W.A. A Review on Electrospun Poly(amino acid) Nanofibers and Their Applications of Hemostasis and Wound Healing. *Biomolecules* **2022**, *12*, 794. [CrossRef]
20. Yang, Y.; Du, Y.; Zhang, J.; Zhang, H.; Guo, B. Structural and Functional Design of Electrospun Nanofibers for Hemostasis and Wound Healing. *Adv. Fiber Mater.* **2022**, *4*, 1027–1057. [CrossRef]
21. Rangel, E.C.; Bento, W.C.A.; Kayama, M.E.; Schreiner, W.H.; Cruz, N.C. Enhancement of polymer hydrophobicity by SF_6 plasma treatment and argon plasma immersion ion implantation. *Surf. Interface Anal.* **2003**, *35*, 179–183. [CrossRef]
22. Vesel, A.; Mozetic, M. New developments in surface functionalization of polymers using controlled plasma treatments. *J. Phys. D-Appl. Phys.* **2017**, *50*, 293001. [CrossRef]
23. Firoozi, M.; Entezam, M.; Masaeli, E.; Ejeian, F.; Nasr-Esfahani, M.H. Physical modification approaches to enhance cell supporting potential of poly (vinyl alcohol)-based hydrogels. *J. Appl. Polym. Sci.* **2022**, *139*, 51485. [CrossRef]
24. Li, J.; Niu, J.; Kim, Y.; Chae, H. Surface wettability control and fluorination modeling of amorphous carbon films fluorinated with CF4 plasma. *Appl. Surf. Sci.* **2023**, *635*, 157668. [CrossRef]
25. Mai, Z.R.; Liu, Q.L.; Bian, Y.S.; Wang, P.; Fu, X.W.; Lin, D.S.; Kong, N.Z.; Huang, Y.Q.; Zeng, Z.J.; Li, D.F.; et al. PCL/Collagen/UA Composite Biomedical Dressing with Ordered Microfiberous Structure Fabricated by a 3D Near-Field Electrospinning Process. *Polymers* **2023**, *15*, 223. [CrossRef]
26. Guo, J.Y.; Liu, Q.; Cheng, J.; Fu, X.; Zhang, Y.C.; Yang, H. Hemostatic cotton based on biocompatible poly(vinyl alcohol)/soluble starch-fish skin collagen composites. *Mater. Today Commun.* **2022**, *32*, 103812. [CrossRef]
27. Laput, O.A.; Vasenina, I.V.; Korzhova, A.G.; Bryuzgina, A.A.; Khomutova, U.V.; Tuyakova, S.G.; Akhmadeev, Y.H.; Shugurov, V.V.; Bolbasov, E.N.; Tverdokhlebov, S.I.; et al. Effect of Nitrogen Arc Discharge Plasma Treatment on Physicochemical Properties and Biocompatibility of PLA-Based Scaffolds. *Polymers* **2023**, *15*, 3381. [CrossRef]
28. Laput, O.A.; Vasenina, I.V.; Shapovalova, Y.G.; Ochered'ko, A.N.; Chernyavskii, A.V.; Kudryashov, S.V.; Kurzina, I.A. Low-Temperature Barrier Discharge Plasma Modification of Scaffolds Based on Polylactic Acid. *ACS Appl. Mater. Interfaces* **2022**, *14*, 41742–41750. [CrossRef]

29. Can-Herrera, L.A.; Oliva, A.I.; Cervantes-Uc, J.M. Enhancement of chemical, physical, and surface properties of electrospun PCL/PLA blends by means of air plasma treatment. *Polym. Eng. Sci.* **2022**, *62*, 1608–1618. [CrossRef]
30. Thongsomboon, W.; Srihanam, P.; Baimark, Y. Preparation of flexible poly(L-lactide)-b-poly(ethylene glycol)-b-poly (L-lactide)/talcum/thermoplastic starch ternary composites for use as heat-resistant and single-use bioplastics. *Int. J. Biol. Macromol.* **2023**, *230*, 123172. [CrossRef]
31. Savoji, H.; Lerouge, S.; Ajji, A.; Wertheimer, M.R. Plasma-Etching for Controlled Modification of Structural and Mechanical Properties of Electrospun PET Scaffolds. *Plasma Process. Polym.* **2015**, *12*, 314–327. [CrossRef]
32. Cai, Z.S.; Qiu, Y.P. The mechanism of air/oxygen/helium atmospheric plasma action on PVA. *J. Appl. Polym. Sci.* **2006**, *99*, 2233–2237. [CrossRef]
33. Ghobeira, R.; Tabaei, P.S.E.; Morent, R.; De Geyter, N. Chemical characterization of plasma-activated polymeric surfaces via XPS analyses: A review. *Surf. Interfaces* **2022**, *31*, 102087. [CrossRef]
34. Paneru, R.; Ki, S.H.; Lamichhane, P.; Nguyen, L.N.; Adhikari, B.C.; Jeong, I.J.; Mumtaz, S.; Choi, J.; Kwon, J.S.; Choi, E.H. Enhancement of antibacterial and wettability performances of polyvinyl alcohol/chitosan film using non-thermal atmospheric pressure plasma. *Appl. Surf. Sci.* **2020**, *532*, 147339. [CrossRef]
35. Rao, X.; Zhou, Q.; Wen, Q.; Ou, Z.; Fu, L.; Gong, Y.; Du, X.; Huo, C. High-Performance and Water Resistant PVA-Based Films Modified by Air Plasma Treatment. *Membranes* **2022**, *12*, 249. [CrossRef]
36. Tifui, G.; Dobromir, M.; Sirghi, L. Stiffening of polydimethylsiloxane surface as result of exposure to low-pressure argon discharge plasma. *Plasma Process. Polym.* **2023**, *20*, 2300014. [CrossRef]
37. Rufato, K.B.; Veregue, F.R.; Medeiro, R.d.P.; Francisco, C.B.; Souza, P.R.; Popat, K.C.; Kipper, M.J.; Martins, A.F. Electrospinning of poly(vinyl alcohol) and poly(vinyl alcohol)/tannin solutions: A critical viewpoint about crosslinking. *Mater. Today Commun.* **2023**, *35*, 106271. [CrossRef]
38. Rogojanu, A.; Rusu, E.; Dorohoi, D.O. Characterization of Structural Modifications Induced on Poly(Vinyl Alcohol) Surface by Atmospheric Pressure Plasma. *Int. J. Polym. Anal. Charact.* **2010**, *15*, 210–221. [CrossRef]
39. Szewczyk, P.K.; Ura, D.P.; Metwally, S.; Knapczyk-Korczak, J.; Gajek, M.; Marzec, M.M.; Bernasik, A.; Stachewicz, U. Roughness and Fiber Fraction Dominated Wetting of Electrospun Fiber-Based Porous Meshes. *Polymers* **2019**, *11*, 34. [CrossRef]
40. Ojah, N.; Saikia, D.; Gogoi, D.; Baishya, P.; Ahmed, G.A.; Ramteke, A.; Choudhury, A.J. Surface modification of core-shell silk/PVA nanofibers by oxygen dielectric barrier discharge plasma: Studies of physico-chemical properties and drug release behavior. *Appl. Surf. Sci.* **2019**, *475*, 219–229. [CrossRef]
41. Das, P.; Ojah, N.; Kandimalla, R.; Mohan, K.; Gogoi, D.; Dolui, S.K.; Choudhury, A.J. Surface modification of electrospun PVA/chitosan nanofibers by dielectric barrier discharge plasma at atmospheric pressure and studies of their mechanical properties and biocompatibility. *Int. J. Biol. Macromol.* **2018**, *114*, 1026–1032. [CrossRef]
42. Ojah, N.; Borah, R.; Ahmed, G.A.; Mandal, M.; Choudhury, A.J. Surface modification of electrospun silk/AMOX/PVA nanofibers by dielectric barrier discharge plasma: Physiochemical properties, drug delivery and in-vitro biocompatibility. *Prog. Biomater.* **2020**, *9*, 219–237. [CrossRef]
43. Paneru, R.; Lamichhane, P.; Adhikari, B.C.; Ki, S.H.; Choi, J.; Kwon, J.S.; Choi, E.H. Surface modification of PVA thin film by nonthermal atmospheric pressure plasma for antifogging property. *AIP Adv.* **2019**, *9*, 075008. [CrossRef]
44. Watson, S.; Pattyn, C.; Garakani, M.M.; Ajji, A.; Reuter, S.; Wertheimer, M.R. Wetting and aging behaviors for several O_2 and NH_3 plasma treated nanofibrous and flat film polymer materials: A comparison. *Plasma Process. Polym.* **2023**, *20*, e2200224. [CrossRef]
45. Fu, Q.; Jin, Y.; Song, X.F.; Gao, J.Y.; Han, X.B.; Jiang, X.Y.; Zhao, Q.; Yu, D.P. Size-dependent mechanical properties of PVA nanofibers reduced via air plasma treatment. *Nanotechnology* **2010**, *21*, 095703. [CrossRef]
46. Zhao, C.; Zhang, X.; Jiang, A.; Pan, Z. Adhesion and protective properties of polyvinyl alcohol nanofibrous composite fabrics. *J. Text. Inst.* **2018**, *109*, 1263–1269. [CrossRef]
47. Ye, T.; Yang, Z.Y.; Hao, R.L.; Guo, J.N.; Dou, G.F.; Meng, Z.Y.; Liu, S.C.; Gu, R.L.; Wu, Z.N.; Sun, Y.B.; et al. Highly Efficient Hemostatic Cross-Linked Polyacrylate Polymer Dressings for Immediate Hemostasis. *Polymers* **2024**, *16*, 863. [CrossRef]
48. Wang, Y.; Zhai, W.; Cheng, S.; Li, J.; Zhang, H. Surface-functionalized design of blood-contacting biomaterials for preventing coagulation and promoting hemostasis. *Friction* **2023**, *11*, 1371–1394. [CrossRef]
49. Cheng, C.; Peng, X.; Xi, L.; Wan, C.; Shi, S.; Wang, Y.; Yu, X. An agar-polyvinyl alcohol hydrogel loaded with tannic acid with efficient hemostatic and antibacterial capacity for wound dressing. *Food Funct.* **2022**, *13*, 9622–9634. [CrossRef]
50. John, J.V.; McCarthy, A.; Karan, A.; Xie, J.W. Electrospun Nanofibers for Wound Management. *Chemnanomat* **2022**, *8*, e202100349. [CrossRef]

polymers

MDPI

Article

The Incorporation of Sulfonated PAF Enhances the Proton Conductivity of Nafion Membranes at High Temperatures

Kun Cai [1,*], Jinzhu Yu [1], Wenjun Tan [1], Cong Gao [1], Zili Zhao [2], Suxin Yuan [1], Jinghui Cheng [1], Yajie Yang [3,*] and Ye Yuan [4]

1 Key Laboratory of Micro-Nano Materials for Energy Storage and Conversion of Henan Province, Institute of Surface Micro and Nano Materials, College of Chemical and Materials Engineering, Xuchang University, Xuchang 461000, China; yjz020914@outlook.com (J.Y.); tanwj@dhu.edu.cn (W.T.); galigiaogiaoBBB@163.com (C.G.); y1061582635@163.com (S.Y.); jinghui871016@163.com (J.C.)
2 XuJue Electric Co., Ltd., Xuchang 461000, China; 13513748576@163.com
3 Key Laboratory of Automobile Materials of Ministry of Education, School of Materials Science and Engineering, Jilin University, Changchun 130022, China
4 Key Laboratory of Polyoxometalate and Reticular Material Chemistry of Ministry of Education, Northeast Normal University, Changchun 130024, China; yuany101@nenu.edu.cn
* Correspondence: caikun2010@gmail.com (K.C.); yangyajie@jlu.edu.cn (Y.Y.)

Abstract: Nafion membranes are widely used as proton exchange membranes, but their proton conductivity deteriorates in high-temperature environments due to the loss of water molecules. To address this challenge, we propose the utilization of porous aromatic frameworks (PAFs) as a potential solution. PAFs exhibit remarkable characteristics, such as a high specific surface area and porosity, and their porous channels can be post-synthesized. Here, a novel approach was employed to synthesize a PAF material, designated as PAF-45D, which exhibits a specific surface area of 1571.9 $m^2 \cdot g^{-1}$ and possesses the added benefits of facile synthesis and a low cost. Subsequently, sulfonation treatment was applied to PAF-45D in order to introduce sulfonic acid groups into its pores, resulting in the formation of PAF-45DS. The successful incorporation of sulfonic groups was confirmed through TG, IR, and EDS analyses. Furthermore, a novel Nafion composite membrane was prepared by incorporating PAF-45DS. The Nyquist plot of the composite membranes demonstrates that the sulfonated PAF-45DS material can enhance the proton conductivity of Nafion membranes at high temperatures. Specifically, under identical film formation conditions, doping with a 4% mass fraction of PAF-45DS, the conductivity of the Nafion composite membrane increased remarkably from 2.25×10^{-3} $S \cdot cm^{-1}$ to 5.67×10^{-3} $S \cdot cm^{-1}$, nearly 2.5 times higher. Such promising and cost-effective materials could be envisioned for application in the field of Nafion composite membranes.

Keywords: Nafion; composite membrane; proton conduction; PAFs

1. Introduction

Proton exchange membrane fuel cells (PEMFCs) are widely acknowledged for their high energy conversion efficiency and environmental friendliness [1]. Proton exchange membranes (PEMs), a crucial component of fuel cells, facilitate the migration and transport of protons while simultaneously serving as a barrier for gas reactants. Nafion, developed by DuPont, is a perfluorosulfonic acid (PFSA) membrane. It is the most commercialized proton exchange membrane and exhibits high proton conductivity under moderate temperature conditions [2]. However, Nafion membranes exhibit certain drawbacks, such as slightly elevated methanol permeability and reduced proton conductivity due to the evaporation of water molecules responsible for proton conduction at high temperatures [3–5]. Consequently, numerous mixed matrix membranes incorporating Nafion have been attracting attention. In recent years, novel porous materials such as metal organic frameworks (MOFs) [6–10] and covalent organic frameworks (COFs) [11–14] have been employed as fillers. As members of the porous material family, porous aromatic frameworks (PAFs) can also serve as fillers due to their exceptional resistance to acid and heat. However, to the best of our knowledge, there are limited reports on the synthesis of PAF-doped Nafion composite membranes.

PAFs, constructed via robust C–C covalent bonds, exhibit numerous exceptional properties, including a high specific surface area and porosity, and have been extensively studied in gas storage or separation [15–17], catalysis [18–20], and other domains [21–27]. Furthermore, they demonstrate notable thermal and chemical stability [16], capable of enduring post-synthetic modification under harsh conditions, such as treatment with chlorosulfonic acid [28–31]. Due to these advantageous characteristics, this paper proposes using PAF materials as novel fillers to address the dehydration issue of Nafion membranes at high temperatures and low humidity. The pores of PAFs have the ability to adsorb water molecules and preserve the proton transport channel. By sulfonating the framework, the hydrophilicity of the pores is enhanced, which also facilitates the substantial supply of protons. However, some PAFs still have certain disadvantages, such as difficulties in monomer synthesis and the need for expensive catalysts. Therefore, PAF-45 was selected.

PAF-45, a member of the PAF family [32], possesses the added benefits of facile synthesis and a low cost. Among its raw materials, the biphenyl monomer and the aluminum chloride catalyst are comparatively economical. This attribute is particularly significant for Nafion membrane fillers as Nafion itself is highly expensive. Therefore, PAF-45 was selected as the dopant. According to the literature [32,33], a novel approach was employed in synthesizing PAF-45 by substituting trichloromethane with dichloromethane, resulting in a material with a larger surface area (1571.9 $m^2 \cdot g^{-1}$), designated as PAF-45D (D = dichloromethane). Then, PAF-45D was treated with chlorosulfonic acid to introduce sulfonic acid groups into its pores, resulting in the product marked PAF-45DS [29,34]. This sulfonation process effectively enhances the hydrophilicity of the PAF-45D pores. These modifications are expected to enhance the conductivity of Nafion films at elevated temperatures. Subsequently, the impact of doping PAF-45DS with varying mass fractions on the proton conductivity of Nafion composite membranes was primarily investigated. Under identical film formation conditions, the conductivity of the Nafion composite membrane increased significantly from 2.25×10^{-3} $S \cdot cm^{-1}$ to 5.67×10^{-3} $S \cdot cm^{-1}$ at 80 °C ~78% RH (relative humidity), nearly 2.5 times, when only doping with a 4% mass fraction of PAF-45DS. This represents a substantial enhancement in membrane performance.

2. Materials and Methods

All commercial chemical reagents were used without further purification. The main reagents were biphenyl (AR, Aladdin Reagents, Shanghai, China), $AlCl_3$ (AR, Aladdin Reagents, Shanghai, China), Nafion (D520, DuPont, Wilmington, DE, USA), chlorosulfonic acid (AR, Aladdin Reagents, Shanghai, China), and dichloromethane (AR, Sinopharm Group SCR Reagents, Beijing, China). Thermogravimetric analyses (TGA) were performed under an atmosphere with a heating rate of 10 °C min^{-1} using a STA 409 PC from 30 °C to 800 °C. Before testing, the sample was cleaned with acid and water. Powder X-ray-diffraction (XRD) analyses were performed on a Bruker D8 Advance with Cu Kα (λ = 1.5418 Å) at 40 kV, 40 mA. Fourier-transform infrared spectra (FT-IR) were obtained on a Thermo Nicolet Nexus 470 FT-IR (Thermo Fisher Scientific, Waltham, MA, USA) spectrometer in the 4000–400 cm^{-1} range using KBr pellets. The N_2 gas sorption measurements were acquired on a BELSORP-max (MicrotracBEL Japan Inc., Osaka, Japan) surface area analyzer at 77 K. The water uptake was analyzed on a JWGB Instrument JW-ZQ100 (JWGB Instrument, Beijing, China) at 298 K. The SEM and EDS pictures were taken on a Nova NanoSEM 450 (FEI Company, Hillsboro, OR, USA).

2.1. Synthesis of PAF-45D

In the experiment, the catalyst was activated first. Anhydrous aluminum chloride (1 g) and dichloromethane (20 mL) were added to a 100 mL round-bottom flask, and the catalyst was stirred at 65 °C for 3 h. After activation, 200 mg of biphenyl was dissolved in 20 mL of dichloromethane and poured into the round-bottom flask. After the reaction system was stirred in an oil bath at 65 °C for 24 h, the power was turned off, and the crude product was obtained after the temperature of the reaction system had dropped to room temperature. Ethanol (50 mL) was added to the product, which was filtered with a pump. To remove the raw materials and catalysts, the resulting material was added to the beaker with 50 mL 2 mol·L^{-1} HCl and stirred for 3 h, followed by ultrasonication for 30 min, and then filtrated. Subsequently, distilled water and ethanol were used successively. The agitation, ultrasonication, and filtration process was repeated. The cleaning procedure was iterated three times. The final product was placed in an oven at 80 °C to dry, and it was named PAF-45D.

2.2. Synthesis of PAF-45DS

PAF-45D (250 mg) and dichloromethane (20 mL) were added to a 100 mL flask, which was filled with N_2 gas and cooled in an ice-water bath. Chlorosulfonic acid (4.0 mL) was slowly dropped into the flask. The mixture was stirred at room temperature for three days; then, the product was poured into 50 mL ice water under stirring conditions. Then, ethanol (50 mL) was added to the water; it was filtered with a pump and then washed with distilled water and ethanol three times successively. The treatment method was the same as above. The final product was placed in an oven at 80 °C to dry, and it was named PAF-45DS.

2.3. Preparation of Nafion Composite Membrane

Nafion D520 is available for commercial use and its solvent is an alcohol–water mixture. However, it is necessary to convert the solvent into a DMF solution in order to achieve the better dispersion of PAF. A certain amount of Nafion D520 was placed in a glass culture dish and positioned in an 80 °C oven for drying. Then, the dry Nafion film was cut into pieces and a certain amount of DMF was added to prepare a 5% (wt.%) Nafion–DMF solution. The conventional solution casting technique was carried out to prepare the composite membranes. Different masses of PAF-45DS (0 wt.%, 2 wt.%, 4 wt.%, 6 wt.%, 8 wt.%, and 10 wt.%) were added to the Nafion–DMF solution, which was then subjected to ultrasonic dispersion for half an hour to obtain the film-forming liquid. Then, it was dropped onto a flat Teflon surface and left at 80 °C for 12 h until the Nafion composite films were obtained.

2.4. Proton Conductivity

The proton conductivity of the membranes was measured via the AC impedance spectroscopy technique with a quasi-four-probe method on a Solartron ModuLab XM (AMETEK SI, Berwyn, PA, USA). A frequency from 1 MHz to 1 Hz, 5 mV AC perturbations, and a 0.0 V DC rest voltage were used during the test. Saturated solutions of KCl were used to act as a humidity source with about 78% relative humidity (~78%RH) at 80 °C. The solution was placed into a sealed bottle for 3 days to ensure that the air reached an equilibrium state. The testing temperature of the impedance studies was generated by water vapor at 80 °C. The Nafion composite films were cut to round pieces with a diameter of 3 mm, and the thickness of the films was tested with a micrometer caliper. Prior to the measurement, the samples were placed in a bottle at the required temperature of 80 °C for a certain time. The conductivity (σ, S cm^{-1}) was calculated using the relation $\sigma = L/(RA)$, where R (Ω) is the sample resistance estimated by the extrapolation of the high-frequency arc crossing to the axis [35]. L (cm) is the thickness and A (cm^2) is the face area.

3. Results and Dicussion

3.1. X-ray Diffraction Analysis

The X-ray diffraction patterns of the samples and the Nafion composite films with different doping amounts were tested. Both the powder sample and the membrane sample were tested on a glass plate with grooves. It can be seen from Figure 1 that the peaks of PAF-45D and PAF-45DS are basically the same, and it shows typical amorphous polymer peaks. This is consistent with the characteristics of the previously reported material [32]. There are no weak peaks for the PAF-45D material, which suggests that the catalyst $AlCl_3$ was basically removed. The curve of PAF-45DS is an amorphous material curve, indicating that no impurities were introduced during the sulfonation.

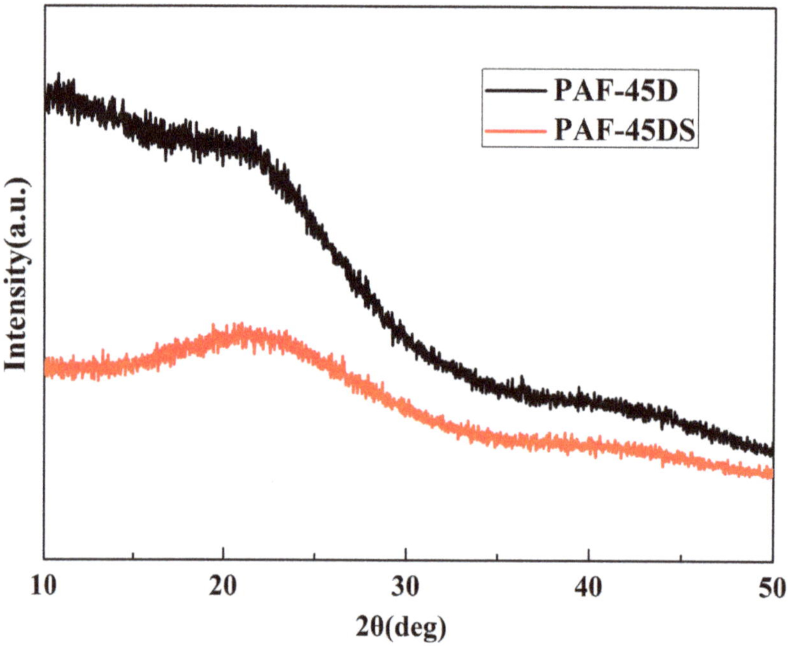

Figure 1. XRD patterns of the PAF powder.

Figure 2 illustrates the XRD patterns of the Nafion composite films doped with different mass fractions of PAF-45DS. It can be observed that as the mass fraction of PAF-45DS increases, the peak position of the Nafion composite films shifts to the right and approaches the amorphous peak of PAF-45DS, indicating the gradual intensification of the amorphous nature of the Nafion composite film.

3.2. TGA and IR

The thermal stability of PAF-45D and PAF-45DS was investigated through thermogravimetric analysis (TGA) experiments. The TG curve was recorded in an air atmosphere from 35 to 800 °C (Figure 3). PAF-45D has less weight loss before 100 °C. After a platform, another weight loss is observed starting from ~350 °C, and the mass begins to drop sharply, which is attributed to the decomposition of the polymer frameworks. A weight loss of 11.2% is observed from 35 to 126 °C in the curve of PAF-45DS, which is attributed to the release of adsorbed water in the pores. After a brief plateau, the mass begins to decrease continually from 200 °C, indicating the collapse of the skeleton. The TGA demonstrates that the introduction of sulfonic acid groups reduces the thermal stability of the material. More importantly, sulfonation enhances the water absorption capacity of PAF-45D.

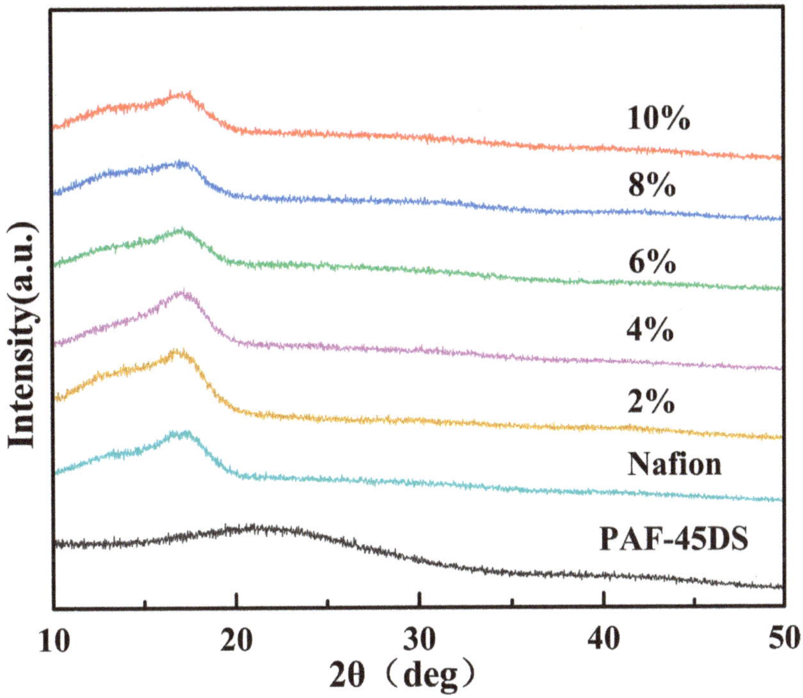

Figure 2. XRD patterns of Nafion composite films with various content of PAF-45DS.

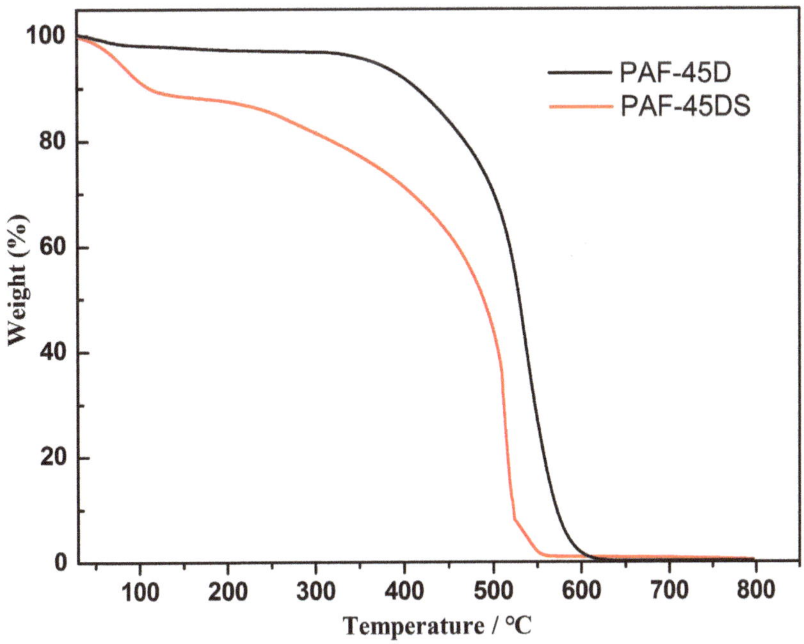

Figure 3. TGA thermograms of PAF-45D and PAF-45DS.

To determine whether the sulfonic acid group was introduced, Fourier transform infrared spectroscopy (FT-IR) was performed on PAF-45D and PAF-45DS. As shown in Figure S1, the band associated with the S–O vibration is detected at 1176 cm^{-1} in the spectrum of PAF-45DS, which is similar to the literature reports [29]. The range 1660~1373 cm^{-1} in the spectrum denotes the C=C stretching vibrations for the benzene ring. The spectroscopy confirms the existence of alkyl groups in the structure by C-H stretching vibrations near 2929 cm^{-1}. This is consistent with the literature report [32,33]. Moreover, the C-H absorption peaks of benzene extend from 730 to 910 cm^{-1}.

3.3. Gas Adsorption

The N_2 adsorption isotherm of the activated sample was measured at 77 K to explore the porosity of these PAF materials. All samples were pumped to a vacuum at 423 K for 8 h before the measurement. According to Figure 4, the two materials exhibit typical I type isotherms, indicating substantial and rapid N_2 uptake at the low-pressure region ($P/P_0 < 0.1$). The adsorption and desorption curves of PAF-45D exhibit distinct hysteresis, a phenomenon commonly observed in many PAF networks, attributed to the interaction between nitrogen and the pore channels. After sulfonation, the hysteresis of the adsorption and desorption curves of PAF-45DS is significantly diminished, indicating that the interaction is reduced.

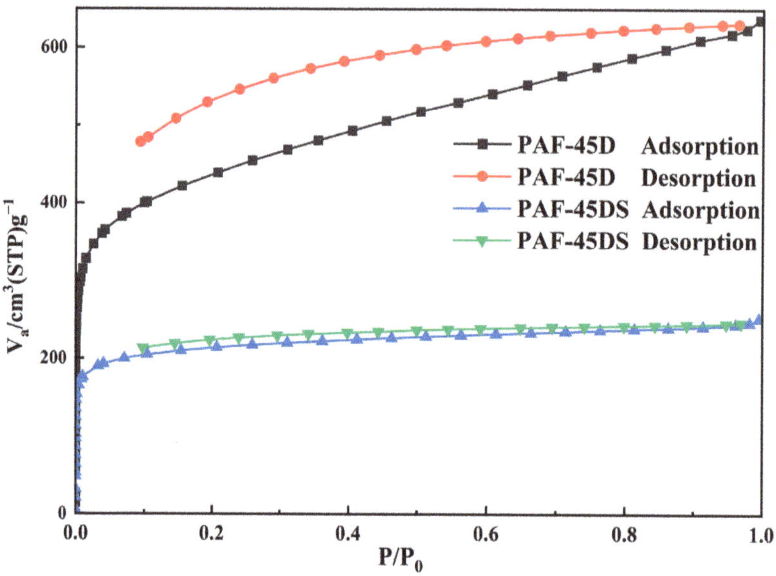

Figure 4. The N_2 adsorption isotherms of PAF-45D and PAF-45DS.

For PAF-45D, the Brunauer–Emmett–Teller (BET) surface area was calculated to be 1571.9 $m^2 \cdot g^{-1}$, and that of PAF-45DS was 806.1 $m^2 \cdot g^{-1}$. The pore size distribution can be obtained by the nonlocal density functional theory (NLDFT) method. The pore sizes of PAF-45D and PAF-45DS are 1.04 nm and 0.44 nm, respectively. These results clearly indicate that the sulfonic acid group occupies the pores of PAF-45D, which reduces the specific surface area and pore size. It also proves the successful introduction of the sulfonic acid group. The water vapor adsorption at 298K was also investigated (Figure S2), demonstrating that the sulfonated material exhibited an enhanced water adsorption capacity due to the presence of hydrophilic sulfonic groups on the frameworks. Despite the reduction in the pore size and surface area of PAF-45DS, it still demonstrates higher water uptake (177 cm^3/g, STP), surpassing that of PAF-45D (128 cm^3/g, STP).

3.4. Morphological Characteristics

The morphological characteristics of the PAF powder and Nafion composite film were obtained by SEM. According to Figure 5, most of the PAF-45D and PAF-45DS particles display irregular spherical shapes, indicating that sulfonation has no significant effect on the morphology of the material. The SEM and EDS images for PAF-45DS were employed to confirm the S elemental content. Figure S3 displays the EDS spectrum obtained from the surface of the sulfonated material. The quantitative analysis results are shown in Table S1. The results indicate the distinct presence of the S element in the PAF-45DS material. The presence of water molecules in the material leads to an excess of oxygen, resulting in an increased O/S ratio.

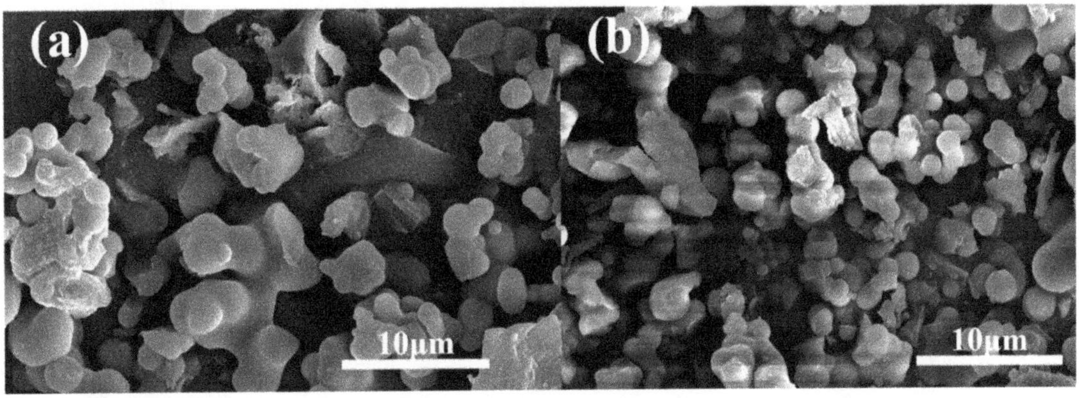

Figure 5. The SEM images of (**a**) PAF-45D and (**b**) PAF-45DS.

The morphology of the Nafion composite membranes can be observed in Figure 6. The SEM images captured from the top surfaces of different membranes reveal that as the PAF-45DS mass fraction increases, the solid particles are packed more closely and the composite membranes become progressively rougher. When the quantity increases to 10 wt.%, the Nafion becomes scarcer and crevices can be seen on its surface. This phenomenon is normal when the dispersing phase increases. As the crystal content increases, it may cause the composite membranes to become more brittle and less flexible. An increase in the solid particle content of PAF-45DS in a mixed matrix membrane also results in enhanced brittleness and decreased flexibility.

3.5. Proton Conduction of Composite Membranes

The proton-conductive properties of the Nafion composite membranes were evaluated by using AC impedance spectroscopy at 80 °C and ~78% RH (Figure 7). The detailed Nyquist curves are shown in Figure S4. The proton conductivity of the pure Nafion membrane (0 wt.% PAF-45DS) was 2.245×10^{-3} S·cm^{-1} at the same film-forming conditions. The proton conductivity of the composite membranes significantly increased with the addition of PAF-45DS. It can be observed that varying the loading amounts of PAF-45DS from 2 wt.% to 4 wt.% leads to an increase in proton conductivity from 4.077×10^{-3} S·cm^{-1} to 5.667×10^{-3} S·cm^{-1}. These results indicate that the addition of PAF-45DS can enhance the proton conductivity of Nafion membranes at high temperatures.

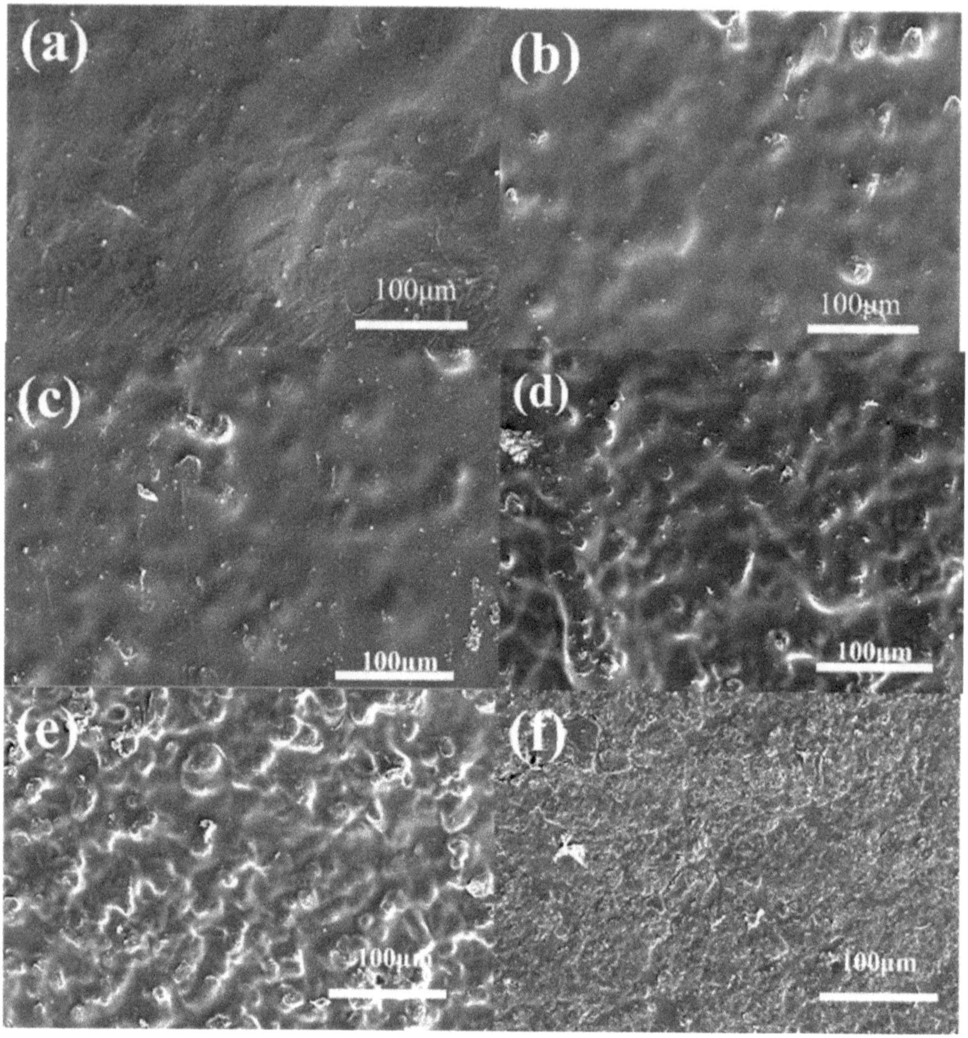

Figure 6. The SEM images of the Nafion composite membranes with different content of PAF-45DS: (**a**) pure Nafion (0 wt.% PAF-45DS) and (**b–f**) the content of 2 wt.%, 4 wt.%, 6 wt.%, 8 wt.%, 10 wt.%, respectively.

Under the same film formation conditions, the conductivity of the Nafion film can be increased by nearly 2.5 times. This may be due to the fact that PAF-45DS provides some sulfonic acid groups and its abundant pores can reduce the loss of water molecules at high temperatures. The proton conductivity of the composite membranes reached a peak value at 4 wt.% PAF-45S. Afterwards, as the content increased, the conductivity began to decrease. The conductivity was 3.310×10^{-3} S·cm^{-1} (6 wt.%), 2.832×10^{-3} S·cm^{-1} (8 wt.%), and 2.528×10^{-3} S·cm^{-1} (10 wt.%), respectively. This may be due to the increased doping mass, which reduces the continuity of the Nafion film and leads to a decrease in conductivity. When the temperature was around 80 °C, we observed a decrease in the proton conductivity of the Nafion composite membrane as the temperature increased. Therefore, we were unable to determine the activation energy for proton conduction under the same test conditions.

Author Contributions: Conceptualization, Y.Y. (Ye Yuan) and K.C.; methodology, K.C., W.T. and Z.Z.; validation, Y.Y. (Yajie Yang) and K.C.; formal analysis, Y.Y. (Ye Yuan); investigation, J.Y., C.G., J.C. and S.Y.; writing—original draft preparation, K.C. and J.Y.; writing—review and editing, Y.Y. (Ye Yuan), Y.Y. (Yajie Yang) and K.C.; visualization, K.C.; supervision, Y.Y. (Ye Yuan); project administration, K.C. All authors have read and agreed to the published version of the manuscript.

Funding: This research was supported by the Natural Science Foundation of Henan (212300410237); the Henan Provincial Department of Science and Technology (222102230052); the Henan Province Key Research and Development Project (221111230400); the Fundamental Research Funds for the Central Universities (2412020ZD008 and GFPY202309); the Excellent Youth Team Program (2412023YQ001); and the CNNC Key Laboratory on Uranium Extraction from Seawater (KLUES202202).

Institutional Review Board Statement: Not applicable.

Data Availability Statement: Data are contained within the article or Supplementary Materials.

Conflicts of Interest: Author Zili Zhao was employed by the company XuJue electric Co., LTD. The remaining authors declare that the research was conducted in the absence of any commercial or financial relationships that could be construed as a potential conflict of interest.

References

1. Peighambardoust, S.J.; Rowshanzamir, S.; Amjadi, M. Review of the proton exchange membranes for fuel cell applications. *Int. J. Hydrogen Energy* **2010**, *35*, 9349–9384. [CrossRef]
2. Karimi, M.B.; Mohammadi, F.; Hooshyari, K. Recent approaches to improve Nafion performance for fuel cell applications: A review. *Int. J. Hydrogen Energy* **2019**, *44*, 28919–28938. [CrossRef]
3. Haider, R.; Wen, Y.; Ma, Z.F.; Wilkinson, D.P.; Zhang, L.; Yuan, X.; Song, S.; Zhang, J. High temperature proton exchange membrane fuel cells: Progress in advanced materials and key technologies. *Chem. Soc. Rev.* **2021**, *50*, 1138–1187. [CrossRef] [PubMed]
4. Ying, J.; Liu, T.; Wang, Y.; Guo, M.; Shen, Q.; Lin, Y.; Yu, J.; Yu, Z. Perspectives on Membrane Development for High Temperature Proton Exchange Membrane Fuel Cells. *Energy Fuels* **2024**, *38*, 6613–6643. [CrossRef]
5. Schmidt-Rohr, K.; Chen, Q. Parallel cylindrical water nanochannels in Nafion fuel-cell membranes. *Nat. Mater.* **2008**, *7*, 75–83. [CrossRef] [PubMed]
6. Wang, Q.; Shen, D.; Tu, Z.; Li, S. Improved performance of lanthanide-doped UIO-66/Nafion hybrid proton exchange membrane for water electrolyzer. *Int. J. Hydrogen Energy* **2024**, *56*, 1249–1256. [CrossRef]
7. Huang, S.Z.; Liu, S.S.; Zhang, H.J.; Han, Z.; Zhao, G.; Dong, X.Y.; Zang, S.Q. Dual-Functional Proton-Conducting and pH-Sensing Polymer Membrane Benefiting from an Eu-MOF. *ACS Appl. Mater. Interfaces* **2020**, *12*, 28720–28726. [CrossRef]
8. Gao, Q.; Zhang, L.; Zhang, H.; Zhang, D.; Xiao, W. Preparation and performance of UIO-66-NH_2 enhanced proton exchange membranes for vanadium redox flow batteries. *J. Solid State Electrochem.* **2024**. [CrossRef]
9. Lim, D.W.; Kitagawa, H. Proton Transport in Metal–Organic Frameworks. *Chem. Rev.* **2020**, *120*, 8416–8467. [CrossRef]
10. Wang, Y.; Gao, H.; Wu, W.; Zhou, Z.; Yang, Z.; Wang, J.; Zou, Y. Nafion-threaded MOF laminar membrane with efficient and stable transfer channels towards highly enhanced proton conduction. *Nano Res.* **2021**, *15*, 3195–3203. [CrossRef]
11. Chen, G.; Ge, L.; Lee, J.H.; Zhu, Z.; Wang, H. Porous coordination polymer-based composite membranes for high-temperature polymer exchange membrane fuel cells. *Matter* **2022**, *5*, 2031–2053. [CrossRef]
12. Yang, Y.; He, X.; Zhang, P.; Andaloussi, Y.H.; Zhang, H.; Jiang, Z.; Chen, Y.; Ma, S.; Cheng, P.; Zhang, Z. Combined Intrinsic and Extrinsic Proton Conduction in Robust Covalent Organic Frameworks for Hydrogen Fuel Cell Applications. *Angew. Chem. Int. Ed.* **2020**, *59*, 3678–3684. [CrossRef] [PubMed]
13. Sahoo, R.; Mondal, S.; Pal, S.C.; Mukherjee, D.; Das, M.C. Covalent-Organic Frameworks (COFs) as Proton Conductors. *Adv. Energy Mater.* **2021**, *11*, 2102300–2102334. [CrossRef]
14. Geng, K.; He, T.; Liu, R.; Dalapati, S.; Tan, K.T.; Li, Z.; Tao, S.; Gong, Y.; Jiang, Q.; Jiang, D. Covalent Organic Frameworks: Design, Synthesis, and Functions. *Chem. Rev.* **2020**, *120*, 8814–8933. [CrossRef] [PubMed]
15. Tian, Y.; Zhu, G. Porous Aromatic Frameworks (PAFs). *Chem. Rev.* **2020**, *120*, 8934–8986. [CrossRef] [PubMed]
16. Li, L.; Ren, H.; Yuan, Y.; Yu, G.; Zhu, G. Construction and adsorption properties of porous aromatic frameworks via $AlCl_3$ triggered coupling polymerization. *J. Mater. Chem. A* **2014**, *2*, 11091–11098. [CrossRef]
17. Wang, Z.; Ma, R.; Meng, Q.; Yang, Y.; Ma, X.; Ruan, X.; Yuan, Y.; Zhu, G. Constructing Uranyl-Specific Nanofluidic Channels for Unipolar Ionic Transport to Realize Ultrafast Uranium Extraction. *J. Am. Chem. Soc.* **2021**, *143*, 14523–14529. [CrossRef] [PubMed]
18. Dai, A.; Li, S.; Wang, T.; Yang, Y.; Tian, Y.; Jing, X.; Zhu, G. Frustrated Lewis pairs in situ formation in B-based porous aromatic frameworks for efficient o-phenylenediamine cyclization. *Chin. Chem. Lett.* **2023**, *34*, 107559. [CrossRef]
19. Yang, Y.; Deng, D.; Zhang, S.; Meng, Q.; Li, Z.; Wang, Z.; Sha, H.; Faller, R.; Bian, Z.; Zou, X.; et al. Porous Organic Frameworks Featured by Distinct Confining Fields for the Selective Hydrogenation of Biomass-Derived Ketones. *Adv. Mater.* **2020**, *32*, 1908243. [CrossRef]

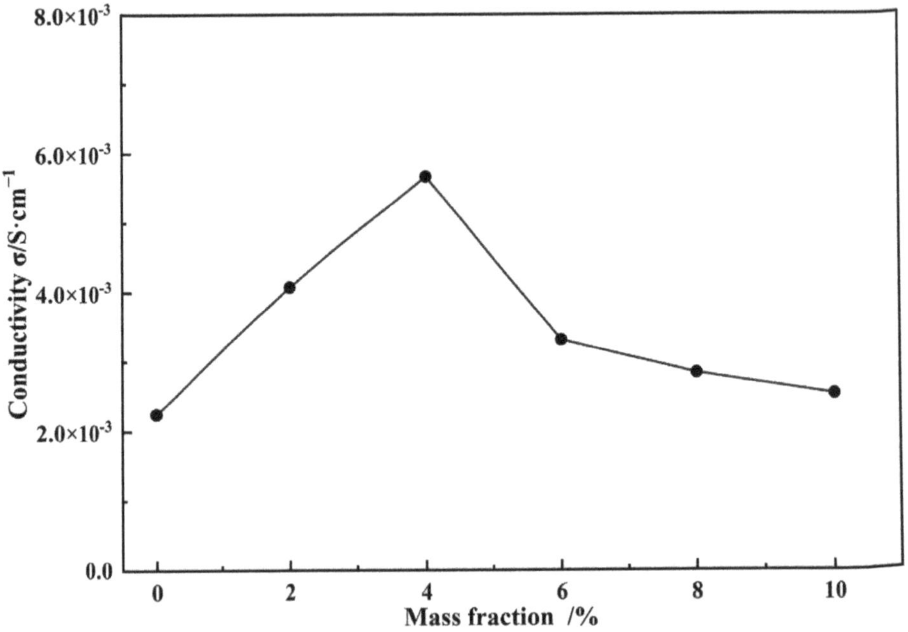

Figure 7. The conductivity of the Nafion composite membranes with different content of PAF-45DS at 80 °C, ~78% RH.

4. Conclusions

In this study, PAF-45D with 1571.9 $m^2 \cdot g^{-1}$ BET surface area was successfully synthesized with $AlCl_3$, dichloromethane and biphenyl. After this, PAF-45D was treated by sulfonation, recorded as PAF-45DS. The successful incorporation of sulfonic groups was confirmed through TG, IR, and EDS analyses. Then, a novel Nafion composite membrane was prepared by introducing PAF-45DS. The SEM images show that as the PAF-45DS mass fraction increases, the solid particles are packed more closely and the composite membranes become progressively rougher. The proton conductivity of the Nafion composite membranes increased from 2.245×10^{-3} $S \cdot cm^{-1}$ (0 wt.% PAF-45DS) to 5.667×10^{-3} $S \cdot cm^{-1}$ (4 wt.%), followed by a gradual decrease to 2.528×10^{-3} $S \cdot cm^{-1}$ (10 wt.%). The results show that the conductivity of the Nafion composite membrane can be significantly enhanced by nearly 2.5 times when doping with only a 4% mass fraction of PAF-45DS. To the best of our knowledge, this is an impressive statistic, indicating that the sulfonated PAF-45DS material can improve the proton conductivity of Nafion membranes in high-temperature environments. Such promising materials not only improve the conductivity but also offer a cost-effective solution for advanced composite membranes. They hold great potential for widespread application in the composite film industry.

Supplementary Materials: The following supporting information can be downloaded at https://www.mdpi.com/article/10.3390/polym16152208/s1, Figure S1: The IR spectra of the PAF-45D and PAF-45DS samples; Figure S2: The water uptake of the PAF-45D and PAF-45DS samples at 298 K; Figure S3: SEM image and EDS mapping of PAF-45DS; Table S1: The quantitative analysis results of the test sample; Figure S4: Nyquist curves of PAF-45DS composite films with different doping amounts at ~78%, 80 °C conditions.

20. Guo, X.X.; Cai, Z.T.; Muhammad, Y.; Zhang, F.L.; Wei, R.P.; Gao, L.J.; Xiao, G.M. Silver-anchored porous aromatic framework for efficient conversion of propargylic alcohols with CO_2 at ambient pressure. *Chin. Chem. Lett.* **2023**, *34*, 107740. [CrossRef]
21. Yuan, Y.; Zhu, G. Porous Aromatic Frameworks as a Platform for Multifunctional Applications. *ACS Cent. Sci.* **2019**, *5*, 409–418. [CrossRef] [PubMed]
22. Bazhenova, M.A.; Kulikov, L.A.; Bolnykh, Y.S.; Maksimov, A.L.; Karakhanov, E.A. Palladium catalysts based on porous aromatic frameworks for vanillin hydrogenation: Tuning the activity and selectivity by introducing functional groups. *Catal. Commun.* **2022**, *170*, 106486–106496. [CrossRef]
23. Zhang, C.; Dong, J.; Zhang, P.; Sun, L.; Yang, L.; Wang, W.; Zou, X.; Chen, Y.; Shang, Q.; Feng, D.; et al. Unique fluorophilic pores engineering within porous aromatic frameworks for trace perfluorooctanoic acid removal. *Natl. Sci. Rev.* **2023**, *10*, nwad191. [CrossRef] [PubMed]
24. Carey, C.; Díaz, J.C.; Kitto, D.; Espinoza, C.; Ahn, E.; Kamcev, J. Interfacial interactions between polymers and selective adsorbents influence ion transport properties of boron scavenging ion-exchange membranes. *J. Membr. Sci.* **2023**, *669*, 121301–121311. [CrossRef]
25. Liu, X.; Zhu, C.; Yin, J.; Li, J.; Zhang, Z.; Li, J.; Shui, F.; You, Z.; Shi, Z.; Li, B.; et al. Installation of synergistic binding sites onto porous organic polymers for efficient removal of perfluorooctanoic acid. *Nat. Commun.* **2022**, *13*, 2132–2141. [CrossRef] [PubMed]
26. Ruan, X.; Zhang, C.; Zhu, Y.; Cai, F.; Yang, Y.; Feng, J.; Ma, X.; Zheng, Y.; Li, H.; Yuan, Y.; et al. Constructing Mechanical Shuttles in a Three-dimensional (3D) Porous Architecture for Selective Transport of Lithium Ions. *Angew. Chem. Int. Ed.* **2023**, *62*, e202216549. [CrossRef] [PubMed]
27. Yuan, Y.; Yang, Y.; Zhu, G. Molecularly Imprinted Porous Aromatic Frameworks for Molecular Recognition. *ACS Cent. Sci.* **2020**, *6*, 1082–1094. [CrossRef] [PubMed]
28. Bazhenova, M.A.; Kulikov, L.A.; Makeeva, D.A.; Maximov, A.L.; Karakhanov, E.A. Hydrodeoxygenation of Lignin-Based Compounds over Ruthenium Catalysts Based on Sulfonated Porous Aromatic Frameworks. *Polymers* **2023**, *15*, 4618. [CrossRef]
29. Li, J.; Huang, L.; Zou, X.; Zheng, A.; Li, H.; Rong, H.; Zhu, G. Porous organic materials with ultra-small pores and sulfonic functionality for xenon capture with exceptional selectivity. *J. Mater. Chem. A* **2018**, *6*, 11163–11168. [CrossRef]
30. Bildirir, H. Post-synthetic sulfonation of a diphenylanthracene based porous aromatic framework. *Org. Commun.* **2022**, *15*, 346–355. [CrossRef]
31. Klumpen, C.; Gödrich, S.; Papastavrou, G.; Senker, J. Water mediated proton conduction in a sulfonated microporous organic polymer. *Chem. Commun.* **2017**, *53*, 7592–7595. [CrossRef] [PubMed]
32. Li, L.; Cai, K.; Wang, P.; Ren, H.; Zhu, G. Construction of Sole Benzene Ring Porous Aromatic Frameworks and Their High Adsorption Properties. *ACS Appl. Mater. Interfaces* **2014**, *7*, 201–208. [CrossRef] [PubMed]
33. Wang, S.; Zhang, C.; Yu, S.; Jiang, S.; Xia, Q.; Chen, L.; Jin, S.; Hussain, I.; Cooper, A.I.; Tan, B. Layered microporous polymers by solvent knitting method. *Sci. Adv.* **2017**, *3*, e1602610. [CrossRef] [PubMed]
34. Kalinina, M.A.; Kulikov, L.A.; Cherednichenko, K.A.; Maximov, A.L.; Karakhanov, E.A. The Effect of Sulfonate Groups in the Structure of Porous Aromatic Frameworks on the Activity of Platinum Catalysts Towards Hydrodeoxygenation of Biofuel Components. *Petrol. Chem.* **2021**, *61*, 1061–1070. [CrossRef]
35. Jaimez, E.; Hix, G.B.; Slade, R.C.T. The titanium(I) salt of N, N-(diphosphonomethyl)glycine: Synthesis, characterisation, porosity and proton conduction. *J. Mater. Chem.* **1997**, *7*, 475–479. [CrossRef]

polymers

MDPI

Article

Inhibition and Mechanisms of Isothiazolinone and Layered Double Hydroxide–Sodium Pyrithione with Modified Hydrophobic Resin Membranes Against Pipeline Moss Fouling

Rudong Zhou [1], Qifeng Zhao [1,2], Aomen Liu [1,2], Hui Xu [1,2], Yang Xu [1], Weijie Wang [1,2], Jicheng Han [1,2], Bo Liu [1,2], Muli Wang [1], Zhigang Di [1,2], Lei Miao [1,2], Yanhui Liu [3,*] and Lei Guo [3,*]

1 CNOOC Changzhou Paint and Coatings Industry Research Institute Co., Ltd., Changzhou 213016, China; zhourd@cnooc.com.cn (R.Z.); zhaoqf4@cnooc.com.cn (Q.Z.); liuaom@cnooc.com.cn (A.L.); xuyang12@cnooc.com.cn (Y.X.); wangwj53@cnooc.com.cn (W.W.); hanjch4@cnooc.com.cn (J.H.); liubo48@cnooc.com.cn (B.L.); wangml9@cnooc.com.cn (M.W.); dizhg@cnooc.com.cn (Z.D.); miaolei3@cnooc.com.cn (L.M.)

2 Key Laboratory of Corrosion Protection for Offshore Oil Industry, Changzhou 213016, China

3 Research Center for High-Value Utilization of Waste Biomass, Institute of Biomedical Engineering, College of Life Science, Qingdao University, Qingdao 266071, China

* Correspondence: 15020033032@163.com (Y.L.); guoleibio@qdu.edu.cn (L.G.); Tel.: +86-152-7532-3063 (L.G.)

Abstract

To address pipeline blockages and corrosion caused by moss, this study evaluates the effectiveness of two treatments, Isothiazolinone (IS) and layered double hydroxide–sodium pyrithione (LDH-SPT) modified hydrophobic resin membranes, in preventing moss growth. Furthermore, we closely examined how IS works at a molecular level to stop moss growth. The sequencing results revealed that the predominant algae identified in the pipeline moss community was a norank species of *Trebouxiophyceae*, accounting for 75.79%. Tests show that IS has strong moss inhibition. It works at low doses (0.2%) and becomes even more effective as the concentration increases. Furthermore, IS remains highly effective at inhibiting moss within a modified hydrophobic resin membrane, but its corrosion resistance is poor. The LDH-SPT@IS composite modified hydrophobic resin membrane addresses the corrosion problem of using IS alone and still works very well at inhibiting moss. Finally, the mechanism of IS's inhibition of moss was elucidated based on experiments and existing literature. It functions by disrupting moss cellular DNA structure and interfering with the mitochondrial electron transport chain. This research provides the basis for developing efficient, durable, and eco-friendly solutions to prevent pipeline corrosion and moss growth, paving the way for new technologies and materials.

Keywords: moss; green algae; isothiazolinone; LDH-SPT; membrane; inhibition mechanism

1. Introduction

Pipeline corrosion is a widespread issue, resulting in serious accidents such as casing perforation, failure of drilling and production equipment, and pipeline fractures [1]. Coastal installations are long exposed to the dual corrosive environment of marine and industrial atmospheres, facing not only intense ultraviolet radiation but also high salt and high humidity. The air is rich in water vapor, oxygen, and acidic pollutants, and the severe corrosion resulting from this greatly threatens the mechanical properties of structural materials [2–5]. However, there are still some problems and challenges in the technology

for anti-corrosion of water supply and drainage pipelines, both domestically and internationally. Therefore, researching and developing efficient, economical, and environmentally friendly pipeline corrosion prevention technologies has significant practical importance and broad application prospects [6].

Moss, characterized as a type of lower plant community widely distributed in humid environments [7], tends to proliferate, adhere, and spread within various water conveyance, drainage, and industrial circulation pipeline systems. In particular, in the humid regions of southern China, due to the climate characterized by high temperatures, abundant rainfall, and high air humidity throughout the year, moss tends to grow easily on the surface of pipelines. Moss not only damages the pipeline structure through physical erosion, but also engages in complex chemical interactions with the pipeline material, accelerating aging and significantly reducing their actual service life [8,9]. During their metabolic processes, mosses continuously release acidic substances such as organic acids, which corrode metal components and induce rust formation, exacerbating the wear and tear of facilities.

In order to effectively address pipeline corrosion challenges in humid environments and prevent moss attachment, and to avoid problems such as reduced transportation efficiency and a sharp increase in maintenance costs caused by pipeline damage and moss blockage, it is urgent to develop an efficient and lasting inhibition technology. Among the current solutions, the use of inhibitors and functional membranes is the focus of current research. Particularly, IS represents a class of highly effective broad-spectrum heterocyclic biocides. Its molecules exhibit strong electrophilic characteristics, enabling specific interactions with sulfhydryl groups within microbial cells. By oxidizing key functional groups in cytoplasmic and membrane proteins, it rapidly suppresses metabolic activity and inhibits cellular growth and reproduction [10,11]. Furthermore, Huang et al. [12] systematically investigated the individual and combined effects of sodium dichloroisocyanurate (NaDCC) and IS on cyanobacteria, Vibrio nattani, and associated microbial communities. The two compounds demonstrated synergistic effects in suppressing cyanobacterial photosynthesis and disrupting microbial cell membrane integrity, providing crucial insights for IS applications in pipeline moss prevention and control.

The LDH-SPT material, formed by loading sodium pyrithione (SPT) onto layered double hydroxides (LDHs), has been extensively applied in corrosion protection and biofouling prevention in recent years [13–17]. LDH possesses a unique layered structure and anion exchange capacity, enabling it to serve as a carrier for stable SPT loading and controlled release. SPT, as a compound with both antibacterial and corrosion inhibition functions, can effectively inhibit microbial activity. Zhang et al. [18] prepared Mg-Al LDH thin films, and the results showed that the films exhibit excellent corrosion protection performance. In addition, research in related fields has also provided theoretical support for the performance optimization of LDH-SPT materials. Cao et al. [19] studied Zn-Al LDHs with different intercalated anions and found that a variety of interlayer anions exhibit a synergistic effect in terms of corrosion protection; Zhang et al. [20] prepared aspartic acid-modified self-healing Li-Al LDHs on 6N01 aluminum alloy, and the membrane possesses long-term anti-corrosion ability. These studies provide valuable references for the structural design and performance enhancement of LDH-SPT materials.

Therefore, this study takes moss from a certain pipeline in Hainan Province as the research object. Its purpose is to systematically investigate the inhibitory effects of IS, LDH-SPT composite materials, and the modified hydrophobic resin membranes prepared from them on pipeline corrosion and moss adhesion, and to conduct an in-depth analysis of the mechanism of action based on experiments and existing literature, so as to provide a theoretical basis and practical references for the optimization of pipeline moss prevention and corrosion protection technologies in hot and humid areas.

https://doi.org/10.3390/polym18050611

Figure 1 illustrates the application process of IS in metal protection. Initially, tinplate, paint, and IS are used as raw materials, which are combined to prepare an IS membrane. The mechanism of action is that IS can not only block electron transport but also form a hydrogen-bonded network with microbial DNA, thereby inhibiting the growth of microorganisms such as moss. Ultimately, this material can be applied in scenarios including metal protection and pipeline protection.

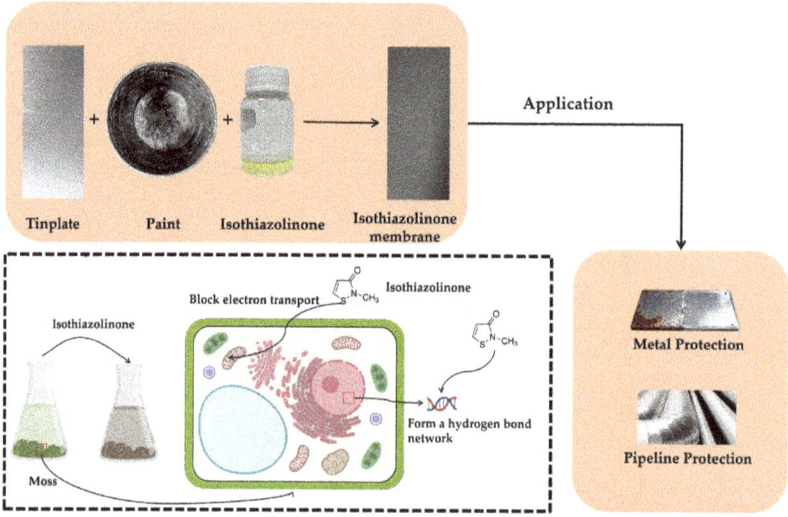

Figure 1. Preparation and application of isothiazolinone (IS) membranes.

2. Materials and Methods

2.1. Chemicals and Materials

The moss sample was collected from a pipeline in Hainan Province. BG-11 culture medium were purchased from Qingdao Hi-Tech Industrial Park Haibo Biotechnology Co., Ltd. (Qingdao, China) Synthetic hydrotalcite ($Mg_6Al_2(CO_3)(OH)_{16}{\cdot}4H_2O$), sodium 2-mercaptopyridine-N-oxide (C_5H_6NNaOS), and IS were procured from Shanghai Aladdin Biochemical Technology Co., Ltd. (Shanghai, China). SPU9560 hydrophobic resin was supplied by CNOOC Changzhou Paint (Changzhou, China) and Membranes Industry Research Institute Co., Ltd. (Nanjing, China).

2.2. Instruments and Equipment

The morphology of LDH-SPT was characterized using scanning electron microscopy (SEM, Thermo Scientific Apreo 2S, Waltham, MA, USA). Elemental analysis of LDH-SPT was performed with an energy dispersive spectrometer (EDS, GENESIS, Offenbach, Germany). Fourier transform infrared spectroscopy (FT-IR, PerkinElmer Spectrum TWO, Waltham, MA, USA) was employed for structural characterization. The moss adhesion and corrosion conditions on the membrane surface were examined using a metallographic microscope (HJ1, Shenyang, China).

2.3. Preparation of SPT-Intercalated LDH Nanomaterials

A certain amount of synthetic hydrotalcite ($Mg_6Al_2(CO_3)(OH)_{16}{\cdot}4H_2O$) was weighed into a crucible and calcined in a muffle furnace at 350 °C for 5 h. Then, 600 mL of CO_2-free deionized water and 100 mL of deprotonated SPT solution were weighed into a three-necked flask. Under N_2 protection, the mixture was stirred at 60 °C, and 10 g of calcined

https://doi.org/10.3390/polym18050611

LDH was weighed and added to it. After ultrasonic treatment for 1 h, the reaction was continued for 24 h. After the reaction, the product was centrifuged, washed with deionized water and absolute ethanol, respectively, and then dried to obtain LDH-SPT.

2.4. Preparation of Hydrophobic Algae-Inhibiting Membranes and Samples

In total, 10 g of hydrophobic resin was added to a mixing tank. Subsequently, 2 g of curing agent, 0.24 g of defoamer, and 1.2 g of thinner were added in sequence, and the mixture was stirred until uniform. Hydrophobic anti-algal membranes with concentrations of 0.2%, 0.5%, and 1% were prepared using the following algal inhibitors: LDH-SPT, IS, and LDH-SPT@IS, respectively.

2.5. Method for Identification of Moss

Metagenomic microbial taxonomic sequencing was employed to analyze 18S rRNA or specific functional gene sequences. Utilizing high-throughput sequencing platforms, the variable regions of these genes were determined to investigate microbial diversity, community composition, and abundance variations.

High-throughput sequencing on the Illumina platform requires the ligation of sequencing adapters before the sequencing process can proceed. To achieve the dual objectives of providing identical sequencing primers for different DNA molecules and enabling sample demultiplexing for distinct users, the adapters are designed with index and barcode sequences. Through the combinatorial use of different indices and barcodes, high-throughput sequencing of numerous samples can be performed simultaneously.

To meet the requirements of high-throughput sequencing library construction, a two-round PCR method was optimized and designed, with the library construction process integrated into the PCR process. This enables efficient and rapid completion of sample detection.

(1) DNA Extraction: DNA was extracted from 2 samples using the E.Z.N.A™ Mag-Bind Soil DNA Kit [21].
(2) PCR Amplification: During the second round of PCR amplification, Illumina bridge PCR-compatible primers were introduced.
(3) Library Quality Control and Pooling: Library fragment size was verified by 2% agarose gel electrophoresis, while library concentration was quantified using a Qubit 3.0 Fluorometer. All samples were mixed in equal amounts at a 1:1 ratio.

2.6. Method for Culturing Moss

In the experiment, the standard BG-11 medium was used as the basic nutrient substrate for algal cultivation. The specific preparation method is as follows: Accurately weigh 1.70 g of BG-11 medium powder, dissolve it in 1 L of deionized water, and stir thoroughly until completely dissolved. Place it in an autoclave and sterilize at 121 °C for 15 min to ensure the medium is sterile. After sterilization, the medium is cooled to room temperature for later use. The inoculation procedure was carried out in a laminar flow hood under sterile conditions. Using a pipette, the algal inoculum was transferred into the sterilized BG-11 medium at an inoculation volume ratio of 5–10% [22,23].

The cultivation process was conducted in an artificial climate incubator with strictly controlled environmental parameters: LED lighting maintained at 2000 lux with a 12 h:12 h light–dark cycle; constant temperature at 25 °C; and intermittent manual shaking performed three times daily for 30 s each time to enhance gas exchange and ensure adequate carbon dioxide supply.

During the cultivation period, the growth status of algae was observed regularly at a fixed time every day. The medium was replaced every two weeks. Meanwhile, regular observations were conducted to monitor the growth status of algae and potential contamination.

https://doi.org/10.3390/polym18050611

2.7. Method for Evaluating Inhibitor Performance

We evaluated the moss inhibition efficacy of two inhibitors, IS and LDH-SPT. The initial concentration of the algal solution was set to 3–5 green algal aggregates with a diameter of approximately 1–2 mm per cubic centimeter. IS and LDH-SPT were tested at three concentrations: 0.2%, 0.5%, and 1%. Three experimental groups were established by combining 1% LDH-SPT with each concentration of IS. Each experimental group was set with two parallel samples, and a blank moss culture solution was established as the control.

The experimental samples were placed in an intelligent climate chamber with the following conditions: temperature at 25 °C, light intensity at 2000 lux (light–dark cycle of 12 h:12 h). The samples were cultured continuously for 10 days, and the experimental phenomena were recorded by taking photos every day to observe the survival status of the moss.

2.8. Method for Evaluating Membrane Inhibition Performance

Membrane samples containing IS and LDH-SPT inhibitors were prepared at concentrations of 0.2%, 0.5%, and 1%, respectively. Three experimental groups were established by combining 1% LDH-SPT with each of the three IS concentrations to prepare composite membranes. The anticorrosion and moss inhibition performance of these membranes was evaluated. Each experimental group was set with three parallel samples, and SPU9560 hydrophobic resin samples were used as controls to eliminate the influence of the substrate itself on moss adhesion.

The coated samples were immersed separately in Petri dishes containing moss culture solution, and the Petri dishes were placed in an intelligent artificial climate incubator. The set cultivation conditions were identical to those in the inhibitor evaluation experiment. Experimental phenomena were recorded by taking photos every day to observe the survival status of moss and assess the adhesion amount of moss on the membrane samples.

3. Results

3.1. Characterization of LDH-SPT

The morphology of LDH-SPT is shown in Figure 2a, which is mainly composed of irregular flakes with rough surfaces, and some of them overlap with each other. Figure 2b shows that the sample has a relatively high oxygen content, with a mass fraction of 46.060%. The mass fractions of other elements are as follows: C at 31.404%, N at 15.279%, Mg at 4.639%, Al at 2.464%, and S at 0.154%. The infrared spectrum of LDH-SPT in Figure 3a shows that the wave number at 3451.2 cm^{-1} is attributed to O-H stretching vibration; 1368.5 cm^{-1} to the asymmetric stretching vibration of NO_3^-; 768.2 cm^{-1} to M-O-H bending vibration (M=Mg, Al) or organic residual vibration; 660.3 cm^{-1} to M-O-M lattice vibration (M=Mg, Al); and 446.9 cm^{-1} to M-O bending vibration or lattice vibration. The infrared spectrum of IS in Figure 3b shows that the wave number at 3397.35 cm^{-1} is attributed to N-H stretching vibration; 1626.4 cm^{-1} to the stretching vibration of C=O (carbonyl group); 1348.3 cm^{-1} to the skeletal vibration of aromatic rings or the stretching vibration of C−N bonds; and 708.6 cm^{-1} to the stretching vibration of C-S-C (thiazole ring).

https://doi.org/10.3390/polym18050611

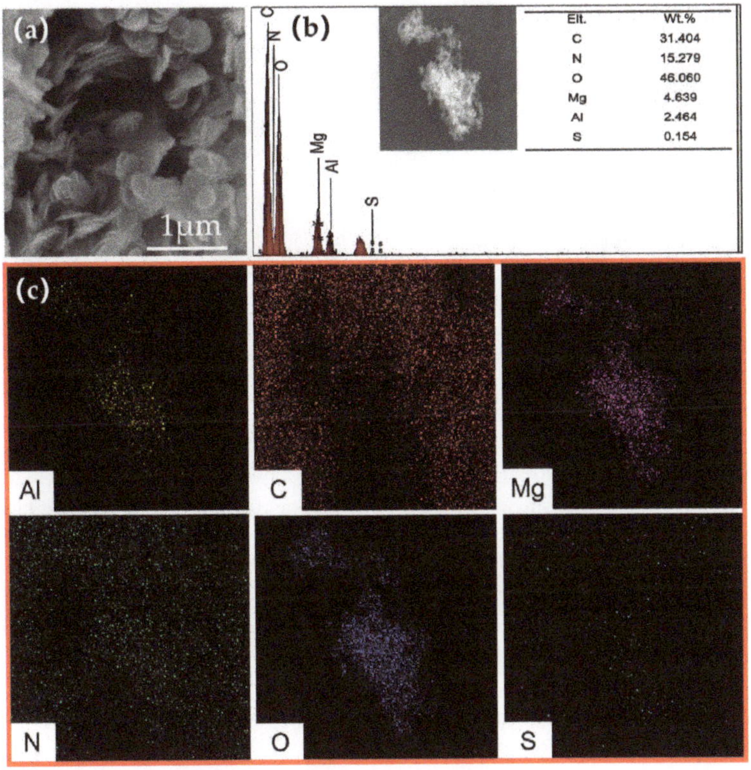

Figure 2. SEM image of LDH-SPT (**a**). EDS elemental mapping and corresponding spectrum analysis of LDH-SPT (**b**). Distribution maps of Al, C, Mg, N, O, and S (**c**).

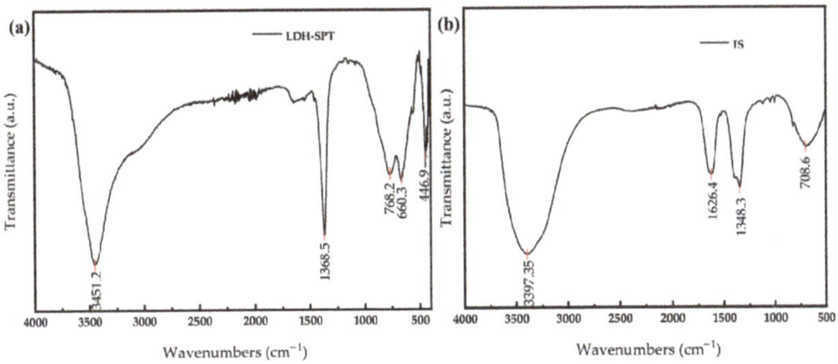

Figure 3. FT-IR spectra of different inhibitors: (**a**) LDH-SPT and (**b**) IS.

3.2. Results of Moss Identification

Based on 18S rRNA sequencing analysis, the species abundance of the moss samples is shown in Table 1. The most dominant algal group was norank *Trebouxiophyceae* (75.79%) within the Chlorophyta phylum. *Jaagichlorella* represented the second most abundant genus at 14.76%, which also belongs to the Chlorophyta phylum. These results indicate that the sample constitutes a community overwhelmingly dominated by green algae.

https://doi.org/10.3390/polym18050611

Table 1. Species abundance of 18S rRNA samples.

Species	Percentage (%)
norank *Trebouxiophyceae*	75.79
Adineta	0.73
Jaagichlorella	14.76
Jenufa	0.04
Tripylina	0.01
Klebsormidium	0.01
Amblydorylaimus	0.01
Orbilia	7.0×10^{-3}
Bradymyces	2.54
Pichia	1.34
Rigidohymena	3.5×10^{-3}
Sterkiella	1.8×10^{-3}

3.3. Investigation of Inhibitor Performance

In the inhibitor screening experiment, the comparison shown in Figure 4(b_1–b_3) demonstrates that IS exhibits significant inhibition effects on common moss species even at low concentrations. Obvious chlorophyll degradation was observed after 1 day of treatment, which was manifested as the color of the moss changing from green to brown. The inhibitory effect improved with the increase in concentration. In Figure 4(c_1–c_3), the three concentrations of LDH-SPT showed no obvious inhibitory effect on the moss. In Figure 4(d_1–d_3), the combination of LDH-SPT and IS exhibited an inhibitory effect on the moss, but its action rate was slower than that of IS alone. For the composite inhibitors of LDH-SPT (1%)@IS (0.5%, 1%), the color change of the moss from green to brown was observed after 4 days of the experiment, while for the composite inhibitor of LDH-SPT (1%)@IS (0.2%), this color change was only observed after 8 days of the experiment.

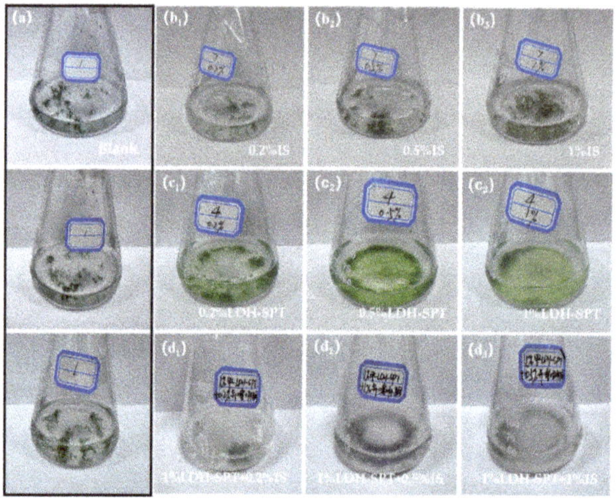

Figure 4. Moss growth under different treatment conditions: (**a**) control group; (**b_1–b_3**) experimental groups with 0.2%, 0.5%, and 1% IS; (**c_1–c_3**) experimental groups with 0.2%, 0.5%, and 1% LDH-SPT; (**d_1–d_3**) experimental groups with LDH-SPT (1%)@ IS (0.2%, 0.5%, 1%) combinations.

IS can rapidly disrupt cellular metabolism, leading to rapid degradation of chlorophyll, thus taking effect quickly even at low concentrations. LDH-SPT itself has weak moss-inhibiting activity, so its effect is not obvious when used alone. When compounded with IS,

https://doi.org/10.3390/polym18050611

the carrier structure of LDH-SPT may slow down the transfer rate of IS to cells through adsorption or sustained-release effects, thereby reducing its action speed. This results in the characteristic that the inhibitory effect still exists but takes effect more slowly.

3.4. Investigation of Membrane Inhibition Performance

Adhesion testing was conducted in accordance with the GB/T 9286-2021 standard [24]. Using a 1 mm grid spacing and tape peel-off method, no peeling of the paint membrane was observed, with smooth and regular cutting edges and no issues such as flaking or chipping. For impact resistance testing, a 1 kg hammer with a 10 mm diameter was dropped from a height of 50 cm onto the composite membrane. After the test, the paint membrane remained intact, with no cracks, damage, or detachment. This demonstrates good mechanical properties of the composite membrane.

In the membrane inhibition experiment, Figure 5(b_1–b_3) demonstrates that compared to the blank membrane, all three IS-containing membranes effectively prevented moss attachment. However, they exhibited poor corrosion resistance with noticeable membrane detachment. Figure 5(c_1–c_3) reveals that the LDH-SPT membranes at all three concentrations showed moss attachment levels similar to the control group, but displayed excellent corrosion protection without any membrane deterioration. Figure 5(d_1–d_3) indicates that the composite membranes containing 1% LDH-SPT combined with three different IS concentrations showed no signs of membrane detachment. Side rust is corrosion caused by incomplete coverage of the membrane. Specifically, the LDH-SPT (1%)@IS (0.2%, 0.5%, 1%) composite membranes exhibited virtually no moss attachment. Further observation via the metallographic microscope in Figure 6 reveals that no moss adhered to the IS membrane, while a trace amount of moss adhered to the LDH-SPT (1%)@IS (1%) composite membrane, which was far less than that on the LDH-SPT membrane. When observing membranes using the upright light source of a metallographic microscope, it was found that purple and green colors are usually caused by light interference effects (especially thin-film interference) if the membrane is thin and has a smooth surface. The unevenness of membrane thickness or changes in microstructure lead to different light interference conditions in different regions, thus presenting different colors.

Figure 7(a_1,a_2) show the SEM images of the cross-section of the membrane before and after the experiment, respectively. The cross-section of the membrane maintained an overall continuous structure, with no large-scale cracking or detachment observed; only slight interfacial changes were noted. Combining Figure 7(b_1,b_2), the EDS mapping results show that the distribution and content of key elements such as Al, Mg, N, and S did not undergo significant changes before and after the evaluation. This phenomenon indicates that during the evaluation process, the chemical composition and elemental distribution of the composite membrane did not exhibit noticeable degradation or loss, nor did the bonding interface between the membrane and the substrate suffer structural damage. These results confirm the excellent stability of the membrane at the microscopic level, providing strong evidence for the long-term anti-corrosion and sustained algae inhibition performance of the membrane.

EIS tests were conducted on three different types of composite membranes after experimentation. Figure 8a exhibits an impedance magnitude of 10^7 ohm, which is the highest among the three. This indicates that the membrane has the strongest electrolyte barrier property and relatively optimal anti-corrosion performance. However, based on previous experimental findings, when the IS concentration is 0.2%, its algae inhibition effect is comparatively the weakest among the three. Figure 8c shows an impedance magnitude of 10^5 ohm, the lowest among the three. This suggests that the membrane has the weakest electrolyte barrier property and relatively the poorest anti-corrosion performance. Yet,

https://doi.org/10.3390/polym18050611

when the IS concentration is 1%, its algae inhibition effect is the strongest among the three. Figure 8b displays an impedance magnitude of 10^6 ohm, lying between that of 8a and 8c. Membranes with impedance below 10^6 ohm are generally considered to have failed. Therefore, with an IS concentration of 0.5%, this formulation represents a relatively balanced ratio between "anti-corrosion" and "algae inhibition" performance.

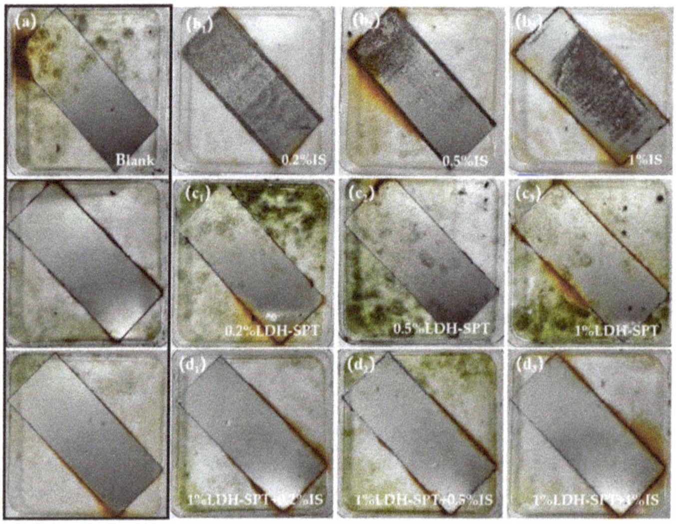

Figure 5. Moss attachment and corrosion conditions on membrane surfaces under different treatments: (**a**) control group; (**b$_1$–b$_3$**) membranes with 0.2%, 0.5%, and 1% IS; (**c$_1$–c$_3$**) membranes with 0.2%, 0.5%, and 1% LDH-SPT; (**d$_1$–d$_3$**) composite membranes with LDH-SPT (1%)@IS (0.2%, 0.5%, 1%).

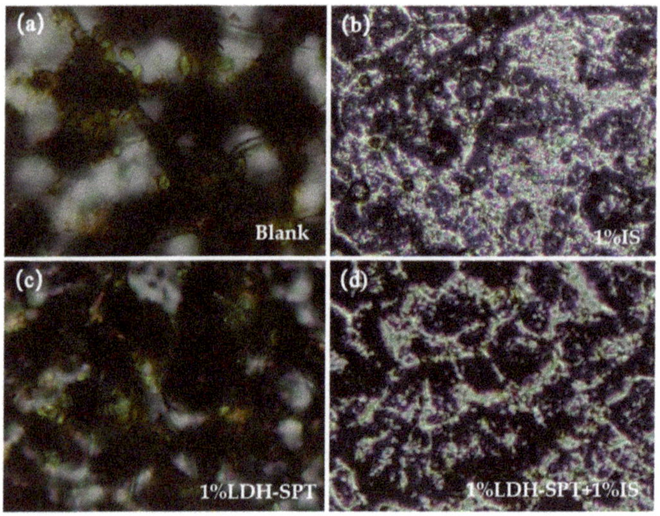

Figure 6. Metallographic microscope images (20 × 10 magnification) of membranes under different treatments: (**a**) control group; (**b**) 1% IS treatment; (**c**) 1% LDH-SPT treatment; (**d**) LDH-SPT (1%)@IS (1%) treatment.

https://doi.org/10.3390/polym18050611

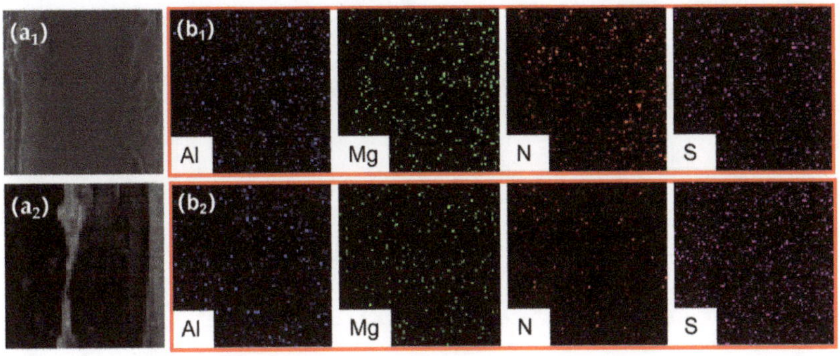

Figure 7. SEM morphology and EDS elemental mapping of the cross-section of the LDH-SPT(1%)@IS(1%) composite membrane: (**a1**) SEM image before experimentation; (**b1**) distribution maps of Al, Mg, N and S before experimentation; (**a2**) SEM image after experimentation; (**b2**) distribution maps of Al, Mg, N and S after experimentation.

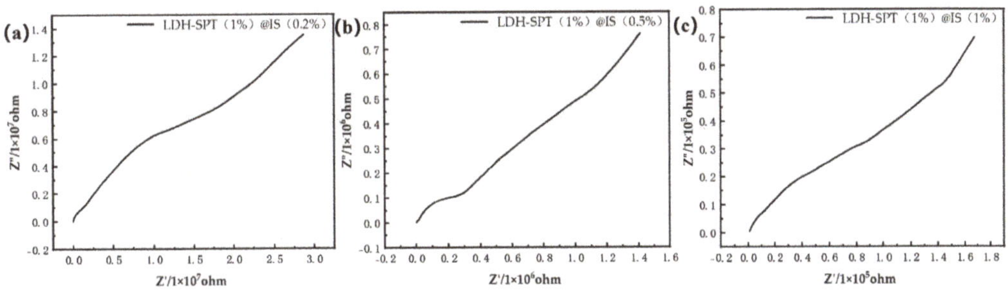

Figure 8. EIS diagrams of the 1% LDH-SPT composite coating at different IS concentrations: (**a**) IS concentration of 0.2%; (**b**) IS concentration of 0.5%; (**c**) IS concentration of 1%.

3.5. Inhibition Mechanism of IS-Based Membranes

The IS molecule has strong electrophilic properties, enabling it to specifically interact with sulfhydryl groups in microbial cells. By oxidizing key functional groups in cytoplasm and membrane proteins, it rapidly inhibits cellular metabolic activity and growth as well as reproduction [12,25].

In terms of molecular action mechanism, IS is a typical electrophilic biocide. Due to the active heterocyclic ring in its molecule, it can specifically bind to intracellular biological macromolecules. The mechanism diagram of IS inhibition in Figure 9 shows that the active groups of IS can form a stable hydrogen bond network with the bases in DNA molecules. This intermolecular force leads to the selective adsorption of IS on the cell surface, which further attacks the nucleophilic centers in the cell and ultimately destroys the secondary structure of DNA [12]. Such structural damage severely disrupts the normal replication and transcription processes of genetic material, resulting in the loss of fundamental physiological functions and metabolic activity in cells.

Furthermore, IS exhibits strong cell membrane penetration capability [26], enabling it to rapidly traverse cell wall and membrane barriers to directly target the electron transport chain on the mitochondrial inner membrane, as illustrated in Figure 9. By inhibiting the activity of key components such as the NADH dehydrogenase complex and cytochrome oxidase, IS disrupts electron transfer processes, consequently interfering with oxidative phosphorylation and ultimately leading to cellular energy metabolism dysfunction [27].

https://doi.org/10.3390/polym18050611

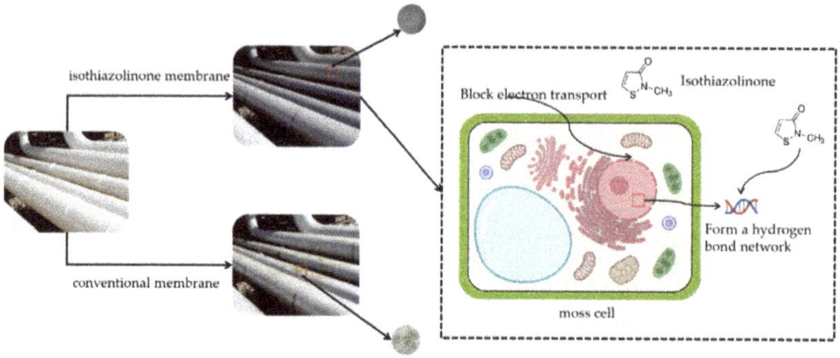

Figure 9. Mechanism diagram of IS inhibition.

After compounding IS with LDH-SPT, from the perspective of the carrier structure of LDH-SPT, LDH materials possess a typical layered two-dimensional nanostructure [28], with regulable ion exchange channels and adsorption sites between layers. This structure enables it to not only serve as a loading carrier for SPT but also form multiple interactions with IS. On one hand, the metal cations between LDH layers can form coordination bonds with polar groups such as nitrogen and oxygen in IS molecules, anchoring part of IS in the interlayer gaps; on the other hand, the layered stacking of LDH creates a physical barrier to the diffusion path of IS, preventing it from being released into the system as rapidly as when used alone.

4. Conclusions

This study systematically investigated the inhibition performance of IS and LDH-SPT inhibitors and their modified hydrophobic resin membranes against pipeline-adhered moss, along with the moss inhibition mechanism of IS. Through high-throughput sequencing of pipeline moss samples from Hainan Province, the dominant species was identified as a norank species of *Trebouxiophyceae*, with a relative abundance as high as 75.79%.

In the study on moss-inhibiting performance, IS exhibited significant inhibitory effects, while LDH-SPT showed no obvious moss-inhibiting effect. However, the LDH-SPT@IS composite inhibitor demonstrated a distinct moss-inhibiting effect. In modified hydrophobic resin membrane applications with three concentrations (0.2%, 0.5%, and 1%), the IS membrane displayed excellent anti-adhesion performance with no moss attached to its surface, but its anti-corrosion effect was poor. LDH-SPT membrane showed excellent anti-corrosion performance but had no obvious moss-inhibiting effect. The experiment confirmed that the LDH-SPT@IS composite membrane significantly improved the anti-corrosion effect of the membrane.

Through its strong electrophilic properties, IS disrupts the secondary structure of cellular DNA and inhibits the electron transport chain on the inner mitochondrial membrane, thereby interfering with genetic information transmission and energy metabolism to achieve the inhibition of moss. This study provides theoretical support and material design directions for the development of efficient and environmentally friendly pipeline moss-prevention technologies, and offers guidance for the selection and optimization of moss-prevention materials in practical applications.

Author Contributions: Conceptualization, Y.L. and L.G.; Data curation, R.Z., Q.Z., A.L. and H.X.; Formal analysis, R.Z., Y.X., W.W., J.H., B.L. and M.W.; Funding acquisition, Y.L. and L.G.; Investigation, R.Z., Q.Z. and A.L.; Methodology, Y.X. and W.W.; Project administration, Y.L. and L.G.; Resources, J.H., Bo Liu, Z.D. and L.M.; Supervision, L.G.; Validation, H.X., M.W., Z.D. and L.M.; Writing—

https://doi.org/10.3390/polym18050611

original draft, R.Z., J.H., Y.L. and L.G.; Writing—review and editing, Y.L. and L.G. All authors have read and agreed to the published version of the manuscript.

Funding: This research received no external funding.

Institutional Review Board Statement: Not applicable.

Data Availability Statement: The original contributions presented in this study are included in the article. Further inquiries can be directed to the corresponding authors.

Acknowledgments: We appreciate Mengqi Zhang and Qingni Yan for their assistance in part of the experiments. They are graduate students at the School of Life Sciences, Qingdao University.

Conflicts of Interest: Authors Rudong Zhou, Qifeng Zhao, Aomen Liu, Hui Xu, Yang Xu, Weijie Wang, Jicheng Han, Bo Liu, Muli Wang, Zhigang Di, and Lei Miao were employed by CNOOC Changzhou Paint and Coatings Industry Research Institute Co., Ltd. The remaining authors declare that the research was conducted in the absence of any commercial or financial relationships that could be construed as a potential conflict of interest.

Abbreviations

The following abbreviations are used in this manuscript:

IS	Isothiazolinone

References

1. Zhang, Y.; Zhong, Z.; Guo, D.; Xu, C.; Du, X.; Wang, H. Experimental investigation of ductile iron pipeline joints rehabilitated with corrosion-protection linings under combined bending and internal pressure. *Eng. Struct.* **2025**, *343*, 121234. [CrossRef]
2. Firoozi, A.; Firoozi, A.; Oyejobi, D.O.; Avudaiappan, S.; Flores, E. Enhanced durability and environmental sustainability in marine infrastructure: Innovations in anti-corrosive coating technologies. *Results Eng.* **2025**, *26*, 105144. [CrossRef]
3. Han, W.; Nie, S.; Wang, Y.; Liu, M.; Chen, Z.; Chen, J.; Yang, B.; Elchalakani, M. Experimental and numerical investigation on cyclic mechanical properties of high-strength steel (HSS) after corrosion damage. *J. Build. Eng.* **2024**, *97*, 110900. [CrossRef]
4. Hocine, A.; Kara Achira, F.S.; Habbar, G.; Levent, A.; Medjdoub, S.M.; Maizia, A.; Dhaou, M.H.; Bezazi, A. Structural integrity assessment of corroded pipelines repaired with composite materials—Literature review. *Int. J. Press. Vessels Pip.* **2024**, *210*, 105253. [CrossRef]
5. Zhang, J.; Yang, P.; Xia, Z.; Wang, C.; Li, Y. Experimental study on the influence of CO_2 corrosion on hydrogen permeation and hydrogen embrittlement sensitivity of X52 pipeline steel. *Eng. Fail. Anal.* **2025**, *176*, 109622. [CrossRef]
6. Mahmoud, A.A.K.L.; Hasan, R. A Comprehensive Survey on Pipeline Monitoring Technologies: Advancements, Challenges, Market Opportunities and Future Directions. *J. Pipeline Sci. Eng.* **2025**, 100353. [CrossRef]
7. Anglana, C.; Barozzi, F.; Capaci, P.; Migoni, D.; Rojas, M.; Fanizzi, F.P.; Di Sansebastiano, G.-P. Characterization of three species of aquatic mosses in axenic culture for biomonitoring and biotechnological applications. *Aquat. Bot.* **2024**, *193*, 103762. [CrossRef]
8. Bai, F.-Y.; Chen, X.-P.; Huang, J.-Z.; Lu, Y.-S.; Dong, H.-Y.; Wu, Y.-H.; Song, S.-L.; Yu, J.; Bai, S.; Chen, Z.; et al. Microbial biofilms on a giant monolithic statue of Buddha: The symbiosis of microorganisms and mosses and implications for bioweathering. *Int. Biodeterior. Biodegrad.* **2021**, *156*, 105106. [CrossRef]
9. Wang, X.; Chen, L.; Liu, J.; Zhou, Y. Lubrication and corrosion inhibition mechanisms of Rauiella Fujisana extracts as lipid soluble additives. *Ind. Crops Prod.* **2025**, *225*, 120459. [CrossRef]
10. Jeschke, M.; Kiehl, K. Effects of a dense moss layer on germination and establishment of vascular plants in newly created calcareous grasslands. *Flora* **2008**, *203*, 557–566. [CrossRef]
11. Li, X.; Zhao, X.; Chen, D.; Guo, G.; Wu, J.; Long, M.; Wu, Q.; Wang, D.; Jiang, H.; Long, L. Uniformity and variance in the effects of moss crusts on soil properties, enzyme activities, and bacterial communities along a subtropical karst degradation gradient. *Catena* **2025**, *255*, 109064. [CrossRef]
12. Huang, H.; Huang, K.; Chen, Y.; Huang, S.; Wang, J.; Wu, H.; Zheng, Z. Individual and combined effects of sodium dichloroisocyanurate and isothiazolinone on the cyanobacteria-Vallisneria natans-microbe aquatic ecosystem. *J. Hazard. Mater.* **2024**, *480*, 136229. [CrossRef] [PubMed]
13. Huang, M.; Lu, G.; Pu, J.; Qiang, Y. Superhydrophobic and smart MgAl-LDH anti-corrosion coating on AZ31 Mg surface. *J. Ind. Eng. Chem.* **2021**, *103*, 154–164. [CrossRef]
14. Jirón-Lazos, U.; Corvo, F.; De la Rosa, S.C.; García-Ochoa, E.M.; Bastidas, D.M.; Bastidas, J.M. Localized corrosion of aluminum alloy 6061 in the presence of Aspergillus niger. *Int. Biodeterior. Biodegrad.* **2018**, *133*, 17–25. [CrossRef]

https://doi.org/10.3390/polym18050611

15. Karami, Z.; Aghazadeh, M.; Jouyandeh, M.; Zarrintaj, P.; Vahabi, H.; Ganjali, M.R.; Torre, L.; Puglia, D.; Saeb, M.R. Epoxy/Zn-Al-CO_3 LDH nanocomposites: Curability assessment. *Prog. Org. Coat.* **2020**, *138*, 105355. [CrossRef]
16. Karami, Z.; Jouyandeh, M.; Ali, J.A.; Ganjali, M.R.; Aghazadeh, M.; Paran, S.M.R.; Naderi, G.; Puglia, D.; Saeb, M.R. Epoxy/layered double hydroxide (LDH) nanocomposites: Synthesis, characterization, and Excellent cure feature of nitrate anion intercalated Zn-Al LDH. *Prog. Org. Coat.* **2019**, *136*, 105218. [CrossRef]
17. Mohammadi, I.; Shahrabi, T.; Mahdavian, M.; Izadi, M. Chemical modification of LDH conversion coating with diethyldithiocarbamate as a novel anti-corrosive film for AA2024-T3. *J. Ind. Eng. Chem.* **2021**, *95*, 134–147. [CrossRef]
18. Zhang, F.; Sun, M.; Xu, S.; Zhao, L.; Zhang, B. Fabrication of oriented layered double hydroxide films by spin coating and their use in corrosion protection. *Chem. Eng. J.* **2008**, *141*, 362–367. [CrossRef]
19. Cao, Y.; Zheng, D.; Luo, J.; Zhang, F.; Wang, C.; Dong, S.; Ma, Y.; Liang, Z.; Lin, C. Enhanced corrosion protection by Al surface immobilization of in-situ grown layered double hydroxide films co-intercalated with inhibitors and low surface energy species. *Corros. Sci.* **2020**, *164*, 108340. [CrossRef]
20. Zhang, C.; Luo, X.; Pan, X.; Liao, L.; Wu, X.; Liu, Y. Self-healing Li-Al layered double hydroxide conversion coating modified with aspartic acid for 6N01 Al alloy. *Appl. Surf. Sci.* **2017**, *394*, 275–281. [CrossRef]
21. Qian, Q.; Wang, Z.; Chen, G.; Zhang, J.; Xu, D.; Ali, H.; Wang, X. Comparison of root and inter-root soil microbial communities of plants infected with crown gall disease of Yinhong plum (*Prunus salicina* Lindl.) based on metagenomes. *Physiol. Mol. Plant Pathol.* **2025**, *140*, 102874. [CrossRef]
22. Ma, J.; Dong, Q.; Li, S.; Ding, H.; Lu, Z.; Xie, Y.; Yong, K.; Xu, J. Effect of Ulva prolifera fragment size on the biomass of developing green algae. *Mar. Environ. Res.* **2025**, *206*, 107037. [CrossRef] [PubMed]
23. Wu, W.-Z.; Li, R.; Ma, Z.-Q.-L.; Liu, C.-X.; Zhao, W.-T. Biochar-algae microspheres based on sodium alginate for the highly efficient adsorption of malachite green dye: Kinetics, isotherms, and mechanism of adsorption. *J. Contam. Hydrol.* **2025**, *272*, 104547. [CrossRef] [PubMed]
24. *GB/T 9286-2021/ISO 2409:2020*; State Administration for Market Regulation (SAMR). Standardization Administration of China (SAC). Paintsandvarnishes—Cross-Cuttest. Standards Press of China (SPC): Beijing, China, 2021.
25. Fernández-Calviño, D.; Rousk, J.; Bååth, E.; Bollmann, U.E.; Bester, K.; Brandt, K.K. Isothiazolinone inhibition of soil microbial activity persists despite biocide dissipation. *Soil Biol. Biochem.* **2023**, *178*, 108957. [CrossRef]
26. Daulisio, M.d.C.Z.; Schneider, R.P. Inactivation of Pseudomonas aeruginosa MDC by isothiazolones and biocide stabilizing agents. *Int. Biodeterior. Biodegrad.* **2020**, *155*, 105090. [CrossRef]
27. Kim, D.; Shin, Y.; Kim, E.-H.; Lee, Y.; Kim, S.; Kim, H.S.; Kim, H.-C.; Leem, J.-H.; Kim, H.R.; Bae, O.-N. Functional and dynamic mitochondrial damage by chloromethylisothiazolinone/methylisothiazolinone (CMIT/MIT) mixture in brain endothelial cell lines and rat cerebrovascular endothelium. *Toxicol. Lett.* **2022**, *366*, 45–57. [CrossRef]
28. Cao, J.; Wu, Y.; Zhao, W. Review of Layered Double Hydroxide (LDH) Nanosheets in Corrosion Mitigation: Recent Developments, Challenges, and Prospects. *Materials* **2025**, *18*, 1190. [CrossRef]

https://doi.org/10.3390/polym18050611

polymers

MDPI

Article

Regulation of MXene Membranes with β-Lactoglobulin Nanofiber-Templated CuS Nanoparticles for Photothermal Antibacterial Effect

Zhuang Liu [1,2], Chenxi Du [2,3], Xin Zhou [1,2,*] and Gang Wei [3,*]

1 School of Basic Medicine, Qingdao University, Qingdao 266071, China; zhuangliu001@outlook.com
2 College of Chemistry and Chemical Engineering, Qingdao University, Qingdao 266071, China; duchenxi5@outlook.com
3 School of Polymer Science and Engineering, Qingdao University of Science and Technology, Qingdao 266042, China
* Correspondence: zhouxin@qdu.edu.cn (X.Z.); wei-lab@qust.edu.cn or wei@uni-bremen.de (G.W.); Tel.: +86-15066242101 (G.W.)

Abstract: Developing advanced antimicrobial agents is critically imperative to address antibiotic-resistant infection crises. MXenes have emerged as a potential nanomedicine for antibacterial applications, but they suffer from suboptimal photothermal conversion efficiency and inherent cytotoxicity. Herein, we report the synthesis of MXene (Ti_3C_2)-based nanohybrids and hybrid membranes through firstly interfacial conjugation of self-assembled β-lactoglobulin nanofibers (β-LGNFs)-inspired copper sulfide nanoparticles (CuS NPs) onto MXene nanosheets, and subsequent vacuum filtration of the created β-LGNF-CuS/MXene nanohybrids. The constructed β-LGNF-CuS/MXene nanohybrids exhibit excellent photothermal conversion performances and satisfactory biocompatibility and minimal cytotoxicity toward mammalian cells, ascribing to the introduction of highly biocompatible β-LGNFs into the hybrid system. In addition, the fabricated β-LGNF-CuS/MXene hybrid membranes demonstrate high efficiency in antibacterial application through the synergistic photothermal and material-related antibacterial effects of both MXene and CuS NPs. Therefore, the ideas and findings shown in this study are useful for inspiring researchers to design and fabricate functional and biocompatible 2D material-based hybrid membranes for antimicrobial applications.

Keywords: MXene; hybrid membrane; CuS nanoparticles; photothermal effect; antibacterial materials

1. Introduction

Antimicrobial-resistant infections have become one of the most pressing public health concerns, primarily driven by the excessive use of antibiotics in clinical practice [1,2]. Bacteria frequently develop resistance through evolutionary adaptation during antibiotic treatment, thereby compromising the clinical utility of traditional antimicrobial therapies [3]. To address this challenge, researchers have been working on identifying and designing the next generation of highly active antimicrobial agents and biomaterials with low propensity for resistance development over the past few decades [4]. Advanced antimicrobial technologies currently being considered as alternatives to antibiotics include photodynamic therapy (PDT) using photosensitizers [5], chemodynamic therapy (CDT) using nanocatalysts [6,7], and photothermal therapy (PTT) using photothermal nanomaterials [8,9]. Two-dimensional (2D) transition metal carbides, carbonitrides, and nitrides (MXenes) are an emerging class of multifunctional nanomaterials that show unique advantages in PTT antimicrobial therapy [10].

MXenes exhibit an extraordinary photothermal conversion efficiency under near-infrared (NIR) laser excitation, enabling localized hyperthermia that induces simultaneous bacterial membrane disruption and protein denaturation [11]. This physical sterilization strategy rapidly destroyed pathogens and avoided the development of antimicrobial resistance [12,13]. However, MXene-based single PTT modalities showed inherent limitations, such as unsatisfying antimicrobial efficacy, and often fail to meet clinical needs [14,15]. Besides, the antibacterial mechanisms of MXenes may be less or even not effective against specific bacteria that have demonstrated high thermoresistance (e.g., *S. aureus*, which can tolerate temperatures as high as about 55 °C) [16]. Another limitation is the inevitable cytotoxicity of MXene, resulting from mechanical and oxidative stress, depending on the dose and duration of exposure [17]. To address these issues, Researchers have proposed synergistic strategies by combining MXene-mediated PTT with other therapeutic modalities or antimicrobial agents to reduce the dose of MXene [18,19]. For instance, MXene has been reported to act as both a photothermal agent and a photosensitizer carrier for synergistic PTT/PDT, with synergistic antibacterial activity against methicillin-resistant *S. aureus* [20]. This multimodal combination therapy can overcome the limitations of monotherapies and effectively suppress the resistance evolution of bacteria through synergistic interplay between distinct bactericidal modalities, providing a novel and promising approach for anti-infective therapy.

As a representative metal-based antimicrobial nanomaterial, copper (Cu)-based nanoparticles (NPs), such as Cu, CuO, Cu_2O, and CuS, have attracted much attention for their broad-spectrum antimicrobial properties and have been applied in clinical scenarios such as wound dressings [21], orthopedic implant coatings [22], and antiviral fabrics [23]. Cu-based NPs exert bactericidal effects through dual synergistic mechanisms: (i) cationic surface interactions that compromise microbial membrane integrity via electrostatic disruption, and (ii) ROS-mediated oxidative cascade through the catalysis of hydrogen peroxide [24,25]. This coordinated action achieves potent microbicidal activity at low concentrations while circumventing conventional pathways of bacterial resistance. However, the cytotoxicity of Cu-based NPs remains an urgent issue, and the required dosage of Cu-based NPs for antibacterial applications is still higher than that of conventional antibiotics. In addition, Cu-based NPs exhibit a morphology-dependent photothermal activity and antibacterial efficiency, requiring a complicated synthesis procedure to enhance photothermal efficiency [26]. To enhance the antimicrobial performance, researchers have developed various functionalization strategies. For example, Cu-based NPs were conjugated with polymer biomaterials to construct hybrid systems or establish slow-release platforms through secondary carrier encapsulation [27,28]. Therefore, MXene-mediated combination therapy of PTT with Cu-based NPs holds great promise for antimicrobial applications. However, the clinical translation of Cu-based NPs still confronts two key barriers: (i) unselective interaction of Cu-based NPs with human cells causes off-target cytotoxicity [29], and (ii) the high surface energy promotes colloidal aggregation and thermodynamic instability [30,31], ultimately compromising antimicrobial efficiency and limiting the scope of clinical applications.

β-Lactoglobulin (β-LG) exhibits distinctive supramolecular assembly capabilities for optimizing metal-ion-based antimicrobial systems. Its sulfhydryl (-SH) and carboxyl (-COOH) groups enable strong coordination chelation of divalent metal ions, forming stable protein–metal supramolecular aggregates [32]. The protein shell of protein–metal nanocomposites blocks free Cu ion exposure and reduces the oxidative stress damage to mammalian cells, thereby enabling cytotoxicity regulation of Cu-based NPs [33]. Additionally, the amphiphilic structure of β-LG forms a dense stabilization layer on the surface of nanocomposites and generates a spatial site-blocking effect that effectively blocks the interparticle van der Waals force interaction, thus contributing to the long-term stability of the colloidal dispersion system [34,35]. Compared with previously reported MXene–CuS and MXene–protein composites, our strategy employs β-lactoglobulin nanofibrils to facilitate spatially controlled CuS growth and hybridization with MXene, yielding membranes with enhanced PTT/CDT antibacterial efficacy and biocompatibility [36,37]. This interfacial engineering strategy based on protein conformation manipulation provides a new approach to enhance the biocompatibility and stability of Cu-based nanomaterials.

In this work, we fabricate antimicrobial MXene-based nanohybrids and hybrid membranes using vacuum filtration, which exhibit rapid bactericidal activity and sustained antibacterial efficacy through photothermal and material-based bacterial killing (Scheme 1). We first prepare β-LG nanofibers (β-LGNFs) by protein self-assembly to enhance the biocompatibility of the antimicrobial membrane and suppress CuS NP aggregation. Next, the pre-incubated β-LGNFs are used to synthesize CuS NPs by the adsorption of Cu^{2+} and the reaction with S^{2-}, forming β-LGNF-CuS nanohybrids. As shown in Scheme 1A, glutaraldehyde (GA) is used as a cross-linking agent to conjugate MXene with β-LGNF-CuS to form β-LGNF-CuS/MXene nanohybrids, which are then vacuum filtered and dried to obtain a membrane-like structure (Scheme 1B). The coordinated antimicrobial effect of CuS NPs and MXene nanosheets endows the fabricated β-LGNF-CuS/MXene membranes with excellent antimicrobial activity against *S. aureus* and *S. epidermidis* via photothermal treatment while maintaining excellent biocompatibility due to the effect of β-LGNFs.

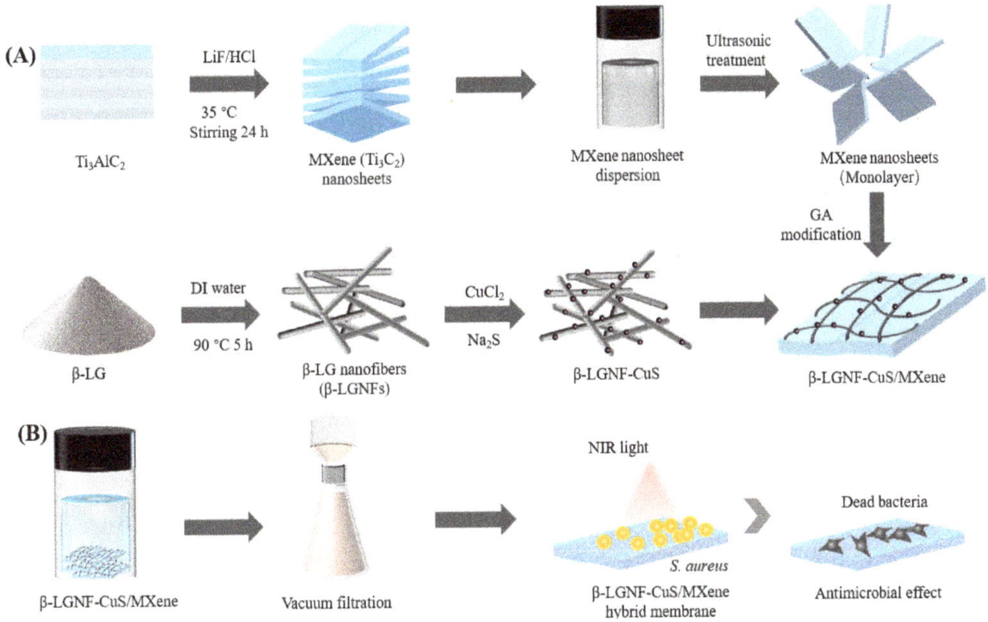

Scheme 1. Schematics of preparations of (**A**) β-LGNF-CuS/MXene nanohybrids and (**B**) β-LGNF-CuS/MXene hybrid membranes for synergistic photothermal and material-related antibacterial application.

2. Materials and Methods

2.1. Materials

β-LG, copper chloride ($CuCl_2$), sodium sulfide (Na_2S), GA, dimethyl sulfoxide (DMSO), and L-ascorbic acid (AA) were purchased from Shanghai Macklin Biochemical Technology Co. (Shanghai, China). Ethanol, hydrochloric acid, hydrofluoric acid (HF), and sodium hydroxide (≥96%) were provided by Sinopharm Chemical Reagent Co. (Shanghai, China). Lithium fluoride (LiF), titanium aluminum carbide (Ti_3AlC_2), and sodium bicarbonate were purchased from Shanghai Yien Chemical Technology Co. (Shanghai, China).

2.2. Synthesis of Amyloid β-LGNFs and β-LGNF-CuS Nanohybrids

The synthesis of β-LGNFs was achieved by protein self-assembly. In brief, 0.04 g of β-LG powder was first dissolved in deionized (DI) water. Then, the solution pH was adjusted to 2.0 with 1 M HCl and diluted to 2% (w/w) using pH 2-adjusted DI water. The solution was then heated to 90 °C for 5 h with continuous stirring, at rates of 120 rpm for the first 3 h and 170 rpm for the last 2 h. The solution (β-LGNFs) was then cooled with ice and stored at 4 °C.

For the bioinspired synthesis of CuS NPs, $CuCl_2$ (10 mM) was added to the β-LGNF solution in a 1:5 ratio (v/v) and stirred for 12 h. An equal volume of Na_2S solution (10 mM) was then added to the stirred mixture to react with the Cu^{2+}, resulting in the formation of β-LGNF-CuS nanohybrids.

2.3. Synthesis of MXene Nanosheets

MXene (Ti_3C_2) monolayered nanosheets were synthesized by HF etching of Ti_3AlC_2 and DMSO intercalation of multilayered MXene nanosheets based on a previous study [38]. Briefly, 1 mg of LiF was dissolved in 20 mL of HCl (12 M) and stirred for 10 min. 1 mg of Ti_3AlC_2 was then added to the mixture and stirred for 24 h at 35 °C. The solution was washed with 1 M HCl to remove unreacted LiF and impurities, followed by repeated washing with deionized water and centrifugation (8000 rpm, 5 min) until the pH of the solution reached 6.0–7.0. Monolayered MXene nanosheets were obtained by sonicating the suspension for 30 min. Subsequently, the supernatant was centrifuged, and the middle layer was collected for morphology characterization. Clay-like multilayer MXene, obtained through centrifugal precipitation, was added to DMSO and intercalated overnight under stirring. The solution was dialyzed with sodium ascorbate-containing dialysis buffer for 24 h to remove DMSO, and then freeze-dried to obtain monolayered MXene for further experiments.

2.4. Synthesis of β-LGNF-CuS/MXene Nanohybrids and Hybrid Membranes

A total of 10 mg of MXene powder was dispersed in 5 mL of sodium ascorbate solution. To further functionalize the MXene surface, 0.1% GA was added into the solution at a volume ratio of 1:10, followed by 12 h of stirring to yield GA-modified MXene. After centrifugation and thorough washing with deionized water (3 times) to remove free GA, 2 mg/mL GA-modified MXene dispersion was then mixed with an equal volume of β-LGNF-CuS nanohybrid solution. After 12 h of stirring, β-LGNF-CuS/MXene nanohybrids were created. The hybrid membranes were constructed through permeating β-LGNF-CuS/MXene nanohybrid solution on the surface of polyethersulfone (PES) membrane via vacuum filtration (0.08 MPa) for 12 h at room temperature.

2.5. Characterization Techniques

The prepared β-LGNFs and MXene nanosheets were characterized using atomic force microscopy (AFM) (FM-Nanoview 6800 AFM, FSM-Precision, Suzhou FSM Precision

Instrument Co., Ltd., Suzhou, China) and transmission electron microscopy (TEM) (Tecnai G2 F20, FEI, Hillsboro, OR, USA). The microscopic structure of nanocomposites and membranes was observed via scanning electron microscope (SEM) (Regulus 8100, Hitachi, Tokyo, Japan). X-ray spectroscopy (XPS) (PHI 5000 VersaProbe III spectrometer, UlVAC-PHI, Chigasaki, Japan) was used to characterize the elemental composition of the samples.

To evaluate the structural durability of the β-LGNF-CuS/MXene hybrid membrane under physiologically relevant conditions, a cyclic stability test was performed. The membrane was immersed in 0.9% NaCl solution and subjected to near-infrared (NIR) laser irradiation (808 nm, 1.0 W/cm^2) for 5 min per cycle, once per day, over a total period of 7 days. After each cycle, the membrane was gently rinsed with deionized water and air-dried at room temperature before the next cycle [39]. Mechanical properties of the hybrid membranes after thermal cycling were assessed via uniaxial tensile testing using a universal testing machine (AGS-X, SHIMADZU, Shanghai, China) at a strain rate of 5 mm/min.

2.6. Controlled Release of Cu^{2+}

Cu^{2+} ion release was quantified by ICP-MS. Hybrid membranes (1 cm^2) were immersed in 5 mL of 0.9% NaCl solution at 37 °C, with or without NIR irradiation (808 nm, 1.0 W/cm^2, 5 min per cycle). At predetermined time points (1, 3, 5, and 7 days), aliquots of the supernatant were collected and filtered through a 0.22 μm membrane before analysis using inductively coupled plasma-mass spectrometry (ICP-MS, Agilent 7900, Tokyo, Japan).

2.7. In Vitro Biocompatibility of β-LGNF-CuS/MXene Nanohybrids

MC3T3 cells were cultured in α-MEM medium supplemented with 10% fetal bovine serum. To assess the biocompatibility of the materials, 2×10^5 cells were seeded in 24-well plates and incubated for 12 h. β-LGNF-CuS/MXene nanohybrids were added to the cell medium at a concentration of 10 μg/mL. Cells were incubated for an additional 24 h and 72 h and then stained using Calcein AM/PI Assay Kit (Beyotime, Shanghai, China) according to the manufacturer's protocol.

2.8. In Vitro Antibacterial Assay

S. aureus and *S. epidermidis* were cultured in Luria–Bertani (LB) agar plates at 37 °C. Control, MXene, and hybrid membranes were added into the medium and cocultured with bacterial suspension overnight to detect the inhibitory effect of the materials on bacterial growth. For the PTT group, a mixture of samples and *S. aureus* or *S. epidermidis* was exposed to NIR irradiation (808 nm, 1.5 W/cm^2) for 30 min. The colonies were counted to quantify the antimicrobial effect of the materials.

3. Results and Discussions

3.1. Synthesis and Characterizations of β-LGNFs

With its abundant functional amino acid residues, amyloid β-LGNFs serve as both reducing agents—capable of directly reducing alloy nanoparticles—and as protective agents that prevent NP aggregation and enhance the dispersive stability of CuS NPs. These fibers also impart excellent biocompatibility and multifunctionality to the composite system, serving as an efficient carrier for antimicrobial materials. Thermal treatment under low pH and low ionic strength conditions is the most commonly employed method to induce fibrogenesis in β-LG, significantly accelerating the rate of protein self-assembly (Figure 1A) [40,41].

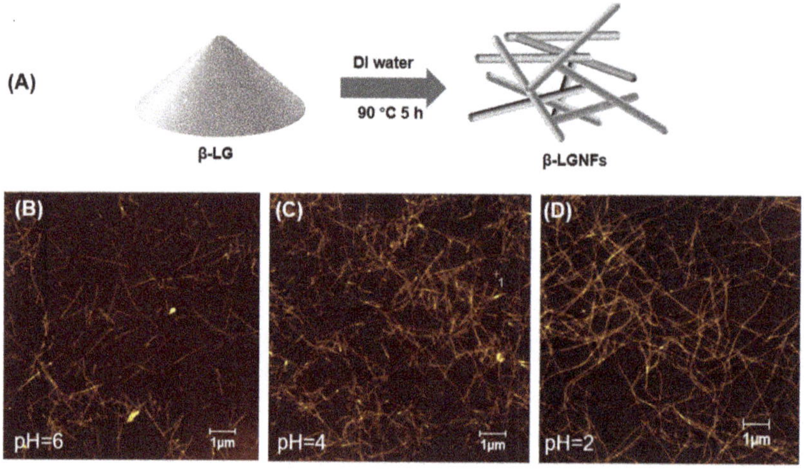

Figure 1. Synthesis and AFM characterization of β-LGNFs. (**A**) Schematic synthesis. (**B–D**) AFM height images of β-LGNFs incubated at different pH values: (**B**) pH 6, (**C**) pH 4, and (**D**) pH 2.

We then compared the morphological features of β-LGNFs that were synthesized under varying pH conditions (Figure 1B–D). The proteins exhibited markedly different aggregate morphologies depending on pH. As the pH decreased, the β-LGNFs underwent distinct morphological transformations. The fiber diameter increased, the fibers elongated, and the degree of entanglement and aggregation intensified, resulting in a denser and more complex network structure. These transformations are likely associated with pH-induced conformational changes in the protein and modifications in intermolecular interactions. At pH 2, β-LGNFs showed a high degree of aggregation, with notable thickening of fiber diameter, elongation, and intertwining, ultimately forming a stable network structure. These results suggest that enhanced intermolecular forces under these highly acidic conditions induced extensive protein molecular aggregation and structural reorganization of the protein nanofibers.

3.2. Synthesis and Characterizations of β-LGNF-CuS Nanohybrids

Subsequently, β-LGNFs prepared by incubation at pH 2 were utilized as templates for the synthesis of β-LGNF-CuS nanohybrids via a bioinspired synthesis method (Figure 2A) [42]. Their morphological characteristics were subsequently analyzed using TEM. The created β-LGNF exhibited elongated and morphologically uniform fiber structures with relatively smooth surfaces (Figure 2B). This homogeneous morphology and surface smoothness may enhance the mechanical properties and dispersibility of the fibers in practical applications. In β-LGNF-CuS hybrids, most CuS NPs were uniformly distributed along the surface of the β-LGNFs, although minor agglomeration was observed (Figure 2C). The deposition of CuS NPs increased the surface roughness of the nanofibers, resulting in an increased specific surface area and consequently offering more active sites for catalytic processes.

To further confirm the formation of CuS NPs on the synthesized β-LGNFs, XPS characterization was performed. The XPS spectrum of the β-LGNF-CuS nanohybrids (Figure 2D) revealed that the nanohybrids contain signals corresponding to Cu 2p, O 1s, N 1s, C 1s, and S 2p. In particular, the detailed XPS spectra of Cu 2p (Cu $2p_{1/2}$ at 952.34 eV and Cu $2p_{3/2}$ at 932.28 eV) and S 2p (at 163.19 eV) further confirmed the successful nucleation and growth of CuS NPs onto the β-LGNF (Figure 2E,F) [43,44]. Based on these results, it can be concluded that β-LGNF-CuS nanohybrids were successfully synthesized.

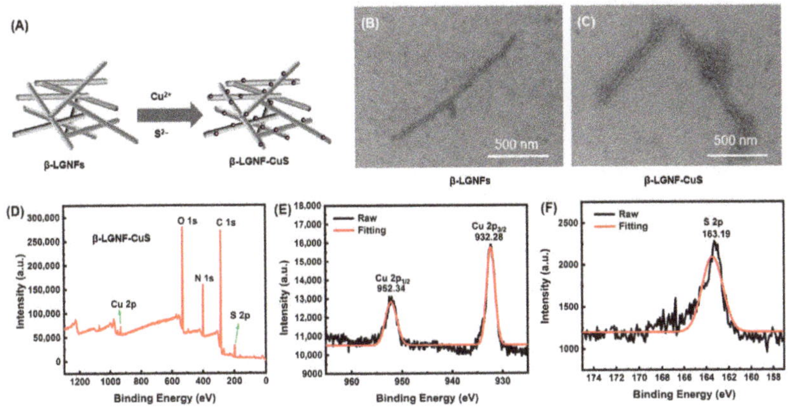

Figure 2. Synthesis and characterization of nanofibers and β-LGNF-CuS nanohybrids. (**A**) Schematic synthesis process. (**B**,**C**) TEM images of β-LGNFs and β-LGNF-CuS nanohybrids. (**D**–**F**) XPS spectrum of (**D**) β-LGNF-CuS, as well as spectra of (**E**) Cu 2P and (**F**) S 2p.

3.3. *Synthesis and Characterization of MXene Membranes*

Two-dimensional materials have great application potential in areas including electronic devices, energy storage, and biomedical devices [45,46]. Two-dimensional monolayered MXene nanosheets were synthesized via HF etching (Figure 3A) [47]. The prepared MXene materials were characterized using AFM and TEM to evaluate their morphology and size in detail. AFM image revealed that the MXene nanosheets exhibit a highly planar surface morphology (Figure 3B). TEM analysis further confirmed the presence of a transparent, film-like single-layer MXene structure with a flat surface and well-defined edges (Figure 3C). These results demonstrate that the MXene monolayer nanosheets prepared by this method possess excellent 2D structural characteristics, with their high transparency and sharp edges indicating high structural integrity and minimal defect density.

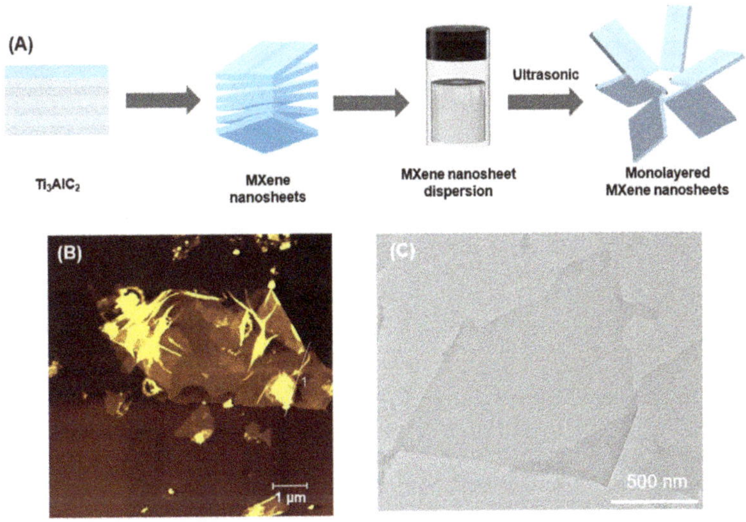

Figure 3. Synthesis and characterization of MXene nanosheets. (**A**) Schematic illustration of MXene synthesis. (**B**,**C**) AFM and TEM images of MXene nanosheets.

β-LGNF-CuS/MXene hybrid membranes were synthesized as illustrated in Figure 4A. After the fabrication of β-LGNF-CuS/MXene hybrid membrane, a photographic image of the resulting membrane shows a uniform black appearance (Figure 4B). The microstructure of the material was further analyzed using SEM and TEM. The MXene nanosheets exhibited a typical layered architecture with a relatively smooth surface and well-defined edges, which is characteristic of 2D materials (Figure 4C). Compared to MXene, SEM and TEM images of the β-LGNF-CuS/MXene hybrid membranes (Figure 4D,E) reveal a uniform distribution of CuS NPs across the MXene nanosheet surface, resulting in the formation of a stable composite structure. Additionally, β-LGNFs promoted the mechanical stability of the final membrane through the interactions with both CuS NPs and MXene nanosheets, we suggest.

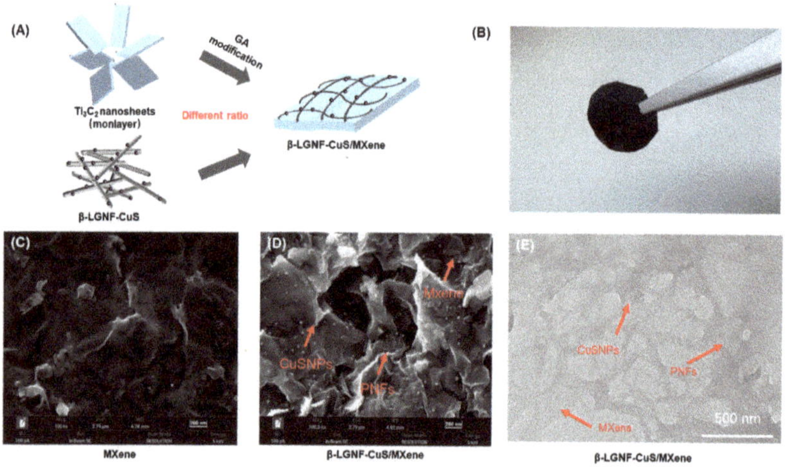

Figure 4. Synthesis and structural characterization of β-LGNF-CuS/MXene hybrid membranes. (**A**) Conjugation of β-LGNF-CuS and GA-MXene. (**B**) Digital photograph of β-LGNF-CuS/MXene membrane. (**C**) SEM image of MXene nanosheets. (**D**) SEM and (**E**) TEM images of β-LGNF-CuS/MXene hybrid membranes.

Considering the importance of membrane stability in practical applications, the stability of the β-LGNF-CuS/MXene composite membrane was investigated under cyclic near-infrared (NIR) irradiation and in physiological saline (Figure S1, Supporting Information). After 7 days of cyclic NIR exposure and immersion in physiological saline, the hybrid membrane maintained its structural integrity without any visible deformation or delamination, indicating excellent resistance to photothermal stress and ionic environments [39]. Uniaxial tensile tests showed that the tensile strength of the composite membranes after photothermal cycling was 473.6 kPa. (Figure S2, Supporting Information).

3.4. *Photothermal Properties and Cu^{2+} Release Profile of β-LGNF-CuS/MXene Nanohybrids*

To evaluate the photothermal properties of β-LGNF-CuS/MXene nanohybrids, a laser with a wavelength of 808 nm and a power density of 1.5 W/cm^2 was used to irradiate suspensions of varying concentrations and compositional ratios. The results demonstrated that the temperature increase of the nanohybrids increased with the suspension concentration, confirming the excellent photothermal conversion capability of the β-LGNF-CuS/MXene hybrids (Figure 5A). Specifically, at a concentration of 0.2 mg/mL, the suspension temperature increased to 58.9 °C within 10 min under NIR light irradiation. Notably, the incorporation of β-LGNF-CuS nanohybrids did not significantly alter the intrinsic photothermal performance of MXene (Figure 5B). Furthermore, the photothermal effect of the β-LGNF-CuS/MXene hybrids remained stable after three consecutive heating-cooling cycles, indicating its excellent photothermal stability (Figure 5C). These favorable photothermal characteristics suggest that the β-LGNF-CuS/MXene hybrids show high potential as efficient PTT reagents, with promising applications in localized thermotherapy, particularly in enhancing antimicrobial efficacy. Additionally, the results from ICP-MS indicated that exposure under NIR significantly promoted the release of Cu^{2+} from the β-LGNF-CuS/MXene membrane (Figure S3, Supporting Information).

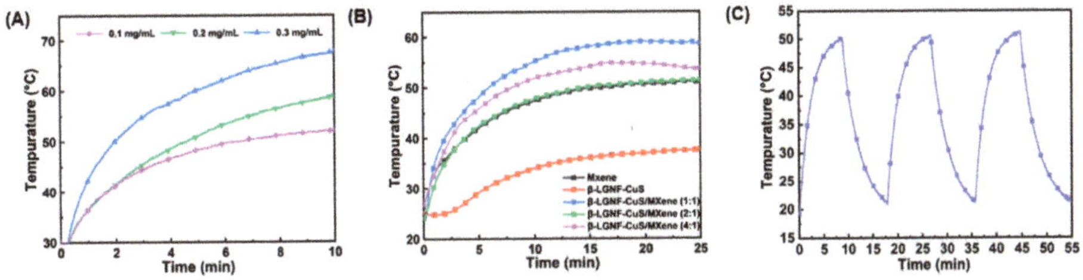

Figure 5. Photothermal performance of β-LGNF-CuS/MXene nanohybrids under NIR irradiation (808 nm, 1.5 W/cm^2). (**A**) Temperature elevation of β-LGNF-CuS/MXene at varying concentrations (1.5 W/cm^2). (**B**) Heating and cooling curves of various samples. (**C**) Photothermal stability of β-LGNF-CuS/MXene nanohybrids (0.2 mg/mL) over three heating-cooling cycles.

3.5. Biocompatibility of β-LGNF-CuS/MXene Nanohybrids

Favorable biocompatibility is imperative for medical devices and clinical translation [48]. Cell live/dead staining was performed to investigate the cytotoxicity of the β-LGNFs-CuS/MXene nanohybrids. As shown in Figure 6A, most materials had no significant effect on cell proliferation after 24 h. However, partial cell death was observed in the MXene group, with a cell survival rate of approximately 86.50% ($p < 0.01$). The toxic effects of MXene may be attributed to apoptosis due to cellular internalization and cell membrane rupture due to MXene-cell contact [49]. Despite minor cell death in some fields of view, most cells retained good morphology and viability, with survival > 90% in the β-LGNFs-CuS and β-LGNFs-CuS/MXene group (Figure 6B). After 72 h, the cell viability in the MXene group decreased to 83.82%, while most of the cells in the β-LGNFs-CuS and β-LGNFs-CuS/MXene groups were still alive. (Figure 6C,D). These results suggest that the protein shell of β-LGNFs-CuS greatly enhances the biocompatibility of CuS NPs. Besides, the interfacial conjugation of β-LGNFs-CuS overcomes the high cytotoxicity of conventional MXene materials. Therefore, we demonstrate that β-LGNFs-CuS/MXene nanohybrids have good biocompatibility, which provides inspiration for future MXene-based membrane materials in biomedical applications.

3.6. Antibacterial Properties of β-LGNF-CuS/MXene Hybrid Membranes

Finally, *S. aureus* was chosen to verify the antibacterial ability of the β-LGNFs-CuS/MXene hybrid membranes. As shown in Figure 7A, antibacterial efficacy was observed in all samples. As previously reported, the sharp outer edges of MXene disrupt bacterial biofilms, leading to bacterial rupture and inactivation [50]. The physical disruption and Cu^{2+} release are the main factors for bacterial killing. However, in the absence of NIR light irradiation, there was no significant difference in the antimicrobial rate (about 75–80%) between all the materials (Figure 7A). In the PTT group (with NIR light irradiation), the antimicrobial properties of all materials except β-LGNF-CuS were enhanced. This is consistent with the poor photothermal performance of β-LGNF-CuS in the above-presented photothermal results. It should be noted that in the presence of NIR light irradiation, the sample of β-LGNF-CuS/MXene (4:1) showed the greatest antibacterial property (nearly

99%) (Figure 6B). While higher β-LGNF-CuS loading reduced photothermal conversion, we demonstrate that the ratio of 4:1 could be the optimal ratio to amplify the synergistic effect of β-LGNFs and MXene for a satisfactory antimicrobial effect. The results support that the synergistic effect of the physical disruption, Cu^{2+} release, and photothermal effect achieves the greatest antimicrobial efficacy. In addition, to demonstrate the efficacy of clearance against other significant pathogens, we also investigated the antimicrobial properties of β-LGNF-CuS/MXene (4:1) against *S. epidermidis* (Figure S4, Supporting Information). β-LGNF-CuS/MXene (4:1) exhibited > 95% antimicrobial efficiency against *S. epidermidis* under NIR light irradiation. Despite the promising antibacterial performance and biocompatibility of β-LGNF-CuS/MXene nanohybrids, potential challenges remain in scaling up this technology for clinical translation. These include the need for precise control of CuS nucleation during large-scale synthesis, and ensuring long-term storage stability. Furthermore, regulatory hurdles for clinical approval of nanomaterials must also be addressed before practical application.

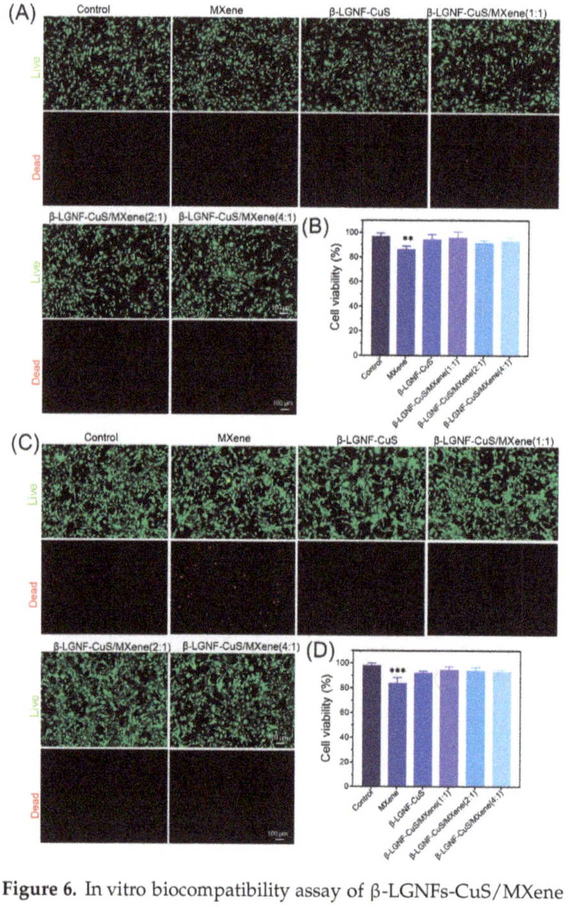

Figure 6. In vitro biocompatibility assay of β-LGNFs-CuS/MXene nanohybrids. (**A**) Fluorescent photographs of live/dead cell staining after 24 h. (**B**) The quantification of cell viability after 24 h. (**C**) Fluorescent photographs of live/dead cell staining after 72 h. (**D**) The quantification of cell viability after 72 h. (n = 3) The data are presented as mean ± standard deviation (SD). ** $p < 0.01$, *** $p < 0.001$.

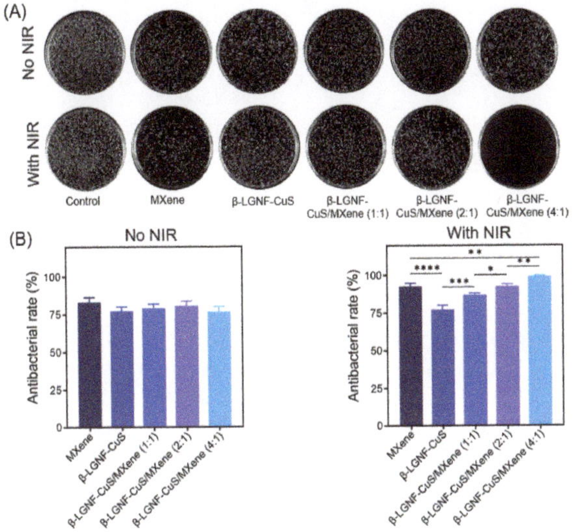

Figure 7. Antibacterial assay of hybrid membranes. (**A**) Agar plate photographs of *S. aureus*. (**B**) The quantification of the antibacterial rate. (n = 3) The data are presented as mean ± standard deviation (SD). * $p < 0.05$, ** $p < 0.01$, *** $p < 0.001$, **** $p < 0.0001$.

4. Conclusions

In summary, we reported the bioinspired, template-assisted synthesis of β-LGNFs-CuS nanohybrids, which were subsequently conjugated with MXene nanosheets to form β-LGNFs-CuS/MXene nanohybrids. Hybrid membranes were then fabricated using vacuum filtration. The resulting β-LGNFs-CuS/MXene nanohybrid exhibited high structural integrity and uniform distribution. As a result, the synthesized nanohybrids presented high photothermal conversion efficiency and excellent biocompatibility, and the fabricated hybrid membranes revealed effective antibacterial activity against *S. aureus* and *S. epidermidis*. Therefore, the proposed organic–inorganic hybrid membranes represent a promising strategy for the design and development of MXene-based antibacterial biomaterials with improved performance. Future research should focus on further optimizing the structure–property relationships of β-LGNFs-CuS/MXene hybrid membranes, especially under in vivo conditions. It is also important to investigate long-term antibacterial efficacy, biodegradation behavior, and immune responses in relevant animal models. In addition, exploring scalable fabrication methods and expanding the hybrid system to other types of pathogenic bacteria could broaden its biomedical applicability.

Supplementary Materials: The following supporting information can be downloaded at https://www.mdpi.com/article/10.3390/polym17141960/s1, Figure S1: β-LGNF-CuS/MXene composite membrane after cyclic near-infrared (NIR) irradiation and physiological saline immersion. Figure S2: Mechanical properties of hybrid membranes after thermal cycling. Figure S3: Release curve of Cu^{2+} in β-LGNF-CuS/MXene composite membrane. Figure S4: Antibacterial effect of hybrid membranes for *S. epidermidis*. (A) Agar plate photographs of *S. epidermidis*. (B) The quantification of the antibacterial rate. (n = 3) The data are presented as mean ± standard deviation (SD).

Author Contributions: Conceptualization, X.Z. and G.W.; data curation, Z.L. and C.D.; formal analysis, Z.L. and C.D.; funding acquisition, G.W.; investigation, Z.L.; methodology, Z.L. and C.D.; project administration, X.Z. and G.W.; software, Z.L. and C.D.; supervision, X.Z. and G.W.; writing—original draft, Z.L.; writing—review and editing, X.Z. and G.W. All authors have read and agreed to the published version of the manuscript.

Funding: This research was funded by the Taishan Scholars Program of Shandong province (No. tsqn201909104).

Institutional Review Board Statement: Not applicable.

Data Availability Statement: Data are contained within the article and Supplementary Materials.

Conflicts of Interest: The authors declare no conflicts of interest.

References

1. Daruka, L.; Czikkely, M.S.; Szili, P.; Farkas, Z.; Balogh, D.; Grézal, G.; Maharramov, E.; Vu, T.-H.; Sipos, L.; Juhász, S.; et al. ESKAPE pathogens rapidly develop resistance against antibiotics in development in vitro. *Nat. Microbiol.* **2025**, *10*, 313–331. [CrossRef] [PubMed]
2. Metsemakers, W.-J.; Moriarty, T.F.; Morgenstern, M.; Marais, L.; Onsea, J.; O'Toole, R.V.; Depypere, M.; Obremskey, W.T.; Verhofstad, M.H.J.; McNally, M.; et al. The global burden of fracture-related infection: Can we do better? *Lancet Infect. Dis.* **2024**, *24*, e386–e393. [CrossRef] [PubMed]
3. Piskovsky, V.; Oliveira, N.M. Bacterial motility can govern the dynamics of antibiotic resistance evolution. *Nat. Commun.* **2023**, *14*, 5584. [CrossRef] [PubMed]
4. Panáček, D.; Belza, J.; Hochvaldová, L.; Bad'ura, Z.; Zoppellaro, G.; Šrejber, M.; Malina, T.; Šedajová, V.; Paloncýová, M.; Langer, R.; et al. Single Atom Engineered Antibiotics Overcome Bacterial Resistance. *Adv. Mater.* **2024**, *36*, 2410652. [CrossRef] [PubMed]
5. Wang, Z.; Liu, X.; Duan, Y.; Huang, Y. Infection microenvironment-related antibacterial nanotherapeutic strategies. *Biomaterials* **2022**, *280*, 121249. [CrossRef] [PubMed]
6. Yu, Y.; Sun, H.; Lu, Q.; Sun, J.; Zhang, P.; Zeng, L.; Vasilev, K.; Zhao, Y.; Chen, Y.; Liu, P. Spontaneous formation of MXene-oxidized sono/chemo-dynamic sonosensitizer/nanocatalyst for antibacteria and bone-tissue regeneration. *J. Nanobiotechnol.* **2023**, *21*, 193. [CrossRef] [PubMed]
7. Sun, G.; Jiang, X.; Liu, C.; Song, S.; Zhang, J.; Shen, J. FeS@LAB-35@Ti_3C_2 as a high-efficiency nanozyme for near infrared light induced photothermal enhanced chemodynamic antibacterial activity and wound healing. *Nano Res.* **2023**, *16*, 2840–2850. [CrossRef]
8. Hu, Y.; Zeng, Q.; Hu, Y.; He, J.; Wang, H.; Deng, C.; Li, D. MXene/zinc ion embedded agar/sodium alginate hydrogel for rapid and efficient sterilization with photothermal and chemical synergetic therapy. *Talanta* **2024**, *266*, 125101. [CrossRef] [PubMed]
9. Hu, P.; Feng, N.; Zhao, S.; Shi, J.; Yang, G.; Guo, W.; Liu, Y.; Fan, K. Microenvironment-Responsive Injectable Thermosensitive Hydrogel Incorporating Nanozymes for Synergistic Breast Cancer Therapy and Postsurgical Adjuvant Treatment. *Adv. Funct. Mater.* **2025**, 2421176. [CrossRef]
10. Huang, H.; Dong, C.; Feng, W.; Wang, Y.; Huang, B.; Chen, Y. Biomedical engineering of two-dimensional MXenes. *Adv. Drug Delivery Rev.* **2022**, *184*, 114178. [CrossRef] [PubMed]
11. He, X.; Qian, Y.; Wu, C.; Feng, J.; Sun, X.; Zheng, Q.; Li, X.; Shen, J. Entropy-Mediated High-Entropy MXenes Nanotherapeutics: NIR-II-Enhanced Intrinsic Oxidase Mimic Activity to Combat Methicillin-Resistant Staphylococcus Aureus Infection. *Adv. Mater.* **2023**, *35*, 2211432. [CrossRef] [PubMed]
12. Zheng, Y.; Yan, Y.; Lin, L.; He, Q.; Hu, H.; Luo, R.; Xian, D.; Wu, J.; Shi, Y.; Zeng, F.; et al. Titanium carbide MXene-based hybrid hydrogel for chemo-photothermal combinational treatment of localized bacterial infection. *Acta Biomater.* **2022**, *142*, 113–123. [CrossRef] [PubMed]
13. Hsu, Y.J.; Nain, A.; Lin, Y.F.; Tseng, Y.T.; Li, Y.J.; Sangili, A.; Srivastava, P.; Yu, H.L.; Huang, Y.F.; Huang, C.C.; et al. Self-redox reaction driven in situ formation of $Cu_2O/Ti_3C_2T_x$ nanosheets boosts the photocatalytic eradication of multi-drug resistant bacteria from infected wound. *J. Nanobiotechnol.* **2022**, *20*, 235. [CrossRef] [PubMed]
14. Wu, F.; Zheng, H.; Wang, W.; Wu, Q.; Zhang, Q.; Guo, J.; Pu, B.; Shi, X.; Li, J.; Chen, X.; et al. Rapid eradication of antibiotic-resistant bacteria and biofilms by MXene and near-infrared light through photothermal ablation. *Sci. China Mater.* **2021**, *64*, 748–758. [CrossRef]
15. Luan, X.; Hu, H.; Sun, Z.; He, P.; Zhu, D.; Xu, Y.; Liu, B.; Wei, G. Assembling Ag_2S quantum dots onto peptide nanosheet as a biomimetic two-dimensional nanoplatform for synergistic near infrared-II fluorescent imaging and photothermal therapy of tumor. *J. Colloid Interface Sci.* **2024**, *663*, 111–122. [CrossRef] [PubMed]
16. Li, J.; Suo, Y.; Liao, X.; Ahn, J.; Liu, D.; Chen, S.; Ye, X.; Ding, T. Analysis of Staphylococcus aureus cell viability, sublethal injury and death induced by synergistic combination of ultrasound and mild heat. *Ultrason. Sonochem.* **2017**, *39*, 101–110. [CrossRef] [PubMed]
17. Lim, G.P.; Soon, C.F.; Ma, N.L.; Morsin, M.; Nayan, N.; Ahmad, M.K.; Tee, K.S. Cytotoxicity of MXene-based nanomaterials for biomedical applications: A mini review. *Environ. Res.* **2021**, *201*, 111592. [CrossRef] [PubMed]

18. Zheng, K.; Li, S.; Jing, L.; Chen, P.-Y.; Xie, J. Synergistic Antimicrobial Titanium Carbide (MXene) Conjugated with Gold Nanoclusters. *Adv. Healthcare Mater.* **2020**, *9*, 2001007. [CrossRef] [PubMed]
19. Zhou, W.; Chen, J.; Liao, T.; Wu, Q.; Guo, N.; Xie, G.; Lin, H.; Li, C.; Liu, Y. Ag Nanoparticles on MXene Nanosheets for Combined Ionic and Photothermal Therapy of Bacterial Infections. *ACS Appl. Nano Mater.* **2024**, *7*, 21261–21274. [CrossRef]
20. Yu, C.; Sui, S.; Yu, X.; Huang, W.; Wu, Y.; Zeng, X.; Chen, Q.; Wang, J.; Peng, Q. $Ti_3C_2T_x$ MXene loaded with indocyanine green for synergistic photothermal and photodynamic therapy for drug-resistant bacterium. *Colloids Surf. B* **2022**, *217*, 112663. [CrossRef] [PubMed]
21. Liu, T.; Ma, M.; Ali, A.; Liu, Q.; Bai, R.; Zhang, K.; Guan, Y.; Wang, Y.; Liu, J.; Zhou, H. Self-assembled copper tannic acid nanoparticles: A powerful nano-bactericide by valence shift of copper. *Nano Today* **2024**, *54*, 102071. [CrossRef]
22. Zhang, Y.; Fu, S.; Yang, L.; Qin, G.; Zhang, E. A nano-structured $TiO_2/CuO/Cu_2O$ coating on Ti-Cu alloy with dual function of antibacterial ability and osteogenic activity. *J. Mater. Sci. Technol.* **2022**, *97*, 201–212. [CrossRef]
23. Chen, G.; Yang, L.; Ma, N.; Yu, S.; Zhou, J.; Cao, R.; Zhang, Q.; Yu, H.; Zhai, M.; Wang, R.; et al. Selection and design principle of efficient antiviral nano-hybrid fiber materials for fighting pandemic viruses: A review. *Nano Today* **2023**, *53*, 102001. [CrossRef]
24. Smith, J.L.; Tran, N.; Song, T.; Liang, D.; Qian, M. Robust bulk micro-nano hierarchical copper structures possessing exceptional bactericidal efficacy. *Biomaterials* **2022**, *280*, 121271. [CrossRef] [PubMed]
25. Chang, T.; Babu, R.P.; Zhao, W.; Johnson, C.M.; Hedström, P.; Odnevall, I.; Leygraf, C. High-Resolution Microscopical Studies of Contact Killing Mechanisms on Copper-Based Surfaces. *ACS Appl. Mater. Interfaces* **2021**, *13*, 49402–49413. [CrossRef] [PubMed]
26. Naskar, A.; Kim, K.-s. Photo-Stimuli-Responsive CuS Nanomaterials as Cutting-Edge Platform Materials for Antibacterial Applications. *Pharmaceutics* **2022**, *14*, 2343. [CrossRef] [PubMed]
27. Han, Q.; Wang, X.; Qiu, L.; Zhou, X.; Hui, Z.; Ni, X.; Xuan, Y.; Lei, X.; Wang, J. Gelatinase Responsive Nanogel for Antibacterial Phototherapy and Wound Healing. *Gels* **2022**, *8*, 397. [CrossRef] [PubMed]
28. Dash, K.K.; Deka, P.; Bangar, S.P.; Chaudhary, V.; Trif, M.; Rusu, A. Applications of Inorganic Nanoparticles in Food Packaging: A Comprehensive Review. *Polymers* **2022**, *14*, 521. [CrossRef] [PubMed]
29. Liu, H.; Lai, W.; Liu, X.; Yang, H.; Fang, Y.; Tian, L.; Li, K.; Nie, H.; Zhang, W.; Shi, Y.; et al. Exposure to copper oxide nanoparticles triggers oxidative stress and endoplasmic reticulum (ER)-stress induced toxicology and apoptosis in male rat liver and BRL-3A cell. *J. Hazard. Mater.* **2021**, *401*, 123349. [CrossRef] [PubMed]
30. Miao, L.; Wang, C.; Hou, J.; Wang, P.; Ao, Y.; Li, Y.; Lv, B.; Yang, Y.; You, G.; Xu, Y. Enhanced stability and dissolution of CuO nanoparticles by extracellular polymeric substances in aqueous environment. *J. Nanopart. Res.* **2015**, *17*, 404. [CrossRef]
31. Fernando, L.; Chen, W.T.; Lai, C.W.; Ye, F.Y.; Lai, P.S.; Lin, J.J.; Yasuda, K.; Song, T.T.; Song, J.M. Biocompatibility and antimicrobial activity of copper(II) oxide hybridized with nano silicate platelets. *Surf. Coat. Technol.* **2022**, *435*, 128253. [CrossRef]
32. Thawari, A.G.; Rao, C.P. Peroxidase-like Catalytic Activity of Copper-Mediated Protein–Inorganic Hybrid Nanoflowers and Nanofibers of β-Lactoglobulin and α-Lactalbumin: Synthesis, Spectral Characterization, Microscopic Features, and Catalytic Activity. *ACS Appl. Mater. Interfaces* **2016**, *8*, 10392–10402. [CrossRef] [PubMed]
33. Rodzik, A.; Pomastowski, P.; Buszewska-Forajta, M.; Railean, V.; Gołębiowski, A.; Buszewski, B.; Niedojadło, K.; Fijałkowski, P.; Robotnik, K.; Rafińska, K. Enhancing wound healing with zinc and silver nanocomposites synthesized with β-lactoglobulin: Antimicrobial properties, collagen deposition, and systemic effects in a C57BL/6J mouse model. *Discover Nano* **2024**, *19*, 150. [CrossRef] [PubMed]
34. Navarra, G.; Tinti, A.; Di Foggia, M.; Leone, M.; Militello, V.; Torreggiani, A. Metal ions modulate thermal aggregation of beta-lactoglobulin: A joint chemical and physical characterization. *J. Inorg. Biochem.* **2014**, *137*, 64–73. [CrossRef] [PubMed]
35. Buszewski, B.; Rodzik, A.; Railean-Plugaru, V.; Sprynskyy, M.; Pomastowski, P. A study of zinc ions immobilization by β-lactoglobulin. *Colloids Surf. A* **2020**, *591*, 124443. [CrossRef]
36. Gao, M.; Zhang, T.; Ng, S.W.L.; Lu, W.; Tian, G.; Ong, W.L.; Kozlov, S.M.; Ho, G.W. Polymeric Layered Films for TiO2-Au/CuS Tandem Photothermal Catalytic H2 Production in Harsh Seawater and Waste Plastic Media. *Adv. Energy Mater.* **2025**, *15*, 2404198. [CrossRef]
37. Mu, R.; Zhu, D.; Abdulmalik, S.; Wijekoon, S.; Wei, G.; Kumbar, S.G. Stimuli-responsive peptide assemblies: Design, self-assembly, modulation, and biomedical applications. *Bioact. Mater.* **2024**, *35*, 181–207. [CrossRef] [PubMed]
38. Zhu, D.; Kong, H.; Yang, G.; He, P.; Luan, X.; Guo, L.; Wei, G. Peptide Nanosheet-Inspired Biomimetic Synthesis of CuS Nanoparticles on Ti_3C_2 Nanosheets for Electrochemical Biosensing of Hydrogen Peroxide. *Biosensors* **2023**, *13*, 14. [CrossRef] [PubMed]
39. Jiang, W.; Wang, Y.; Liu, Y.; Jia, Q.; Zhao, C.; Yang, J.; Ren, W.; Liu, Q. Enhanced water purification through the double regulation of GO/MXene membranes with sodium alginate and KOH. *Sep. Purif. Technol.* **2025**, *363*, 132287. [CrossRef]
40. Bolder, S.G.; Vasbinder, A.J.; Sagis, L.M.C.; van der Linden, E. Heat-induced whey protein isolate fibrils: Conversion, hydrolysis, and disulphide bond formation. *Int. Dairy J.* **2007**, *17*, 846–853. [CrossRef]
41. Lara, C.; Gourdin-Bertin, S.; Adamcik, J.; Bolisetty, S.; Mezzenga, R. Self-assembly of ovalbumin into amyloid and non-amyloid fibrils. *Biomacromolecules* **2012**, *13*, 4213–4221. [CrossRef] [PubMed]

42. Peydayesh, M.; Mezzenga, R. Protein nanofibrils for next generation sustainable water purification. *Nat. Commun.* **2021**, *12*, 3248. [CrossRef] [PubMed]
43. Lendel, C.; Solin, N. Protein nanofibrils and their use as building blocks of sustainable materials. *RSC Adv.* **2021**, *11*, 39188–39215. [CrossRef] [PubMed]
44. Zhang, Y.C.; Qiao, T.; Ya Hu, X. A simple hydrothermal route to nanocrystalline CuS. *J. Cryst. Growth* **2004**, *268*, 64–70. [CrossRef]
45. Lukatskaya, M.R.; Mashtalir, O.; Ren, C.E.; Dall'Agnese, Y.; Rozier, P.; Taberna, P.L.; Naguib, M.; Simon, P.; Barsoum, M.W.; Gogotsi, Y. Cation Intercalation and High Volumetric Capacitance of Two-Dimensional Titanium Carbide. *Science* **2013**, *341*, 1502–1505. [CrossRef] [PubMed]
46. Sarycheva, A.; Polemi, A.; Liu, Y.; Dandekar, K.; Anasori, B.; Gogotsi, Y. 2D titanium carbide (MXene) for wireless communication. *Sci. Adv.* **2018**, *4*, eaau0920. [CrossRef] [PubMed]
47. Naguib, M.; Kurtoglu, M.; Presser, V.; Lu, J.; Niu, J.; Heon, M.; Hultman, L.; Gogotsi, Y.; Barsoum, M.W. Two-Dimensional Nanocrystals Produced by Exfoliation of Ti_3AlC_2. *Adv. Mater.* **2011**, *23*, 4248–4253. [CrossRef] [PubMed]
48. Chen, Z.; Wang, B.; Yang, C.; Lv, Z.; Wei, Y.; Pan, T.; Xuan, F.; Zhou, X.; Chen, H.; Shen, H.; et al. 3D Printed Pedicle Screws with Microarc Oxidation Ceramic Interfaces Enhance Osteointegration and Orthopedic Fixation Feasibility. *ACS Appl. Mater. Interfaces* **2024**, *16*, 31983–31996. [CrossRef] [PubMed]
49. Zhang, S.; Meng, L.; Hu, Y.; Yuan, Z.; Li, J.; Liu, H. Green Synthesis and Biosafety Assessment of MXene. *Small* **2024**, *20*, 2308600. [CrossRef] [PubMed]
50. Rasool, K.; Helal, M.; Ali, A.; Ren, C.E.; Gogotsi, Y.; Mahmoud, K.A. Antibacterial Activity of $Ti_3C_2T_x$ MXene. *ACS Nano* **2016**, *10*, 3674–3684. [CrossRef] [PubMed]

MDPI AG
Grosspeteranlage 5
4052 Basel
Switzerland
Tel.: +41 61 683 77 34

Polymers Editorial Office
E-mail: polymers@mdpi.com
www.mdpi.com/journal/polymers

www.ingramcontent.com/pod-product-compliance
Lightning Source LLC
Chambersburg PA
CBHW040838250926
51850CB00007BA/1345